Laboratory Safety

A Self Assessment Workbook

2e

to my wonderful husband and children

Publishing team
Aimee Algas Alker (editorial/proofreading)
Erik N Tanck (design/production)
Joshua Weikersheimer (publishing direction)

Notice

The views and opinions expressed by the authors are those of the authors only and do not necessarily reflect the views and opinions of the authors' employer or of the American Society for Clinical Pathology.

Trade names for equipment and supplies described are included as suggestions only. In no way does their inclusion constitute an endorsement of preference by the author or the ASCP. The author and ASCP urge all readers to read and follow all manufacturers' instructions and package insert warnings concerning the proper and safe use of products. The American Society for Clinical Pathology, having exercised appropriate and reasonable effort to research material current as of publication date, does not assume any liability for any loss or damage caused by errors and omissions in this publication. Readers must assume responsibility for complete and thorough research of any hazardous conditions they encounter, as this publication is not intended to be all inclusive, and recommendations and regulations change over time.

20 19 18 17 16   5 4 3 2 1

Printed in the United States of America

 ISBN 978-089189-6463

Laboratory Safety

A Self Assessment Workbook

second edition

Diane L Davis, PhD, MT(ASCP)Sc, SLS, CLS(NCA)
Professor of Clinical Laboratory Science
Health Sciences Department, Salisbury University
Salisbury, MD

Contents

ISBN 978-089189-6463

Contents

 ISBN 978-089189-6463

 ISBN 978-089189-6463

An overview of laboratory safety

1. In a teaching laboratory, a 22-year-old medical laboratory science student acquired *Salmonella typhi*, which caused an intestinal abscess. In addition to antibiotic therapy, the abscess required 2 separate surgeries for drainage and creation of a temporary diverting ileostomy. The student also required hyperalimentation but was able to completely recover in 6 months, at which time the ileostomy was reversed. (Hoerl D, Rostowski C, Ross SL et al [1988] Typhoid fever acquired in a medical technology teaching laboratory. *Lab Med* 19(3):166-168)
2. Prior to 1976, sodium azide was used extensively as a preservative in laboratory reagents and was disposed of in ordinary plumbing. Copper and lead pipes cause the formation of metal azides, which are more explosive than nitroglycerin; therefore, when this disposal hazard was unrecognized, serious explosions resulted from the manipulation of pipes and drains in laboratories. (Rose SL [1984] *Clinical Laboratory Safety*, JB Lippincott)
3. "80 mL of diazomethane dissolved in ether detonated in a domestic type refrigerator. The door blew open, the frame bowed out, and the plastic lining ignited, causing a heavy blanket of soot to be deposited far down the adjoining corridor." (National Fire Protection Association [1990] *From Health Care Facilities*, NFPA 99)
4. A research facility hired students for the summer, and 6 of them were given the job to clean out a closed satellite lab. With minimal instruction, these students were told to move equipment and dispose of chemical and biological wastes. When they were cleaning out the cold room, on of them left some chemicals on the benchtop at room temperature, and the students all went to lunch. A short time later, the chemicals exploded and a terrible odor necessitated evacuation of the entire building. The students had been working in the room without protective gear and would have been seriously injured if they had inside the room during the explosion. (Ambrozak D [1996] In my opinion. *Lab Med* 27(5):304)
5. On September 20, 2000, an MRI technician died from a nitrogen gas leak at New York Presbyterian Hospital. Since nitrogen is colorless and odorless, levels incompatible with life can accumulate without detection. (www.healthsafetyinfo.com, accessed September 28, 2000)
6. A worker believed a can of sodium had completely converted to sodium hydroxide, so he decided to flush it with water. The remaining unconverted sodium reacted with the water and caused the can to explode, resulting in a fire that blocked the only door to the lab, so workers had to evacuate via the windows. Fortunately, the windows were on the first floor, and a few weeks previously the security bars over the windows had been removed due to the safety department's insistence. (Furr AK, ed [2000] *CRC Handbook of Laboratory Safety*, 5e. CRC Press)

The incidents above are real, and because all hazards cannot be completely eliminated, laboratory accidents still occur. Laboratory workers are often under pressure to produce results in a hurry and can become careless or tempted to take shortcuts. Even when laboratory workers are not under pressure, familiarity with routine hazards may desensitize even conscientious workers to the need for appropriate caution.

 ISBN 978-089189-6463 ©ASCP 2016

Purpose of this text

The purpose of this text is to heighten the reader's awareness of safety issues and to promote the ability to

1. learn and apply the appropriate regulatory and professional requirements for safety protocols, documentation, and long term records
2. establish standard laboratory operating procedures to incorporate the appropriate safety techniques, abate/manage hazards, and properly dispose of waste
3. recognize potential safety hazards and/or hazards revealed through incidents and take action to prevent future accidents
4. react promptly and correctly once an incident occurs
5. create a safety culture in which all workers regard themselves as important and valued members of a team that prioritizes safety for everyone

This text is not intended to be all inclusive. Many special situations require procedures not discussed herein, and this text will only address the most common hazards in a laboratory. Information cited in the text on the many safety regulations is accurate at the time of publication, but the reader should always check the most current version of any regulations before making changes in the laboratory. Resources from which current and additional information can be obtained are included in Appendix 1.1 at the end of Unit 1. The reader is encouraged to consult these or other sources before embarking on any procedure that is unknown and therefore potentially dangerous.

Learning objectives

This text is designed for self teaching. Most units end with practice questions and a table that summarizes major points. Units 9 and 13 are designed for the reader to apply knowledge gained in other units. A posttest covering all of the material at the end of the text is intended for comprehensive evaluation of learning based on the objectives below.

Following study of the material contained in this text, the reader should be able to:

- explain the importance of each individual responsible for complying with safety rules and regulations, and promoting a safe environment as a member of the entire laboratory team
- discuss aspects of laboratory safety management, such as the appointment of a laboratory safety officer, a safety/quality management committee, and an incident review committee, and how these entities continuously examine new regulations, formulate procedures, train staff, analyze incidents, and reformulate procedures as necessary
- define the following terms and explain their significance to laboratory safety:
    - regulation
    - standard/performance standard
    - accreditation
    - certification
    - license
    - equivalency/reciprocity
    - administrative control
    - engineering control
    - work practice control

Learning objectives

- explain the role of the following governmental bodies/regulations in laboratory safety:
    - US Occupational Safety and Health Administration (OSHA), describing the requirements for employers and employees under the general duty clause and with regard to the following OSHA standards and advisories:
        - Hazard Communication Standard ("Right to Know," "HAZCOM")
        - Hazardous Chemicals in Laboratories Standard ("Lab Standard")
        - Bloodborne Pathogens Standard (including changes mandated by the Needlestick Safety and Prevention Act)
        - Formaldehyde Standard
        - Personal Protective Equipment Standard
        - Control of Hazardous Energy Standard ("lockout/tag out")
        - ergonomics advisories
        - tuberculosis advisories
    - US Environmental Protection Agency (EPA)/Resource Recovery and Conservation Act (RRCA)
    - US Nuclear Regulatory Commission (NRC)
    - US Department of Transportation (DOT)/US Postal Service (USPS)
    - US Department of Homeland Security (DHS)
- explain the role of recommendations from the following professional organizations in laboratory safety:
    - US Centers for Disease Control and Prevention (CDC)/National Institute for Occupational Safety and Health (NIOSH)
    - National Institutes of Health (NIH)
    - National Fire Protection Association (NFPA)
    - Clinical Laboratory Standards Institute (CLSI)
    - voluntary accrediting bodies such as College of American Pathologists (CAP), the Joint Commission (TJC), and Commission on Accreditation of Laboratories (COLA)
    - National Toxicology Program (NTP)
    - International Agency for Research on Cancer (IARC)
- discuss fire hazards with respect to the following:
    - the fire "quadrahedron" and its relevance to preventing fire
    - classes of fires, including examples of each class
        - NFPA graphic symbols for each class of fire
        - precautions for each class of fire and general principles of fire prevention
        - appropriate means to extinguish each class of fire; extinguishing agents that are contraindicated
    - the definition and significance of each of the following with respect to fire safety:
        - flash point
        - upper/lower explosion limit
        - autoignition temperature
        - pyrophoric compounds
    - education and training, including fire drills
    - physical layout of the laboratory, and creation of optimal primary and secondary evacuation routes
    - fire safety equipment, specifically the correct use of
        - fire alarms: visual and auditory
        - sand buckets
        - fire extinguishers
- "PASS" acronym describing general proper use
- when/how to use water, dry chemical, CO_2, Class K and Halotron extinguishers

ISBN 978-089189-6463

- contraindications for fire extinguisher use
  - ☐ fire hoses
  - ☐ fire blankets
  - ☐ respirators
  - ○ evacuation and emergency plans in cases of fire, including the "RACE" acronym
- discuss chemical safety with respect to
  - ○ labeling of chemicals and waste, including interpretation of the following labeling systems:
    - ☐ NFPA
    - ☐ Hazardous Materials Information System (HMIS)
    - ☐ Globally Harmonized System
  - ○ the definition and significance of each of the following with respect to chemical safety:
    - ☐ signal word
    - ☐ target organ
    - ☐ permissible exposure limit
    - ☐ environmental monitoring
    - ☐ licensed waste handler
  - ○ proper storage design, organization in storage, and inventory management of chemicals and waste
  - ○ securing chemical materials against theft and terrorism
  - ○ best practices to safely handle chemicals and chemical waste
  - ○ information on safety data sheets (SDSs) and best practices for staff education
  - ○ classes of chemical hazards, giving examples/definitions as well as special procedures for handling each of the following:
    - ☐ corrosives
    - ☐ ignitibles: flammables and combustibles
    - ☐ health hazards: carcinogens, teratogens, mutagens, sensitizers, irritants, hepatotoxins, nephrotoxins, neurotoxins, antineoplastic drugs, asphixiants
    - ☐ unstable or reactive compounds including explosives and oxidizers
    - ☐ incompatible mixtures
    - ☐ physical hazards—chemicals of concern in histology/autopsy suites: paraffin, formaldehyde, and xylene
  - ○ proper use and maintenance of a chemical fume hood
  - ○ proper methods and regulations for disposing chemicals
  - ○ appropriate protective wearing apparel for handling each chemical hazard
  - ○ chemical spill protocols, including the "CLEAN" acronym and general "HAZWOPER" requirements
  - ○ general principles of minimizing and managing chemical hazards

Learning objectives

- discuss electrical safety with respect to
  - nature of electricity, conductors, insulator,s and circuits
  - effects of electricity on the human body and injuries consistent with electrical exposure
  - circuit requirement for shocks and the physical consequences of shock
  - the 5 essential principles of safe practice with electricity, explaining and give examples of each
    - insulation
    - grounding
    - guarding
    - circuit protection devices to include surge protection devices, fuses, circuit breakers and ground fault interrupters
    - safe work practices to include "lockout/tag out," emergency generators/power sources, uninterruptible power supplies, and proper use of electrical equipment
  - identification and prevention of electrical hazards
  - when/how extension cords may be used
- discuss equipment safety with respect to
  - safe use and disposal of glassware
  - safe use and disposal of sharps, including needles, scalpels, bone saws, cryostats, microtomes
  - safety engineered sharp device selection and injury reporting that complies with the Needlestick Prevention and Safety Act
  - centrifuge safety
  - steam sterilizer (autoclave) safety
- Discuss biological hazards with respect to:
  - the universal biohazard symbol: color and where/how to display
  - types of biological hazards, including
    - classes of microbes: viruses, bacteria, prions, protozoa, and fungi
    - microbial products or components: toxins, DNA, RNA
    - microbe sources such as specimens, including relative risk of specimen types
    - fomites
    - insects, rodents, plants
    - aerosols: how they occur, how they are prevented
    - spills
  - vectors of infection: direct inoculation, inhalation, direct contact, aerosols, splashes, insects
  - essential components of a biosafety program
  - CDC/NIH biosafety levels and essential safety requirements at each level
  - NIH biosafety levels for recombinant DNA and RNA
  - World Health Organization (WHO) biosafety levels
  - important micro-organisms in each CDC/NIH biosafety level
  - identification of critical agents associated with bioterrorism and securing the laboratory
  - reporting and specimen referral in the Laboratory Response Network for suspected bioterrorism
  - CDC Universal Precautions, Standard Precautions, and Transmission Based Precautions
  - examples of common hazards and situations
  - precautions used to reduce risk for each type of biohazard

- precautions unique to autopsy suites and histology labs
- precautions for working with animals in the lab
- special issues regarding individual pathogens such as hepatitis B virus, human immunodeficiency virus, hepatitis C virus, infectious prions (such as Creutzfeldt-Jakob disease), Ebola virus, and *Mycobacterium tuberculosis* to include
  - signs of infection
  - typical means of acquiring infection
  - vaccine availability
  - testing for presence of infection
  - treatment of workers exposed to organism
  - applicable biosafety level
  - special precautions in the laboratory
- physical containment requirements including
  - appropriate manipulation techniques
  - biosafety cabinets: proper usage, maintenance, selection
  - protective wearing apparel
  - design of histology/autopsy suites and microbiology laboratories
- biohazard decontamination to include
  - chemical agents

• low, middle, and high level disinfectants

• sterilants

• antiseptics

• proper use of bleach as mid level disinfectant (concentration, expiration)

• debulking material and sufficient contact time

  - heat, pressure, autoclaving
  - radiation: ionizing and ultraviolet
  - special procedures for infectious prions
- response protocols for spills and accidental exposures
- methods for proper biohazard disposal
- risk assessment and mitigation for tasks involving biohazards

• discuss compressed gases with respect to

- the GHS compressed gas hazard symbol: color, where/how to display
- definition and characteristics of compressed gases
- correct labeling and color coding of cylinders
- hazards associated with compressed gases including asphyxiation, fire, intoxication, chemical reaction, and violent release of pressure
- appearance of patent and properly labeled cylinders, and criteria to reject cylinders from supplier
- methods of safe transport, storage, usage, and inventory
- attaching and reading a regulator on a compressed gas tank and proper cylinder installation for use
- methods for handling empty and full tanks
- protocols for emergencies related to compressed gases
- special precautions with cryogens and gases for medical use

Learning objectives

- discuss radioactive materials with respect to
  - the universal radiation hazard symbol: color, where/how to display
  - definitions of terms and their relevance to radiation safety including
    - α, β, and γ radiation
    - rems
    - Curies and Becquerels
    - half-life
    - "decay in storage"
  - effects of radiation on the human body
  - sources of radiation in laboratories
  - NRC regulations to include
    - "ALARA" principle to keep radiation exposure "as low as reasonably achievable"
    - licensing and documentation requirements
    - required poster, signage, and staff information
  - managing radiation exposure by time, shielding, and distance
  - precautions including correct shielding (lucite for α, β, lead for γ)
  - safe storage, including security against theft and terrorism
  - proper usage, manipulation, and confinement
  - environmental and personnel monitoring
  - methods of disposal, including waste segregation by type and half-life
  - spill confinement and cleanup
- discuss laboratory waste and waste management to include
  - proper waste disposal techniques for every waste category and staff training
  - EPA categories of waste generators, including requirements for central collection and satellite collection areas
  - EPA waste categories, giving examples in each category
    - listed wastes
    - characteristic wastes
    - universal wastes
    - mixed wastes
  - non-EPA waste categories, giving examples in each category
    - infectious and regulated medical waste
    - sharp waste
    - radioactive waste
  - general EPA and RCRA requirements to include
    - GHS labeling
    - importance of using, tracking, and retaining waste manifests per regulations
    - definition of "cradle to grave" responsibility, including use of licensed waste handlers
    - when/how waste is disposed in the sanitary sewer, landfill, or nonhazardous waste streams
  - waste minimization including segregation, planning, reducing, reusing, and recycling
  - protocols for accidental waste release, including general requirements of "HAZWOPER" regulations

 ISBN 978-089189-6463

- list and describe important basic work practices common to all labs, giving examples, to include
  - no eating, drinking, smoking, or other hand to face contact
  - complying with policies on, eg, protective equipment, working alone, unattended operations
  - complying with personal dress policies to include
    - leg coverings
    - closed toe, fluid impermeable shoes
    - avoidance of jewelry, long nails, artificial nails, contact lenses
    - pulling hair back to avoid contamination
    - identification badges worn properly on "breakaway" or retractable lanyards
  - wearing protective equipment properly and removing it before leaving the laboratory
  - refraining from jokes, horseplay, drugs, and alcohol
  - maintaining a neat and clean work area, decontaminating and removing trash as required
  - decontaminating hands, using method(s) appropriate to the expected contamination, to include glove removal, handwashing techniques, alcohol gels, and segregated sinks
  - planning procedures to be executed efficiently without rushing
  - refraining from the use of electronic devices that are not necessary for the work at hand
  - maintaining devices and laboratory furniture in good repair
- explain the purpose of and correct techniques to select, install, maintain, and use safety equipment to include
  - signage, including recognition of common symbols (eg, eyewash, shower)
  - telephones or other means of emergency communication
  - fire safety equipment
  - deluge shower
  - eye wash
  - personal protective apparel/ equipment
    - goggles and safety glasses with side shields
    - face shields
    - lab coats and aprons
    - footwear and shoe covers
    - gloves: latex, nitrile, vinyl, chemical resistant, puncture resistant
    - masks and respirators
    - ear plugs and muffs
  - chemical fume hood
  - biological safety cabinet
  - containers for sharps and broken glass
- discuss the causes, signs, symptoms, and prevention strategies regarding latex allergies
- discuss the importance of filing an accident report, the information that should be included in such a report, and the analysis and action that should take place after an accident
- outline how and when OSHA should receive accident reports and what forms should be used
- describe important considerations for emergency and disaster planning in the laboratory
- list essential contents in a laboratory first aid kit

Learning objectives

- list and describe the "Check, Call, Consent, Care" first steps to take when encountering an accident situation
- discuss basic first aid in the following categories:
  - positioning the victim to maintain an open airway
  - assessing and restoring breathing
  - assessing and restoring circulation: cardiopulmonary resuscitation (CPR) or automated external defibrillator (AED)
  - stopping bleeding or supportive care for internal bleeding
  - treating shock
  - treating injuries to include
    - chemical and thermal burns
    - bone, muscle, and joint injuries
    - eye injuries
    - cuts and punctures
  - recognizing and/or treating heart attacks, strokes, and sudden illness
- identify graphic symbols or universal symbols corresponding to the following:
  - chemical hazards: flammables, oxidizers, corrosives, poisons, explosives, health and environmental hazards
  - biohazard
  - radiation hazard
  - compressed gas hazard
  - laser hazard
  - electrical hazard
  - personal protective equipment, eg, gloves, goggles, glasses, lab coats
  - safety equipment, eg, deluge shower, eyewash, fire extinguisher, fire blanket
- given the location and type of work that a laboratory performs, list and justify the appropriate safety equipment and contents of first aid kits for that lab*
- given the location and type of work that a laboratory performs, formulate laboratory protocols that meet all applicable safety standards*
- given an accident case history, perform root cause analysis and list the precautionary measures that should have been taken to prevent the accident
- given an accident situation, list the appropriate remedial actions that should be taken
- describe what is meant by "a culture of safety" in the laboratory and discuss ways to promote and sustain a safety culture in a given laboratory situation*

**Denotes terminal objective, which indicates what the learner should be able to do upon completion of the text*

ISBN 978-089189-6463

Unit 1

Introduction to safety

1

Safety in the laboratory is the responsibility of every person who uses the facility. This obligation includes not only creating a safe environment for colleagues but also considering the safety of others in the institution and community. Serious accidents in a laboratory can injure other occupants of the same building, and careless disposal of hazardous wastes can contaminate the environment of an entire community. Safety is unequivocally a serious duty for each person who practices laboratory science.

Some studies indicate that "just knowing" an institution's safety policies is not enough to ensure that they will be followed. The best defenses against a job related accident are concentration on the work at hand and an attitude that an accident is always possible, even during the most innocuous procedures. Henry states that "while inexperience may be a cause for some accidents, others may be a result of ignoring known risks, haste, carelessness, fatigue, or mental preoccupation" (*Clinical Diagnosis and Management by Laboratory Methods,* 2001).

An important first step toward safety in the work environment is strict compliance with standards established by regulatory/governmental bodies, research institutions, and professional organizations. Based on proper recommendations/requirements, standard operating procedures are developed and published, followed by adequate training and monitoring of all personnel, equipment, and activities. These measures will greatly decrease the chance of an accident occurring but must be regarded as *minimal* requirements. Each facility has its own unique set of circumstances that may not be adequately addressed by generic recommendations. These circumstances should be scrutinized by safety management staff to create additional necessary measures.

Each facility should have a person or a committee responsible for safety. In smaller laboratories, the laboratory director could likely handle the duties for ensuring safety, but in larger, more complex facilities, it is more useful to have a safety or quality management committee consisting of representatives from each area. Those who are responsible for laboratory safety must monitor current, newly published, and relevant regulations/recommendations, and should have the freedom, authority, and budget to formulate policies based on the most current information. After implementation and personnel training, the procedures must be reviewed at least on an annual basis, particularly if regulations have changed or if incidents have occurred. The safety officer and/or the safety committee should routinely conduct surveys/inspections of the facility (sometimes called "safety audits") to ensure that staff members are properly complying with procedures and no unexpected hazards are evident. "Near miss" events and

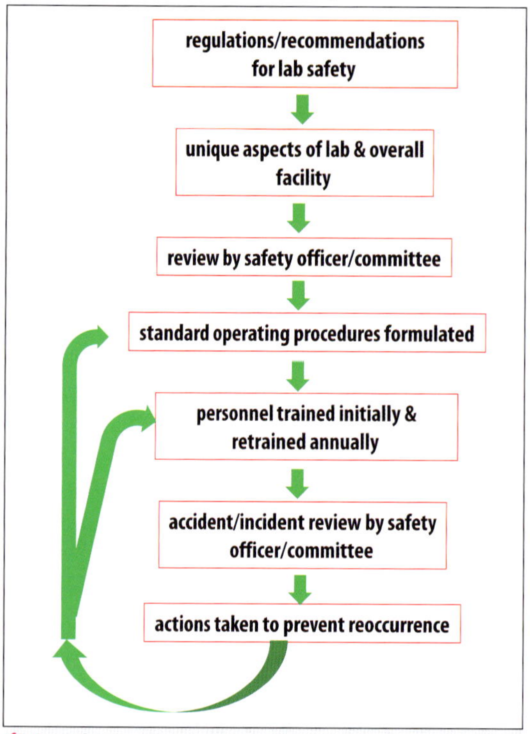

f1.1 safety management process

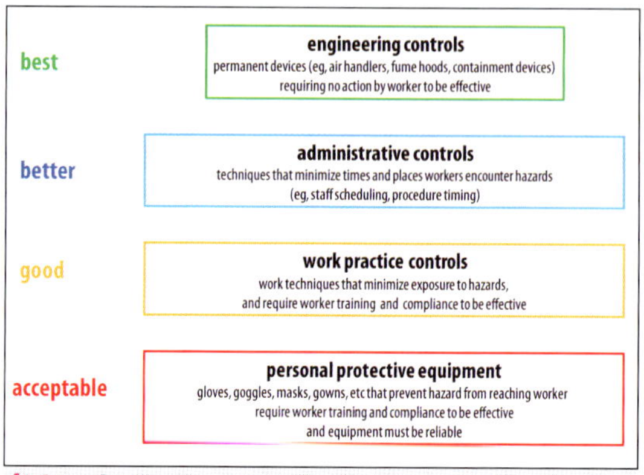

f1.2 safety management tools

accidents, while unfortunate, are excellent opportunities to evaluate the validity of safety procedures and to identify areas that need improvement. As illustrated in **f1.1**, safety must be viewed as a continuous process, rather than an event, as standards evolve and accidents point out flaws in the current system.

The remainder of this Unit outlines some of the many organizations that set standards for safety. Some of these have legal regulatory authority, and others do not. It is helpful at this point to introduce terms that are often used in relation to meeting safety standards.

Local, state, and federal governments pass laws with general principles that must be followed. Many of these laws create and/or authorize regulatory agencies that make the detailed *regulations* or *standards* that ensure compliance with the law. These standards and regulations can be enforced by the agencies through licensing and/or mandatory inspections. Often, these agencies require *licenses* to perform specific operations. Because different levels of government can write laws that overlap in function, agencies may decide to grant *equivalency* or *reciprocity* to other government agencies. For example, the Environmental Protection Agency (EPA) regulates waste, but if a state has stricter waste regulations, then the EPA may grant equivalency to holders of the state license, and therefore only the state regulations must be followed. By contrast, research institutions and professional organizations issue standards (not regulations) that are adopted entirely voluntarily. Instead of a license, these organizations can offer *accreditation* or *certification*. Laboratories can voluntarily conform to the standards, submit to an inspection, and gain the prestige of being accredited by these organizations.

The first strategy in the management of any hazard is to eliminate it altogether. For example, replacing a mercury thermometer with a nontoxic thermometer eliminates the need to manage the mercury hazard. If the hazard cannot be eliminated, management measures are often characterized as engineering, administrative, or work practice controls, or the use of personal protective equipment (PPE), as illustrated in **f1.2**. An engineering control is a physical setup or device that separates workers'

 ISBN 978-089189-6463

exposure to a hazard. A chemical fume hood is an engineering control. This is the preferred method of hazard reduction as it requires no action from the worker to for it to perform its function and it is permanent. Administrative controls are management techniques, such as employee scheduling and workplace protocols, which minimize places and times that workers are exposed to a hazard. It is the second most effective tool. Work practice controls are the third most effective tool, and they include specifications to perform tasks in particular ways that minimize or eliminate hazards. They require worker training, implementation, and compliance to be effective. Pipetting acid inside a sink next to an eye wash is a work practice control. Finally, when PPE is required, this suggests that the hazard cannot reliably be eliminated in any other way. Elimination of the hazard is the most desirable strategy as the use of PPE requires worker compliance and reliability of the equipment to prevent injury.

At the time of publication, every effort was made to ensure the descriptions of the regulations throughout this book were accurate. The reader must consult the regulatory agencies for the most current version of each regulation before substantive changes in policy are made. Many changes in safety policy require considerable financial commitment and inconvenience, and safety officers will be most persuasive to managers if they can cite current regulations and specifications as justification for any changes. In many cases, safety officers may be asked to discriminate between the "nice to do" and "need to do" procedures. Laboratory hazards must be reduced to the least that are reasonably achievable, and safety officers must become skilled in risk assessment to determine the actual level of danger in each situation.

Part I: United States regulatory agencies with legal jurisdiction over laboratories

A. Occupational Safety & Health Administration (OSHA)

The primary federal law dealing with safety in the workplace is the Occupational Safety & Health Act of 1970. Section 5(a)(1) of the act, known as "the general duty clause," states:

> "Each employer – (1) shall furnish to each of his employees employment and a place of employment which are free from recognized hazards that are causing or are likely to cause death or serious physical harm to his employees; (2) shall comply with occupational safety and health standards promulgated under this Act.
>
> Each employee shall comply with occupational safety and health standards and all rules, regulations, and orders issued pursuant to this Act which are applicable to his own actions and conduct."

The Act covers all workers in the United States except self employed farmers who don't employ workers outside their own families, and employees covered by other federal agencies. It does not cover those who are not paid employees (eg, volunteers, visitors, students). It requires that employers minimize or eliminate hazards in the workplace, but it also obligates employees to follow the safety protocols set forth by their employer. The law created a new regulatory agency within the US Department of Labor, the Occupational Safety and Health Administration (OSHA), to oversee the establishment of safety standards for facilities, equipment, and procedures, and to enforce adherence to

1

those standards. An individual state or US territory may form its own occupational safety body if OSHA deems that its requirements are at least as strict as the federal requirements. Approximately half of the states have elected to do this, so safety management staff must be sure to obtain information from the regulatory body—state or federal—under which it operates. Laboratories in a state with an OSHA accepted safety body are directly accountable for state regulations, not federal.

In general, OSHA requires employers to have a comprehensive safety policy that provides employees with appropriate safety equipment, adequate safety training, and free medical care in the event of an incident on the job. Training sessions must be held during paid working hours and must be at an appropriate educational/language level for each employee. Employers must also prominently display OSHA posters that inform employees of their rights. An important required poster, 3165, Appendix 1.1, is easily obtained at www.osha.gov/Publications/osha3165.pdf. Documentation of all activities (hazard correction, topics and attendance roster for safety training, medical follow-up, accident investigation, and so on) is extremely critical. By law, safety training records must be kept for 5 years, and medical and exposure records of must be kept for the duration of employment plus 30 years (with some exceptions that are not likely to apply to laboratories). This is necessary to provide information in the event that an employee experiences a condition in the future that is traced back to the job. OSHA has also established guidelines for which incidents in these records are reportable to OSHA either immediately or within a specified timeframe. Prompt medical care is required at the time of the incident, no matter what hour of the day or night, as well as long term follow-up care, as needed. More information on accidents and reporting is in Unit 12.

If workers have complaints about uncorrected hazardous conditions or an illness they suspect to be job related, they should attempt to work with their employer to resolve them. However, if they are not satisfied with the resolution of the issue, they have the legal right to report it to the local OSHA office. Employer retaliation or discrimination toward employees who report hazardous conditions is prohibited by law. When needed or after complaints, OSHA inspects working areas and issues citations to employers who fail to comply with the minimum safety regulations. OSHA does not accept expense or inconvenience as an adequate reason for failure to adopt an appropriate safety protocol. OSHA fines can be thousands of dollars per violation, so it is essential that laboratories comply with OSHA standards, not only to make the workplace safe but also to avoid devastating financial penalties.

There are 2 general categories of OSHA standards. The first spells out requirements in great detail. Examples include safety showers and eye washes, which must be 55 feet or 10 seconds from any point in the laboratory, and standards for safety glasses specifying the type of material to be used, the thickness of the material, the presence of side shields, and so on. Compliance with such standards is merely a matter of implementing the provisions or purchasing products manufactured to OSHA required standards.

The second type of standard is a performance standard. The Bloodborne Pathogens Standard described later in this Unit is an excellent example of such a standard. It stipulates that employers must

 ISBN 978-089189-6463

prevent employees from being exposed to human blood and body fluids. It does not specify the means by which facilities should accomplish this goal, only that the goal be achieved. Therefore, one facility may elect to purchase work shields behind which employees can work, and another facility may choose to dress employees in gowns, masks, and goggles.

A brief word of caution: strict compliance with all published OSHA standards does not guarantee that OSHA cannot levy fines against an institution. Under its "general duty clause" of ensuring a safe working environment, OSHA can inspect and potentially issue fines based on obvious hazards for which specific standards do not exist. OSHA standards take time to develop into final form, but this does not absolve employers from doing what they can to abate a hazard before the final rule. Most wise employers, especially at high risk facilities, immediately implement many proposed OSHA standards even without a finalized version. For example, in recent years, workplace violence (eg, homicides, assaults, terrorism) has been on the rise. Following OSHA's recommendation, many employers have instituted workplace violence prevention programs that include security provisions to protect against terrorism even though a specific standard for such is not in force.

The 5 OSHA standards described below have significant impact on the medical laboratory. Several years ago, a proposed ergonomics standard was withdrawn, but OSHA can still issue citations for poor ergonomics under the general duty clause. Because laboratory workers can experience ergonomic injuries, an overview of the issues is also presented below. Additional OSHA standards are discussed elsewhere in this text when relevant. Postal and website addresses for OSHA appear in Appendix 1.2. Regional and state OSHA offices should be consulted by those operating under their auspices, and links to these can be easily found at www.osha.gov. In addition, OSHA provides "quick starts" for compliance associated with various industries. The links associated for healthcare and laboratories may be useful:

www.osha.gov/dcsp/compliance_assistance/quickstarts/health_care/index_hc.html

www.osha.gov/Publications/laboratory/OSHA3404laboratory-safety-guidance.pdf

Hazard Communication Standard (CFR 1910.1200)

This standard requires that employers and industries inform employees about the chemicals to which they are exposed on the job, and report for public record the chemicals they are disposing of in the environment. It has been referred to as the "HazCom" and "Right to Know" standard. Every hazardous chemical used in a laboratory must have a safety data sheet (SDS) on file that is accessible to the worker. An SDS contains data on the hazards, composition, handling methods, and disposal of a particular chemical. (A sample SDS is included in Unit 3.) It must be documented that all workers have been informed of what chemicals are in the laboratory, how to access the SDSs (physically or electronically), how to handle the chemicals safely, and what rights they as workers have under the OSHA standard. Chemical mixtures that contain <1.0% of a hazardous chemical or <0.1% of a carcinogen are exempt, but this should be documented. Manufacturers will generally provide letters stating that products are exempt from SDS requirements, and these should be kept with the lab's SDSs for easy access.

 ISBN 978-089189-6463

Hazardous Chemicals in Laboratories (CFR 1910.1450)

The unique nature of laboratory manipulation of chemicals led to the implementation of this standard to address issues not adequately treated by the Hazard Communication Standard or other OSHA standards. At its inception, OSHA focused primarily on regulating large industries, and chemical manipulation in many of these facilities involved large amounts of a limited variety of chemicals. By contrast, laboratories handle smaller amounts of a vast array of chemicals, and many of the OSHA standards that had been written for other industries did not easily apply. This OSHA standard is commonly referred to as the "Lab Standard."

The major requirement of this standard is the development and implementation of a chemical hygiene plan (CHP). A CHP must address virtually every aspect of the procurement, storage, handling, and disposal of chemicals in use in a facility. The plan must be unique to the operations and chemicals in the facility, and the OSHA standard lists categories that must be addressed without being overly prescriptive. Exposure to chemicals must be minimized by establishment of standard procedures, requirements for PPE, engineering controls (eg, fume hoods, air handlers), and waste disposal procedures. For some chemicals, the environment must be monitored for levels that require action or medical attention. Exact procedures to obtain free medical care for work related exposures must be stated. Finally, the means to administer the plan must be specified. Persons responsible for procurement and placement of SDSs, organizing training sessions, monitoring of employee work practices, and annual revision of the CHP must be named. Additional information is provided in Unit 3.

Bloodborne Pathogens (CFR 1910.1030)

This standard requires that a facility establish an exposure control plan (ECP) to minimize "reasonably anticipated" employee exposure to potentially infectious materials such as human blood and body fluids. This plan is similar to the CHP in that every aspect of expected employee contact with potentially infectious materials must be addressed. Many provisions are identical, including requirements for PPE, engineering controls, training, medical care, and administration of the plan.

This standard was implemented because of concerns regarding major bloodborne pathogens including human immunodeficiency virus (HIV), hepatitis B&C viruses, and the Creutzfeldt-Jacob agent. Unique to ECPs is the stipulation that all employees with potential exposure to blood and body fluids be offered at no charge a vaccine against hepatitis B within 10 days of hire. Employees refusing the vaccine must sign a waiver. Although the standard was developed because of particular pathogens, implementation of all provisions essentially protects workers from almost all other infectious organisms typically transmitted through work with human materials.

Since provisions of the standard only apply to employees who are "reasonably anticipated" to have exposure to infectious materials, ECPs must provide for task assessments in which the amount and types of potential exposures are evaluated for every job category. The assessment must also state what measures will be taken to prevent the exposures from occurring. OSHA requires that "Universal Precautions" be used; in other words, every exposure is potentially infectious and must be prevented. Universal, Standard, and Transmission Based Precautions and

 ISBN 978-089189-6463

other information on biohazards will be discussed more thoroughly in Unit 5

Effective 2001, the Needlestick Safety and Prevention Act imposed additional requirements under the OSHA Bloodborne Pathogens Standard. All facilities must examine procedures that use needles and sharps to determine how safety devices such as self sheathing needles can be incorporated to minimize injuries from contaminated sharps. There are documentation requirements for injuries caused by sharps, whether or not they are contaminated with blood and body fluids. More information on safety with sharps is contained in Unit 4.

Formaldehyde Standard (CFR 1910.1048)

Formaldehyde (the major ingredient in formalin) is used as a tissue preservative. It is an irritant and can cause allergic reactions (a sensitizer). In 2011 the National Toxicology Program also classified it as a carcinogen. This standard seeks to minimize formaldehyde exposure and requires monitoring of the environment for formaldehyde fumes. Areas that use formaldehyde containing reagents such as surgical pathology, histology, and autopsy suites must verify that the chemical fume hoods/backdraft vents are in proper working order, and that the ventilation systems provide an adequate number of air exchanges to ensure that the permissible exposure limits of formaldehyde are not exceeded. Additional information is provided in Unit 3.

Personal Protective Equipment Standard (CFR 1910.132)

The previously discussed standards require PPE for chemical and biological hazards, but the Personal Protective Equipment Standard is broader in that it recognizes these hazards as well as physical hazards such as heat (eg, from hot plates), light (eg, lasers), sharps (eg, needles, scalpels), and electricity. It requires a hazard assessment of the workplace, as well as one in writing, and supplying the correct equipment and training employees to use it. In general, an assessment can be placed in the standard operating procedure for any activity. A paragraph describing the expected hazards and the PPE required to perform the activity will suffice. For a given laboratory or section, a master list of activities can also be created, listing the hazards associated with each activity and the PPE required. The risk assessment should be performed at least annually and when changes have occurred. The costs of purchase, maintenance, storage, decontamination, and/or disposal of PPE are the employer's responsibility.

In general, equipment must comply with requirements set forth by OSHA and/or the American National Standards Institute (ANSI). (It is common for OSHA standards to specify ANSI requirements to avoid unnecessary duplication of effort.) Specifications exist for gloves, goggles, face shields, helmets, footwear, respirators, and so on. In addition to training on the protective equipment, employers are required to perform fit testing to insure that the devices are of the proper size and functionality. Additional information is provided in Unit 10.

Ergonomics

OSHA's definition of ergonomics is

> "Ergonomics is a discipline that involves arranging the environment to fit the person in it. When ergonomics is applied correctly in the work environment, visual and musculoskeletal discomfort and fatigue are reduced significantly."

f1.3 proper positioning for using a computer

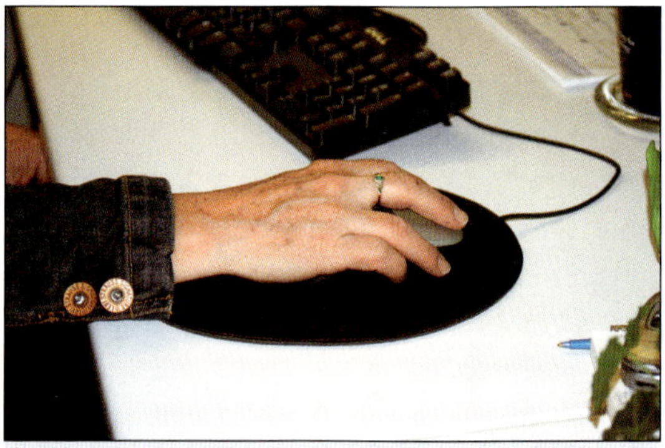

f1.4 use of a wrist pad to maintain proper wrist position

At the time of publication, the formal OSHA ergonomics standard was still withdrawn, but OSHA has made extensive recommendations on ergonomic safety issues on its website and in its publications. Various ergonomic work practices have potential application in the laboratory. Some important issues are as follows.

Effects of long term use of electronic devices with screens

Increased use of computers and other electronic devices with screens in all areas of society makes this an issue to be taken seriously. Antiglare screens with adjustable positions are being used in many workplaces where employees spend many hours of the day using a computer. A computer screen slightly below eye level is easier to read than one above eye level. Users should look away from the screen to a distant spot every 15 minutes to reduce eye strain. In addition, as shown in f1.3, users should be a proper distance away from the VDT (18-30 inches) with heads vertical, not at an angle. Users whose vision does not permit this distance may need to have their eyeglass prescriptions adjusted if the screen display cannot be altered. It is important to have adequate task lighting that is positioned to avoid screen glare.

Hearing loss

The noise from all equipment being operated simultaneously should not exceed 85 decibels over an 8 hour period. A general rule is that at 85 decibels, one must shout to be heard in conversation, so the noise level should be investigated. If noise levels cannot be controlled, a hearing conservation program (Occupational Noise Exposure Standard 29 CFR 1910.95) is required.

Cumulative trauma disorders caused by repetitive motion

Carpal tunnel syndrome may be a familiar example of this type of disorder. Excessive force for repetitive tasks and side to side twisting of the wrist should be avoided. Keyboarding and using a computer mouse are associated with repetitive task injury. There are specific recommendations for constructing workstations. For example, the position of the keyboard and computer mouse must be in the same plane as the hands when the elbows are at a 90° angle to avoid awkward wrist positions. The use of a wrist pad to maintain the proper position is shown in f1.4. Repetitive use of pipetting devices can also cause injury. Several models of pipettes have been developed that fit more comfortably in the hand and require less twisting and/or force to operate. Use of

 ISBN 978-089189-6463

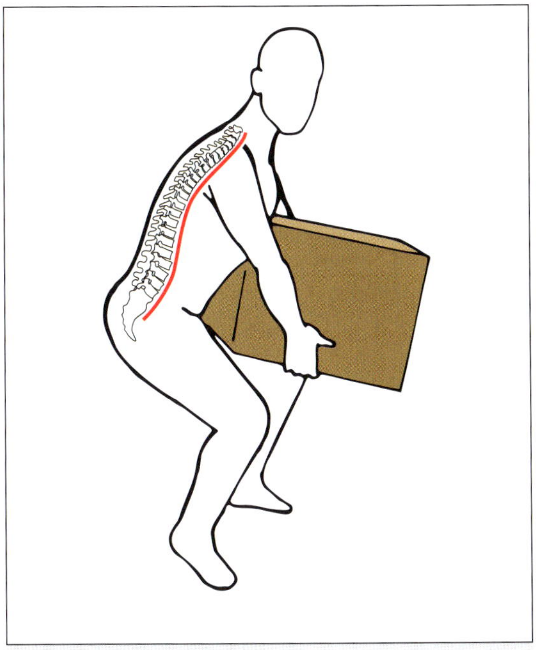

f1.5 proper lifting of a heavy object

f1.6 chair with back support & 5 legs

these types of pipettes should be considered, especially for laboratories performing a high volume of manual pipetting.

Skeletomuscular disorders

Back strain and other skeletomuscular disorders are frequent reasons for job absenteeism, and occupationally induced skeletomuscular injuries must be reported on OSHA Form 300. It is important to minimize moving heavy objects and/or to seek help from other people or assisting devices, such as carts and hand trucks. As shown in f1.5, if heavy objects must be lifted, one should spread the legs shoulder width apart for a wide base of support, face the object directly, bend at the knees, firmly hold the object close to the body with the chin tucked, maintain body symmetry, and lift with the legs, avoiding any twisting motion. It is also important not to lock the knees and to maintain body weight over the feet. The same principles hold when the object is put down. Heavy wheeled objects should be moved using a pushing rather than a pulling motion. Overhead lifting should be minimized, since the lcg muscles cannot be used and the back and arms must support the entire weight. Heavy objects should be stored close to the ground.

Many laboratory tasks and workstations require sustained periods of standing. Well cushioned antifatigue floor mats that can be easily cleaned and decontaminated should be considered. Task analysis should also be performed to determine if workflow can be modified to minimize standing time.

Laboratory chairs and computer workstations should be designed to be comfortable and supportive. Well fitting armrests, back support, and laboratory chairs with 5 wheels on the base are recommended as shown in f1.6. Chairs should be adjustable so that each user may get optimal support and position relative to the work surface. There are microscopes available that prevent hunching over the eyepieces and have focus knobs positioned

 ISBN 978-089189-6463

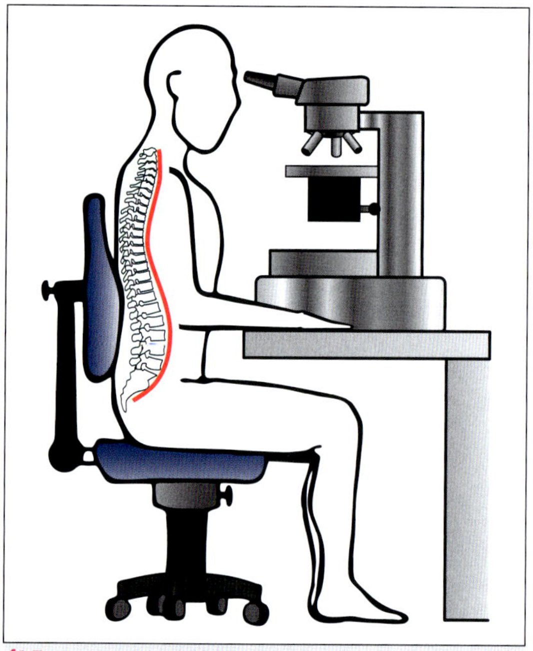

f1.7 proper position when seated

f1.8 improper posture for telephone use

more comfortably for repeated wrist action f1.7. Using the telephone with the awkward posture shown in f1.8 can also cause problems, and if employees must spend significant time on the phone, headsets and/or speakerphones should be considered. (Speakerphones are also recommended to reduce biohazard transmission.) Even if a workstation is designed comfortably, staff should be encouraged to get up periodically and move around to avoid fatigue.

Working with chemical fume hoods and biological safety cabinets can require awkward physical positioning and reaching for objects within the cabinet. Task analysis and optimum positioning of materials are crucial in developing work techniques that minimize physical stress when using these devices. Frequently used objects should be within 6 inches of the opening, as long as proper airflow can be maintained.

In all cases, skeletomuscular strain is reduced with frequent changes of position when doing a task, periodic breaks from any single task, and rotation of tasks within a work period. OSHA recommends standing, stretching, and taking short walks for approximately every 20 minutes of sitting.

Laboratory specific ergonomic recommendations from OSHA can be found at this website: www.osha.gov/Publications/laboratory/OSHAfactsheet-laboratory-safety-ergonomics.pdf

Slips, trips, falls

Workspaces should be arranged so that they are easily navigable and so that supplies, power cords, and furniture do not block the ordinary movement associated with the workflow. Of course, wet spills and other accidents must be cleaned up immediately or the affected area blocked off. In histology labs, it is important to minimize paraffin buildup on the floor by cleaning and by using sticky mats.

B. Environmental Protection Agency (EPA)

Medical waste washing up on US beaches several years ago caused intense scrutiny of how laboratory waste was being discarded. The EPA regulates disposal of

 ISBN 978-089189-6463

waste at a national level, and many state and local governments have additional restrictions on the manner in which laboratory waste can be disposed. Nothing should be flushed down the sink, thrown in the trash, incinerated, or autoclaved without checking with federal (EPA) and local laws. Laboratories should have standard operating procedures that outline protocols for waste disposal to comply with all governmental regulations, and these procedures should be reviewed before an item is discarded.

Some types of waste require removal by licensed waste handlers. Before sending waste out of the laboratory, however, it is important to thoroughly investigate the integrity of a waste handler. According to the Resource Conservation and Recovery Act (RCRA), each facility is responsible for its waste up until its *ultimate* disposal, regardless of the contractual obligations of the waste handler (the so called "cradle to grave" principle). Huge penalties for improper disposal can be levied against institutions because of the damage that can be done to the community and the natural environment. Additional information is provided in Unit 8.

C. Nuclear Regulatory Commission (NRC)

The Nuclear Regulatory Commission (NRC) is responsible for the licensing and inspection of facilities handling radioactive materials. State and local governments may have additional restrictions on the use and handling of radioactivity that should be investigated in addition to compliance with NRC licensure. In general, the amount of radioactivity used in medical laboratory testing is very small and minimally hazardous. This usually qualifies medical laboratories for an NRC general license that is only slightly restrictive. Additional sources of radiation in medical facilities include specimens and blood irradiators, and protocols must be developed for these situations. Further discussion is in Unit 7.

D. US Department of Transportation (DOT) & US Postal Service (USPS)

In the US, transport and shipping of hazardous substances are regulated by the Department of Transportation (DOT) Transportation of Hazardous Materials Regulations and the US Postal Service (USPS). Globally, the International Air Transport Association (IATA) Dangerous Goods Regulations must be followed. In recent years, the United Nations project to globally harmonize warning symbols and hazard protocols has essentially made the regulations of participating countries in substantial agreement, eliminating duplication and confusion. This has increased safety for all, as multiple nomenclatures and processes no longer exist in the nations that have agreed to participate in the Global Harmonization System (GHS). The US participates in the GHS, so in addition to safety gains, there are also gains of efficiency and convenience when preparing materials for transport.

Many laboratory items are chemical, biological, or radioactive hazards that require special shipping, handling, and transport. Dry ice as a preservative material is also hazardous. If there is doubt about how to move any item to or from a laboratory, current DOT/IATA regulations should be consulted and/or help should be requested from the USPS. USPS Publication 52 "Hazardous, Restricted, and Perishable Mail" contains detailed information regarding what substances can be mailed and how they must be labeled and packaged, and is available at pe.usps.com/text/pub52/welcome.htm.

Penalties for noncompliance, including criminal charges, are severe because of

 ISBN 978-089189-6463

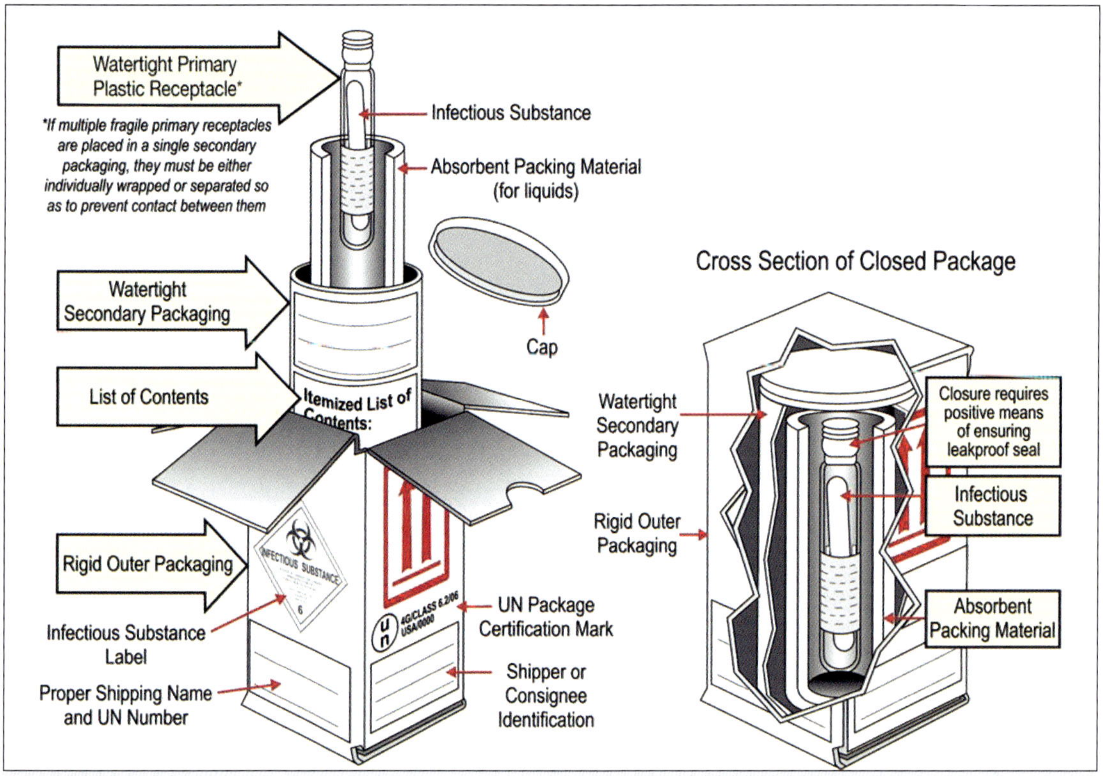

f1.9 specimen packaging for shipment

the risk of injury to innocent people who handle a hazardous package or who travel in the vicinity while hazards are being moved. If a private/commercial company is being used, it must comply with these regulations, so that company's policy may dictate that hazardous material will not be transported. Clear communication with these carriers is essential before using their services for any hazardous material.

In general, nonbreakable, leakproof containers should be used to transport hazards. If it is unavoidable that breakable items must be shipped, they should be thoroughly padded and sealed before shipping. Liquid materials should be packed with enough absorbent material to soak up leakage of the entire package. To ship biohazardous specimens, primary, secondary, and tertiary containers are necessary to guard against leakage. The packages also must be clearly labeled as to the nature of the hazard within. Accompanying paperwork must be placed in the outermost container to avoid contamination and to permit the person who opens the container to know the contents without opening the more dangerous inner containers. DOT has certain GHS codes and symbols for chemical materials, and the universal biohazard and radiation symbols can be used for dangerous biological and radioactive materials, respectively. **f1.9** is an example.

Although USPS is not subject to DOT regulations, the departments do not have conflicting requirements. USPS states, "The basic premise of the postal mailability statutes is that anything 'which may kill or injure another, or injure the mails or other property' is nonmailable." There are some exceptions, outlined in the USPS mailability guideline developed

 ISBN 978-089189-6463

to harmonize with DOT regulations, so following instructions in USPS Publication 52 should be sufficient for mailing hazardous items. USPS is generally *more* restrictive about what can be mailed, compared to what DOT will permit to be transported by other means. Some hazards can never be mailed, such as Category A infectious substances and certain chemicals. If only limited quantities are present, Category B infectious substances and some chemicals can be acceptable, if packaged properly. Published DOT/IATA requirements for transport of regulated waste, sharps, chemicals, infectious materials, etc, are very complicated and include mandates for such details as the size of the package, composition of the packaging materials, labels that must be present, volumes permitted, and information required on the packing slip. DOT/GHS labeling for chemicals will be discussed in Unit 3. A common shipping manifest for waste is used by both DOT and EPA in all 50 states, and is shown in Appendix 1.3. Additional discussion of waste is in Unit 8. The reader is encouraged to become thoroughly trained in the exact regulations before attempting to move any hazardous substance. IATA training is required every 2 years, and DOT training is required every 3 years. The Centers for Disease Control and Prevention (CDC) also provide free online training at www.cdc.gov/labtraining.

There are 3 categories of GHS recognized biohazard materials. An interactive guide to evaluate the shipping category of a potentially infectious material can be found at www.cdc.gov/od/eaipp/shipping/. This guide takes the user through a series of questions that permit the classification of a biological material. The 3 categories are as follows.

Per the CDC: "Category A infectious substance is one that is transported in a form capable of causing permanent disability or life threatening or fatal disease to otherwise healthy humans or animals when exposure to it occurs. An exposure occurs when an infectious substance is released outside of its protective packaging, resulting in physical contact with humans or animals. Examples are the Ebola, Junin, and Nipah viruses. These substances are assigned the identification number of UN 2814, UN 2900, as appropriate.

"Category B infectious substance is one that does not meet the criteria for inclusion in Category A. A Category B infectious substance is not in a form generally capable of causing permanent disability or life threatening or fatal disease to humans or animals when exposure to it occurs. Specimens previously termed 'diagnostic' and 'clinical' belong to Category B substances. These substances are assigned the identification number UN 3373."

Per USPS: "Exempt human or animal specimen means a human or animal sample (including, but not limited to, secreta, excreta, blood and its components, tissue and tissue fluids, and body parts) transported for routine testing not related to the diagnosis of an infectious disease."

Since specimen transport and mailing are common activities, a detailed discussion is provided in Appendix 1.4 for exempt and Category B substances. The reader should refer to the IATA shipping guidelines for other biological materials.

Another common activity is the transportation of specimens and blood products for testing and transfusion, respectively, at another site. DOT does permit transportation of these materials by motor vehicle, including public taxis. Blood products for transfusion are not DOT regulated because they have been thoroughly tested and have no evidence of infectious disease. All other specimens

 ISBN 978-089189-6463

must have 3 packaging layers that prevent breakage and leakage. The exterior container, such as a cooler, must withstand a drop of 4 feet without compromising the specimens. It is important to test the container, as many available for purchase by the public would not meet this criterion. An additional complication is the need to protect personally identifiable health information on specimens, so seals or locks on containers may be necessary, depending on the situation.

E. Clinical Laboratory Improvement Amendments of 1988 (CLIA '88)

The original Clinical Laboratory Improvement Act (CLIA) was passed in 1967. 2 decades later, certain problems with the quality of results from some medical laboratories drove Congress to update CLIA '67 with a new and more stringent package of regulations (CLIA '88). This set of amendments governs all medical laboratories. CLIA '88 empowers the Secretary of Health and Human Services to set laboratory and personnel standards (especially with regard to quality assurance and proficiency testing), to grant licensure, to perform inspections, and to invoke penalties for noncompliance. Small laboratories, such as those in physicians' offices that perform tests that are "so simple and accurate as to render the likelihood of erroneous results negligible (CLIA '88)," may be granted licenses for waived testing only, and license requirements are not as stringent as those for other levels of testing. CLIA licenses for waived testing do not exempt a laboratory from compliance with safety standards, but there may be exemptions to some OSHA provisions for laboratories with <11 employees and minimally hazardous material.

F. US Department of Homeland Security (DHS)

Potential acts of terrorism take many forms, and the Department of Homeland Security (DHS) is the central coordinating authority for antiterrorism measures in the United States. Laboratories can house chemicals and biological agents, so they are potential targets. In 2014, the Protecting and Securing Chemical Facilities from Terrorist Attacks Act of 2014 was passed, and detailed "risk based performance standards" for maintaining site security are constantly evolving. Healthcare sites are specifically mentioned by DHS as potentially housing materials of interest, and security risk assessments of facilities are required. DHS has developed a "chemicals of interest" list, which identifies chemicals and amounts that are likely to be targets of terrorism (further discussed in Unit 3), and it supports bioterrorism standards from entities such as the CDC and the US Department of Agriculture. Laboratories should have protocols for securing physical facilities and screening people to prevent unauthorized access, even by the general public, who could be harmed by hazardous lab materials. Even if laboratories do not possess particularly hazardous materials, many facilities could be targets for bombs, violence, or hostage situations, and the DHS is a good source for facility protocols. As a routine part of safety evaluation, medical laboratories should consider the general security of the facility as well its staff's ability to respond during emergencies and terrorist events with appropriate specimen collection, handling, and testing. Additional discussion regarding emergencies is contained in Unit 12.

ISBN 978-089189-6463

Part II: professional/research bodies for voluntary compliance

A. Centers for Disease Control & Prevention (CDC) & National Institute for Occupational Safety & Health (NIOSH)

The CDC conducts extensive research and monitoring of infectious diseases in the United States and throughout the world. Although it does not directly regulate laboratories, its definitions of minimal accepted practice for biohazard handling are generally recognized by OSHA and other regulatory agencies. Barring other local and special regulations, it is prudent laboratory policy to practice the CDC recommendations for biohazards. A joint publication from the CDC and the National Institutes of Health (NIH) entitled "Biosafety in the Microbiological and Biomedical Laboratories" is an excellent source for the definition of biosafety levels and containment procedures for specific organisms. Extensive materials at www.cdc.gov can be downloaded at no charge, including virtually all of the CDC publications cited in this work. The National Institute for Occupational Safety and Health (NIOSH) is an important division of the CDC with regard to safety in the workplace, and a great deal of useful information can also be obtained at no charge at the NIOSH website. Addresses and contact information for both the CDC and NIOSH are listed in Appendix 1.2, and biohazard handling is further discussed in Unit 5.

The CDC also coordinates the national Laboratory Response Network (LRN), which organizes medical laboratories in emergency events and bioterrorist attacks. In addition to the DHS mentioned earlier, the CDC is a good resource for emergency response protocols for laboratories with respect to biohazards.

B. National Institutes of Health (NIH)

Many of the health issues studied at the NIH relate to occupational illness, including the ergonomics issues discussed above. It is a valuable resource for new information on improving health and safety that may not yet be incorporated in OSHA standards.

C. National Fire Protection Association (NFPA)

The National Fire Protection Association (NFPA) publishes numerous, varied, and extremely useful publications. These publications provide fire prevention information for laboratory problems such as chemicals, electricity, and compressed gases, as well as specifications for facility design and personnel training. Because the NFPA is a private professional organization, there is a charge for many of its publications. Additional information on fire safety is contained in Unit 2.

D. Clinical Laboratory Standards Institute (CLSI)

Clinical Laboratory Standards Institute (CSLI) is an organization that publishes peer reviewed, consensus standards for a wide variety of medical laboratory processes. Because it focuses on medical laboratories, its recommendations are uniquely targeted to those sites in a way that other organizations serving multiple constituencies cannot. This makes its publications particularly useful to medical laboratory professionals. CLSI publishes documents on safety and waste management but is also a private professional organization that must charge for its publications.

E. Voluntary accrediting bodies

Many independent agencies that inspect hospitals and laboratories, such as the College of American Pathologists (CAP), the Joint Commission (TJC), and Commission on Accreditation of Laboratories (COLA), have safety standards that are useful guidelines for formulating policy and require that laboratories have protocols for emergency operations. The guidelines from these organizations are slanted toward healthcare and can offer useful perspectives to hospital laboratories. For example, TJC addresses employee fatigue caused by long work shifts and times that interfere with the natural sleep cycle in some of its recent publications. The shift work associated with healthcare unquestionably can cause fatigue and compromise personal and patient safety.

Part III: training

All staff must undergo thorough training in laboratory protocols upon being hired or when procedures substantively change. Training materials must be at a language and educational level appropriate for each staff category, and trainers must hold appropriate credentials, if required. Generally, there is a common knowledge base for all employees (eg, fire, first aid, accident reporting), and then additional training is tailored to particular job categories. At a minimum, annual updates are required, but they are not generally as detailed as initial training. In all cases, training should be documented, and staff should take a test and/or perform procedures after training to demonstrate competency. Staff must read employee handbooks and safety procedures, and sign a form acknowledging that they have read and understood the material they have received.

Part IV: safety management

As indicated at the beginning, safety management is an ongoing process. Initial risk assessments must be systematic and thorough when formulating standard operating procedures, and safety officers must be vigilant for changes or incidents in the laboratory that may require new procedures or additional training. It is useful to customize a safety checklist from another source such as CAP or CLSI to document both initial and ongoing safety audits and risk assessments.

Continued compliance with standards from many sources is a daunting task. Although multiple individual plans are discussed above (eg, CHP, ECP), there is no prohibition against combining all protocols related to safety into 1 cohesive plan that integrates all requirements. Indeed, bombarding staff with multiple procedures that are duplicative or, even worse, contradictory will reduce effectiveness. In addition, decisions regarding physical laboratory safety are not made in a vacuum. Laboratories are accountable to multiple constituencies, and the needs of all aspects of operations must be balanced. Including laboratory safety as a component of a comprehensive employee handbook that addresses requirements in other areas is not only acceptable but also reasonable. Additional discussion of this topic is in Unit 14.

 ISBN 978-089189-6463

Summary table: safety regulatory & accrediting bodies

Topic	Comments
Compulsory regulatory terminology	law, regulation, standard, license, inspection, reciprocity, equivalency
Voluntary accreditation terminology	standard, inspection, accreditation, certification
Compliance measures	engineering, administrative and work practice controls, personal protective equipment
Regulatory bodies with legally enforceable requirements	state and local jurisdictions may have additional requirements
OSHA	authority to insure safety in the workplace under "general duty clause": employers must provide safe workplace employees must comply with safety protocols 1. Hazard Communication Standard 2. Hazardous Chemicals in Laboratories Standard 3. Bloodborne Pathogens Standard/Needlestick Safety 4. Formaldehyde Standard 5. Personal Protective Equipment Standard 6. Ergonomics recommendations 7. Workplace violence recommendations 8. Portions of many other standards may apply to lab
EPA	authority to regulate waste disposal under RCRA
NRC	authority to regulate radiation handling and disposal
DOT/USPS/IATA	authority to regulate shipping and transportation of hazards
CLIA 1988	law that regulates medical laboratories
US Department of Homeland Security	coordinates general security and standards for at risk facilities
Bodies with nonenforceable, voluntary standards	it is highly advisable to adopt these standards on a voluntary basis to improve safety
CDC	standards for handling infectious micro-organisms
NIH/NIOSH	standards for occupational health and safety
NFPA	standards for fire safety
CLSI	standards for safety and waste management
Others	TJC, COLA, and CAP for medical labs, for example

Self evaluation questions

1 What is the first thing an employee should do if he or she finds an unsafe working condition?

- **a** report it to the supervisor
- **b** report it to the local OSHA office
- **c** assume that the supervisor will take care of it
- **d** conceal it so that OSHA does not issue a citation
- **e** ask other employees what they are doing to correct the problem

2 How does OSHA govern the health and safety of employees?
i. sets minimum standards for performing hazardous jobs
ii. issues citations and fines to employers who do not comply with standards
iii. inspects laboratories to ensure that they meet safety standards
iv. ensures that laboratory personnel are qualified to perform the tests they are doing

- **a** i, ii & iii
- **b** ii, iii & iv
- **c** i & iii
- **d** ii & iv
- **e** all of the above

3 Match the OSHA Standard below to its major provisions. Choose 2 correct answers per item.

Hazard Communication b, d ________

Hazardous Chemicals in Laboratories c, f ________

Bloodborne Pathogens a, e ________

- **a** hepatitis B vaccine
- **b** "Right to Know"
- **c** CHP
- **d** SDSs
- **e** ECP
- **f** environmental chemical monitoring

4 When lifting a heavy object you should

- **a** bend at the waist
- **b** bend at the knees
- **c** hold the object with both arms extended away from the body
- **d** b & c
- **e** a & c

 ISBN 978-089189-6463

5 Which of the following describes the correct way to use a computer?

- **a** use a chair with a minimum of 3 wheeled legs
- **b** position the monitor so that you have to look up and not down
- **c** position the monitor so that it is not any farther away than 20 inches
- **d** maintain your arm at a 90° angle with your wrist straight and supported
- **e** all of the above

6 Your laboratory has added a test that requires the use of a chemical new to your laboratory. Before you can dispose of this chemical, you must check

- **a** EPA regulations
- **b** state and local regulations
- **c** the SDS for the chemical
- **d** the facility's CHP
- **e** all of the above

7 The most important reason to analyze accidents and near misses is to

- **a** discipline employees who are involved
- **b** try to hide the deficiencies from OSHA
- **c** prevent the accident from happening again
- **d** reduce worker's compensation insurance premiums
- **e** justify the purchase of equipment to respond to the accident next time

 ISBN 978-089189-6463

8 Match the entity below to its function in setting standards for laboratories.

CDC __e__

EPA __a__

NRC __g__

NIH/NIOSH b

DOT/USPS __h__

NFPA __d__

CLIA '88 __c__

CAP/TJC/COLA __f__

a Waste disposal
b Research on occupational injury
c Medical laboratory licensure
d Fire prevention information
e Handling biohazards
f Voluntary accrediting bodies
g Licensure for handling radioactive materials
h Shipping of hazardous materials

Which of the entities listed above have legal enforcement authority over medical laboratory activities?

OSHA, CLIA '88, EPA, NRC, DOT/USPS

Which of the entities listed above do laboratories comply with on a voluntary basis?

CAP/TJC/COLA CDC, NIH/NIOSH, NFPA

9 You must mail a serum specimen from a healthy adult to a laboratory for cholesterol testing. What will you do?

a put the specimen into sealed primary and secondary containers
b add enough absorbent material to completely soak up the full volume of the specimen
c clearly label an outside carton with the appropriate cautionary symbols and messages and pack the primary/secondary containers into it
d all of the above
e find another means of transportation, as serum/blood specimens cannot be mailed in the United States

 ISBN 978-089189-6463

10 You must mail a serum specimen from a child who may have chicken pox to test for antibodies. Which statement is correct?

- **a** you must label the specimen as "Category A" infectious substance with a UN 2814 label and mail it in proper packaging
- **b** you must label the specimen as "Category B" infectious substance with a UN 3373 label and mail it in proper packaging
- **c** this is a "Category A" infectious substance and cannot be mailed
- **d** this is a "Category B" infectious substance and cannot be mailed

11 Your laboratory performs a procedure with a hazardous chemical. Which safety management technique below is the BEST of those listed?

- **a** require gloves, goggles, and a fluid resistant lab coat to perform the procedure
- **b** only perform the procedure on night shift
- **c** only perform the procedure in a chemical fume hood
- **d** only allow the most competent staff to perform the procedure

12 Which of the statement(s) below is (are) TRUE?

- **a** hospitals do not use enough chemicals to be considered "at risk" facilities by the Department of Homeland Security
- **b** laboratory access should be restricted to prevent injury of unauthorized people in the laboratory
- **c** since a standard does not exist for workplace violence, OSHA cannot fine facilities for violent incidents that could have been prevented
- **d** all of the above

 ISBN 978-089189-6463

Answers

1 a

2 a

3 Hazard Communication: b, d
Hazardous Chemicals in Laboratories: c, f
Bloodborne Pathogens: a, e

4 b

5 d

6 e

7 c

8 CDC: e
EPA: a
NRC: g
NIH/NIOSH: b
DOT/USPS: h
NFPA: d
CLIA '88: c
CAP/TJC/COLA: f

These entities have legal enforcement authority over medical laboratory activities: OSHA, EPA, NRC, DOT, USPS, CLIA '88
Laboratories comply with these entities on a voluntary basis: CDC, NIH, NIOSH, NFPA, TJC, COLA, CAP

9 d

10 b

11 c

12 b

 ISBN 978-089189-6463

Appendix 1.1: OSHA poster 3165

Source: www.osha.gov/Publications/osha3165.pdf , accessed 11/2015

U.S. Department of Labor

OSHA®
Occupational Safety and Health Administration

Job Safety and Health IT'S THE LAW!

All workers have the right to:

- A safe workplace.
- Raise a safety or health concern with your employer or OSHA, or report a work-related injury or illness, without being retaliated against.
- Receive information and training on job hazards, including all hazardous substances in your workplace.
- Request an OSHA inspection of your workplace if you believe there are unsafe or unhealthy conditions. OSHA will keep your name confidential. You have the right to have a representative contact OSHA on your behalf.
- Participate (or have your representative participate) in an OSHA inspection and speak in private to the inspector.
- File a complaint with OSHA within 30 days (by phone, online or by mail) if you have been retaliated against for using your rights.
- See any OSHA citations issued to your employer.
- Request copies of your medical records, tests that measure hazards in the workplace, and the workplace injury and illness log.

This poster is available free from OSHA.

Employers must:

- Provide employees a workplace free from recognized hazards. It is illegal to retaliate against an employee for using any of their rights under the law, including raising a health and safety concern with you or with OSHA, or reporting a work-related injury or illness.
- Comply with all applicable OSHA standards.
- Report to OSHA all work-related fatalities within 8 hours, and all inpatient hospitalizations, amputations and losses of an eye within 24 hours.
- Provide required training to all workers in a language and vocabulary they can understand.
- Prominently display this poster in the workplace.
- Post OSHA citations at or near the place of the alleged violations.

FREE ASSISTANCE to identify and correct hazards is available to small and medium-sized employers, without citation or penalty, through OSHA-supported consultation programs in every state.

Contact OSHA. We can help.

1-800-321-OSHA (6742) • TTY 1-877-889-5627 • www.osha.gov

OSHA 3165-04R 2015

Appendix 1.2: addresses, phone numbers & websites for safety agencies

Accessed 11/2015

American National Standards Institute (ANSI)
1819 L St NW, 6th floor
Washington, DC 20036
(202) 293-8020
www.ansi.org

Centers for Disease Control (CDC)
1600 Clifton Rd NE
Atlanta, GA 30333
(800) 311-3435, (404) 4989-1515
www.cdc.gov

Clinical Laboratory Managers Association
989 Old Eagle School Rd, Suite 815
Wayne, PA 19087
(610) 995-2640
www.clma.org

College of American Pathologists (CAP)
325 Waukegan Rd
Northfield, IL 60093-2750
(404) 639-3311
www.cap.org

Compressed Gas Association
4221 Walney Rd, 5th floor,
Chantilly, VA 20151-2923
(703) 788-2700
www.cganet.com

Clinical Laboratory Standards Institute (CLSI)
(formerly National Committee for Clinical Laboratory Standards [NCCLS])
940 West Valley Rd, Suite 1400
Wayne, PA 19087-1898
(610) 688-0100
www.clsi.org

National Fire Protection Agency (NPFA)
1 Batterymarch Park
Quincy, MA 02169-9101
(617) 770-3000
www.nfpa.org

National Institute for Occupational Safety and Health (NIOSH)
4676 Columbia Pkwy
Cincinnati, OH 45226
(800) 35-NIOSH (64674)
www.cdc.gov/niosh/homepage.html

National Institutes of Health (NIH)
9000 Rockville Pike
Bethesda, MD 20892
Telephone numbers for each division at www.nih.gov/health/infoline.htm
www.nih.gov

US Department of Homeland Security (DHS)
Washington, DC 20528
(202) 282-8000
www.dhs.gov

US Department of Labor
Occupational Safety and Health Administration (OSHA)
200 Constitution Ave NW
Washington, DC 20210
Hotline: (800) 321-OSHA (6742)
Telephone numbers for each division at www.osha.gov

US Department of Transportation (DOT)
US Department of Transportation
1200 New Jersey Ave SE
Washington, DC 20590
(202) 366-4000
www.dot.gov
hazmat.dot.gov/training/Transporting_Infectious_Substances_Safely.pdf (accessed January 11, 2008)

US Environmental Protection Agency (EPA)
Ariel Rios Bldg
1200 Pennsylvania Ave NW
Washington, DC 20460
Telephone numbers for each division at www2.epa.gov/aboutepa/mailing-addresses-and-phone-numbers
www.epa.gov

US Food and Drug Administration (FDA)
5600 Fishers Lane
Rockville MD 20857-0001
(888) INFO-FDA (463-6332)
www.fda.gov

US Nuclear Regulatory Commission (NRC)
One White Flint North
11555 Rockville Pike
Rockville, MD 20852-2738
(800) 368-5642, (301) 415-7000
www.nrc.gov

US Postal Service General Information
www.usps.gov
Information on Hazardous, Restricted and Perishable Mail (Publication 52)
pe.usps.com/text/pub52/

 ISBN 978-089189-6463

Appendix 1.3: EPA/DOT hazardous waste manifest

Source: www2.epa.gov/sites/production/files/2015-06/documents/newform.pdf , accessed 11/2015

Please print or type. (Form designed for use on elite (12-pitch) typewriter.) Form Approved. OMB No. 2050-0039

UNIFORM HAZARDOUS WASTE MANIFEST | 1. Generator ID Number | 2. Page 1 of | 3. Emergency Response Phone | **4. Manifest Tracking Number**

GENERATOR

5. Generator's Name and Mailing Address | Generator's Site Address (if different than mailing address)

Generator's Phone:

6. Transporter 1 Company Name | U.S. EPA ID Number

7. Transporter 2 Company Name | U.S. EPA ID Number

8. Designated Facility Name and Site Address | U.S. EPA ID Number

Facility's Phone:

9a. HM	9b. U.S. DOT Description (including Proper Shipping Name, Hazard Class, ID Number, and Packing Group (if any))	10. Containers No.	10. Containers Type	11. Total Quantity	12. Unit Wt./Vol.	13. Waste Codes
	1.					
	2.					
	3.					
	4.					

14. Special Handling Instructions and Additional Information

15. **GENERATOR'S/OFFEROR'S CERTIFICATION:** I hereby declare that the contents of this consignment are fully and accurately described above by the proper shipping name, and are classified, packaged, marked and labeled/placarded, and are in all respects in proper condition for transport according to applicable international and national governmental regulations. If export shipment and I am the Primary Exporter, I certify that the contents of this consignment conform to the terms of the attached EPA Acknowledgment of Consent.
I certify that the waste minimization statement identified in 40 CFR 262.27(a) (if I am a large quantity generator) or (b) (if I am a small quantity generator) is true.

Generator's/Offeror's Printed/Typed Name | Signature | Month | Day | Year

INT'L

16. International Shipments ☐ Import to U.S. ☐ Export from U.S. Port of entry/exit:
Transporter signature (for exports only): Date leaving U.S.:

TRANSPORTER

17. Transporter Acknowledgment of Receipt of Materials
Transporter 1 Printed/Typed Name | Signature | Month | Day | Year
Transporter 2 Printed/Typed Name | Signature | Month | Day | Year

DESIGNATED FACILITY

18. Discrepancy
18a. Discrepancy Indication Space ☐ Quantity ☐ Type ☐ Residue ☐ Partial Rejection ☐ Full Rejection
Manifest Reference Number:

18b. Alternate Facility (or Generator) | U.S. EPA ID Number
Facility's Phone:

18c. Signature of Alternate Facility (or Generator) | Month | Day | Year

19. Hazardous Waste Report Management Method Codes (i.e., codes for hazardous waste treatment, disposal, and recycling systems)
1. | 2. | 3. | 4.

20. Designated Facility Owner or Operator: Certification of receipt of hazardous materials covered by the manifest except as noted in Item 18a
Printed/Typed Name | Signature | Month | Day | Year

EPA Form 8700-22 (Rev. 3-05) Previous editions are obsolete.
DESIGNATED FACILITY TO DESTINATION STATE (IF REQUIRED)

Appendix 1.4: shipping instructions for category B infectious substances

Source: www.cdc.gov/nceh/vsp/cruiselines/OPRP/docs_word/biological_substances_shipping_detailed.doc, accessed 11/2015

Biological Substances, Category B Packing and Transportation Requirements

Specimen shipments in the United States and internationally are regulated under either the hazardous materials regulations (US) or dangerous goods regulations (international). Clinical specimens shipped as part an evaluation of an acute gastroenteritis outbreak are classified as Biological Substances, Category B." As such, these specimens require packaging that meets Department of Transportation (DOT) (domestic shipments in the United States) Transportation of Hazardous Materials Regulations (HMR) and International Air Transport Association (IATA) (international air shipments worldwide) Dangerous Goods Regulations (DGR).

The packaging of these materials must be of good quality and be strong enough to withstand leakage of contents, shocks, pressure changes, humidity, vibration and manual or mechanical handling considered incident to ordinary transportation. This is interpreted to mean that the contents should not leak to the outside of the shipping container, even if there should be leakage of the primary receptacle(s) during transit. The packaging should be resilient enough to withstand rough handling, passage through cancellation machines, sorters, conveyors and other similar equipment.

Key Definitions:

1. Infectious substance: a material known to contain or are reasonably expected to contain pathogens. Pathogens are defined as micro-organisms (including bacteria, viruses, rickettsiae, parasites and fungi) and other agents such as prions that cause disease in human or animals.
2. Biological products: products derived from living organisms which are manufactured and distributed in accordance with the requirements of appropriate national authorities, which may have special licensing requirements, and are used either for prevention, treatment, or diagnosis of disease in humans or animals, or for the development, experimental or investigational purposes related thereto. They, include, but are not limited to, finished or unfinished products such as vaccines.
3. Cultures: the result of a process by which pathogens are intentionally propagated. This definition does not include patient specimens as defined below [Note: The VSP does not require the shipment of cultures in support of investigations of cruise ship-associated outbreaks of acute gastroenteritis]
4. Patient specimens: those collected directly from humans or animals, including, but not limited to, excreta, secreta, blood and its components, tissue and tissue fluid swabs, and body parts being transported for purposes such as research, diagnosis, investigational activities, disease treatment and prevention.

DOT Hazardous Materials Division 6.2 materials can include:

- Biological products
- Cultures and stocks
- Diagnostic specimens
- Material of trade
- Regulated medical waste
- Sharps
- Toxin
- Used health care product

The International Air Transport Association (IATA) Dangerous Goods Regulations (DGR) classifies infectious substances into 2 categories, Biological substance, Category A and Biological substances, Category B:

a. Biological Substance, Category A: an infectious substance which in a form that, when exposure to it occurs, is capable of causing permanent disability, life threatening or fatal disease in otherwise healthy humans or animals. The proper shipping names for these substances are: UN2814 (infectious substances, affecting humans) and UN2900 (infectious substances, affecting animals)

b. Biological Substance, Category B: An infectious substance which does not meet the criteria for inclusion in Category A. Infectious substances in category B must be assigned to UN 3373. [Note Category B infectious substance is the classification of clinical specimens requested in support of a VSP acute gastroenteritis outbreak investigation.]

Note: Environmental samples (including food and water samples) which are not considered to pose a significant risk of infection are generally not subject to the IATA or USDOT shipping requirements.

All Category B infectious substances require three components ("triple packaging") for shipment:

 ISBN 978-089189-6463

1. Leakproof Primary Receptacles. Specimen must be placed in a leak-proof container known as a primary receptacle. All primary receptacles (ie, specimen collection containers, Cary-Blair or Para-Pak transport systems) must have positive closures, such as a screw-on cap. The primary receptacle may be glass, metal or plastic and must have a volumetric capacity of not more than 500 mL (16.9 ounces) for liquid diagnostic specimen; or not more than 500 g (1.1 pounds) for a solid diagnostic specimen. When shipped by air, the primary or secondary container must be able to withstand, without leakage, an internal pressure producing a pressure differential of not less than 95 kPa (14 psi) in the range of −40° C to 55° C (−40° F to 130° F). For solid specimens, the primary receptacle must be sift-proof (ie, the solid material does not leak out of the packaging). For Category B infectious substances shipped from cruise ships for in support of outbreak investigations of acute gastroenteritis, the primary receptacle is the stool/vomitus specimen container for viral analysis; or a bacterial transport medium and container system, such as Cary-Blair or Para-Pak C&S. The VSP Operations Manual 2005 requires that all passenger cruise ships within VSP jurisdiction are required to maintain a minimum of ten (10) containers each for viral and bacterial specimens.
2. Leakproof Secondary Packaging. To prevent contact between multiple primary receptacles, each must be individually wrapped or separated and placed inside a leak-proof secondary container. An example of the secondary container is a leakproof biohazard bag.
   a. For a liquid, an absorbent material, such as paper towels, cotton/cotton balls, bubble wrap or cellulose wadding, must also be included and be capable of absorbing the entire contents of the primary receptacle(s) in the event that the primary receptacle(s) are damaged.
   b. When shipped by aircraft, the primary receptacle or secondary container must be capable of withstanding (without leakage) an internal pressure producing a pressure differential of 95 kPa (0.95 bar, 14 psi) or less in a range of −40° C to 55° C (−40° F to 130° F). A biohazard pressure bag is an example of a suitable secondary container for air transit.
   c. An itemized list of contents must be placed between the secondary and outer containers and should be protected by storage in a leakproof plastic bag (eg, a Ziplock bag). For specimens shipped to the CDC labs, the CDC EpiForm satisfies this requirement. Recommend enclosing the forms and documents inside a plastic bag and taping to the inside of the box, preferably on top of the Styrofoam insulated pack (or similar container). When shipping to other labs, check with the receiving lab for documentation requirements.
   d. If a courier such as DHL, FedEx or UPS is used, then you must write the waybill number on the outside of each secondary container.
   e. Do not over pack the secondary container, as this may cause breakage of the primary receptacles. As a rule, if you cannot place a pencil between the primary receptacles after the absorbent material is added, then the secondary container is too full.
3. Outer Packaging. The primary receptacle(s) and the secondary container(s) are then placed inside a sturdy outer container that has a minimum of one rigid side of 4 inches in width (100 mm × 100 mm).
   a. The outer container must consist of corrugated fiberboard, wood, metal or rigid plastic and be appropriately sized for its contents.
   b. Dry ice or ice packs/freezer blocks are placed between the secondary container(s) and the outer container when the specimens require refrigeration.
   c. For liquid specimens, the outer container must not contain more than a total of 4 liters (L) (excluding ice or dry ice used to keep specimens cold).
   d. For solid specimens, the outer container must not contain more than a total of 4 kilograms (kg).Each complete package must be capable of withstanding a 4 foot (1.2 meter) drop test outlined in IATA and DOT regulations.|
4. The minimum required outer container markings and labels include
   a. The UN 3373 label with the words Biological Substances, Category B next to the diamond
   b. The outer container must also have the name, address and telephone number of the shipper, as well as the name, address and telephone number of the receiver/consignee.

 ISBN 978-089189-6463

c. Category B shipments DO NOT require an Infectious Substance label, Shipper's Declaration for Dangerous Goods or emergency response information.

d. For details on shipping Category B infectious substances, visit: www.iata.org/NR/rdonlyres/9C7E382B-2536-47CE-84B4-9A883ECFA040/0/Guidance_Doc62DGR_50.pdf

Comments regarding ice and dry ice shipments

Although clinical specimens can be shipped with ice or dry ice, the VSP discourages using either of these methods for shipping, as ice has been known to leak in transit and dry ice has caused concerns on the part of air transport carriers. The simplest way to maintain clinical specimens cold is to use ice packs as the primary coolant.

Note: Category B infectious substances are not permitted for transport in carryon or checked baggage and must not be carried on a person.

References

1. IATA Packing Instruction 650 – Biological Substances, Category B (www.iata.org/NR/rdonlyres/9C7E382B-2536-47CE-84B4-9A883ECFA040/0/Guidance_Doc62DGR_50.pdf)
2. DOT 49 CFR Parts 171-180 (ecfr.gpoaccess.gov/cgi/t/text/text-idx?c=ecfr&tpl=/ecfrbrowse/Title49/49cfrv2_02.tpl)
3. DOT Pipeline and Hazardous Materials Safety Administration: How to transport infectious substances (hazmatonline.phmsa.dot.gov/services/publication_documents/Transporting%20Infectious%20Substances%20Safely.pdf)
4. Vessel Sanitation Program Operations Manual 2005 (www.cdc.gov/nceh/vsp/operationsmanual/OPSManual2005.pdf)

Biological substances, category B shipment checklist

- ☐ All primary receptacles (ie, specimen collection containers) have positive closures, such as screw on caps
- ☐ All screw on caps are wrapped with Parafilm or adhesive tape
- ☐ Each primary receptacle is labeled with the patient's name (or other unique identifier) and date the sample was collected
- ☐ For liquid specimens, the primary receptacle is leakproof and contains a maximum of 500 mL
- ☐ When shipped by air, the primary or secondary containers are able to withstand, without leakage, an internal pressure producing a pressure differential of not less than 95kPa (14 psi) in the range of –40°C to 55°C (–40°F to 130°F)
- ☐ For solid specimens, the primary receptacle is siftproof and contains a maximum of 500g
- ☐ The primary receptacles are individually wrapped or separated and placed inside a leakproof secondary container
- ☐ The secondary container is certified by the manufacturer prior to use
- ☐ Absorbent material has been placed between the primary receptacle and the secondary container (enough absorbent material to absorb the entire contents of all the primary receptacles)
- ☐ The secondary container is not over packed (a pencil will fit between the primary receptacles after the absorbent material has been added)
- ☐ An itemized list of the contents is included with each shipment; the list includes the telephone number, fax number, and an e-mail address where problems may be reported by the receiving lab (note: a completed EpiForm satisfies this requirement for CDC labs)
- ☐ A sturdy outer package is used to ship the primary receptacle(s) and secondary container(s); the outer packaging consists of corrugated fiberboard, wood, metal, or rigid plastic and is appropriately sized for the contents
- ☐ For liquid, the outer packaging does not contain more than a total of 4 L; each individual primary receptacle contains and maximum of 500 mL
- ☐ For solids, the outer packaging does not contain more than a total of 4 kg; each individual primary receptacle contains a maximum of 500 g
- ☐ If a courier such as DHL, FedEx, or UPS is used, then the waybill number has been written on the outside of each secondary container
- ☐ The minimum package size in the smallest overall external dimension is 4 inches or if using "double mailers", they have been placed in a plastic envelope (pouch) provided by the courier
- ☐ Each completed package is capable of withstanding a 4 foot (1.2 meter) drop test as outlined by IATA and DOT
- ☐ The outermost packaging includes an approved "Biological Substances, Category B" label and all other labels and markings required by DOT and IATA
- ☐ Ice packs and insulated outer packaging is being used to assure specimen integrity during transit

 ISBN 978-089189-6463

Fire safety

4 things are required for a fire to begin: oxygen (or oxidizing agent), fuel, heat, and a self perpetuating chemical chain reaction. This is the so called "fire quadrahedron." It is an important concept because lack of any 1 of the 4 components will prevent a fire or extinguish an existing one.

Fire hazards

A fire hazard is a substance that is easily ignitable, and many materials used in laboratories fit this definition. Since oxygen cannot be eliminated, fire prevention hinges on minimizing heat sources and managing fuels. Heat sources such as hot plates and Bunsen burners should be used sparingly and never left unattended. Sufficient clearance and ventilation must be given to electrical devices that generate heat, and smoking in the laboratory is absolutely forbidden. Fire hazards are classified by fuel type, and the means of extinction, control, and risk management are different for each type **t2.1**. Detailed specifications for fire safe laboratory design and practice are contained in the National Fire Protection Association (NFPA) Standard 45, which is aimed at laboratories that use chemicals. Medical laboratories may also find NFPA Standard 99 for Health Care Facilities and Clinical Laboratory Standards Institute (CLSI) QMS04-A2 "Laboratory Design" useful.

t2.1 Types of fire extinguishers

Class of fire	Traditional NFPA symbol	NFPA pictogram	Water extinguishers	Dry chemical & CO_2 extinguishers
ordinary combustibles (eg, wood, clothing, paper): Class A	A		yes	yes, but CO_2 not considered optimal since gas dissipates
flammable liquids & gases: Class B	B		no (spreads liquid & fire)	yes
energized electrical equipment: Class C	C		no (risk of shock)	yes
combustible metals: Class D	D		no (intensifies fire)	no, sand or special extinguishing agents required
kitchen hot oil cookers: Class K	K		no (can cause oil explosion)	no (can cause oil explosion)

Class A fire: ordinary combustibles

Ordinary combustible solids include paper, wood, fabric, and plastic. (Fuels that burn and leave *ash* are usually Class *A*.) Prevention of a Class A fire in a laboratory is best accomplished by good housekeeping. Combustible material must be kept away from heat sources and stored in its proper place. Unnecessary material must be discarded, and work should be arranged so that clutter is minimized.

Class A fire can be extinguished with water, a dry chemical extinguisher, or sand. A carbon dioxide (CO_2) extinguisher is not hazardous to the user for Class A fires in ventilated spaces, but it is not considered optimal because the fire can continue to smolder when the CO_2 gas dissipates.

Clothing fires are Class A fires. A person with burning clothing can be wrapped in a fire blanket or some other large fabric, such as a curtain or a sheet, and rolled on the floor to eliminate the source of oxygen. A person wrapped in a fire blanket must not be allowed to stand because that might lead to the "chimney effect" of heat rising toward his or her face. If a safety shower is available and there are no chemicals or electricity to contraindicate its usage, a clothing fire can be effectively extinguished in the safety shower. Laboratory staff should be instructed to "stop, drop, and roll" if their clothing catches fire, a person on fire may panic and begin running. This only worsens the fire, and it is best to restrain a victim and roll him or her on the floor until fire equipment can be used.

Class B fire: flammable liquids & gases

Class B fires in the home are typically caused by oil, grease, paint, or gasoline. (Liquids can *boil*, so they are Class *B* fires.) In the laboratory, many organic solvents and compressed gases are potential Class B fire hazards. The main cause of ignition and spread of solvent fires is vaporization of the fuel, which allows it to mix with oxygen in the air. When this occurs, any source of heat can complete the "fire quadrahedron."

Many solvents are highly volatile and form a substantial amount of vapors, even at room temperature. A solvent's flash point is the temperature at which there are enough vapors to form combustible, potentially explosive mixtures with the air at the solvent's surface. Organic solvents should be used and stored with a knowledge of their flash points. **t2.2** gives examples of some common flammable solvents and their flash points. Note that all of the solvents shown except xylene exceed their flash points at normal room temperature. Solvents are classified as hazards based on their flash points. Class I solvents have flash points <100°F (38°C), Class II solvents have flash points between 100°F (38°C) and 140°F (60°C), and Class III solvents have flash points that exceed 140°F (60°C). The most dangerous solvents have the lowest flash points.

In addition to flash point, the hazard of a flammable liquid is assessed by its autoignition temperature and its Lower Explosive Limit (LEL)/Upper Explosive Limit (UEL). At its autoignition temperature, a liquid is hot enough to combust without an additional heat source. The LEL is the minimum concentration of vapor in air required to sustain a flame. At concentrations lower than the LEL of a

 ISBN 978-089189-6463

t2.2 Common flammable solvents & their flammability

Solvent	Flash point (°C)	Flash point (°F)	Lower explosive limit (%)	Upper explosive limit (%)	Autoignition temp °C/°F
acetone	−20	−4	3	13	465/869
ethanol	17	62	3	19	365/689
ethyl ether	−45	−49	2	36	160/320
heptane	−4	25	1	7	204/399
isopropanol	12	53	2	13	399/750
methanol	12	53	7	36	470/878
toluene	4	39	1	7	535/995
xylene	25	77	1	7	463/867

flammable vapor, there is insufficient fuel to combust. The UEL is the maximum concentration of vapor in air above which a flame cannot be ignited. At this level, there is too much fuel in relation to the oxygen to sustain a flame. Therefore, the lower the LEL and the higher the UEL, the more hazardous a flammable is. **t2.2** illustrates that ethyl ether is the most hazardous flammable of those shown, as it has the lowest flash point and autoignition temperature as well as the widest LEL/UEL range. Isopropanol and methanol have identical flash points, but methanol is more hazardous because of its higher UEL. Fortunately, autoignition temperatures of common laboratory solvents are way too high to create risk in most environments. However, in the event of fire, some of those temperatures could result, rendering flammable storage sites important areas to avoid during evacuation.

Because many flash points are below room temperature, it may seem that refrigerators should be used to store solvents. However, many domestic refrigerators have motors, lights, and even heating elements (in frost free models) that could provide heat or sparks, which could ignite any vapors confined in the refrigerators. In addition, many domestic refrigerators have drains that could bring chemical spills in contact with motor components. If refrigerators are used for flammable storage, they should be explosion proof models specially built using NFPA guidelines and hard wired (not plugged in) to electricity. Signage on refrigerators should make it clear whether or not flammables are allowed to be stored inside.

Volatile liquids should be kept away from heat sources, sunlight, and electrical switches. Flammables must only be used around nonsparking equipment and explosion proof heat sources, never open flames. All flammables should be tightly sealed to prevent accumulation of vapors anywhere, but especially when stored in the refrigerator because vapors generally cannot escape a closed unit. Room temperature storage areas must be vented, and vented cabinets or fume hoods that are running should be used to store the liquids. Also recommended are specially designed flammable safety cans with spring closing lids to contain

f2.1 flammable safety can

vapors, valves to relieve pressure, and flame arresters in spouts to dissipate point heat sources **f2.1**.

Flammable storage is designed to isolate materials from ignition sources, especially so that they do not worsen a fire in the vicinity. No materials should be stored on top of or immediately adjacent to flammable cabinets as they could catch fire and increase the temperature inside the cabinet. Bulk supplies of flammables should be kept in specially designed storage at all times, and large metal containers should be grounded to prevent electrical sparks. Workers should only remove the amount needed for the expected short term work volume. The Occupational Safety and Health Administration (OSHA) and NFPA have standards for the use and storage of flammable liquids, including construction standards and maximum permissible amounts based on the site. Per CLSI, if an area lacks a fire protection system such as sprinklers, no more than 1 gallon can usually be kept on open shelving and up to 2 gallons per 100 square feet can be kept in a safety can or cabinet. These limits change based on the flash point of the flammables being stored and the amount of fire protection provided. It is important to note that the amount of flammable waste stored in the lab is included in the total allowable amount. In general, laboratories should only keep the minimum amount of flammable solvents and gases on the premises necessary for normal operation and remove flammable waste promptly.

Very small amounts of water soluble flammables can sometimes be disposed of in the drain with cold water. Generally, however, this should be avoided to avoid accumulation of vapors heavier than air in the pipes, which can ignite ("flashback"). Flammable wastes must be stored in fire safe containers until they can be picked up by a licensed waste disposal firm, which usually incinerates them under carefully controlled conditions.

In Unit 10, alcohol based hand gels will be discussed as a means to decontaminate hands without having to wash them. Many medical facilities have installed numerous gel dispensers to reduce the spread of microbes. Unfortunately, the increased use of these alcohol based gels has also resulted in an increase of small fires associated with them. Gel vapors are usually ignited by a spark from a device or an electrostatic spark, such as that created from walking across a carpet in low humidity. Staff who use the gels must be reminded to let them dry thoroughly before using any electrical devices. Facilities that can maintain humidity over 50% will also reduce the risk.

Class B fires can be extinguished with the use of dry chemical or CO_2

 ISBN 978-089189-6463

extinguishers. Water should not be used directly on the liquid because it can push the fluid around and cause the fire to spread. In addition, many solvents are less dense than water and will float on top of it, continuing to burn. That said, if it can be done safely, water can be used on adjacent burning materials if they are Class A, to reduce the overall fire burden.

Class C fire: energized electrical equipment

Devices connected to a source of electricity are considered Class C fires. (Electrical equipment uses *current* so they are Class *C* fires.) Frayed wires or malfunctioning equipment can provide the spark necessary to ignite flammable vapors or materials. A fire in or around equipment that is plugged in adds the hazard of electrical shock to fighting the fire. A nonelectrical conductor must be used to extinguish Class C fires because conductors could carry the electricity back to the firefighter and cause electrical shock. Ordinary tap water contains ions, which make it an excellent conductor, so water should never be used on a Class C fire. If it is possible, electrical power should be cut off from the area of the fire. If electrical power can be eliminated with certainty, then burning equipment can be considered a Class A fire and treated accordingly. If, however, there is any question that the equipment is energized, only dry chemical or CO_2 extinguishers should be used.

Class D fire: combustible/reactive metals

Although rarely used in most laboratory settings, elemental metals such as magnesium, sodium, potassium, and lithium are not only combustible but also highly reactive, especially with water. For instance, pure sodium is stored in mineral oil because it is capable of reacting with the water in ordinary environmental humidity. Logically, then, water and wet extinguishing agents must never be used on Class D fires, making them difficult to control. In addition, they spread easily and could explode. Class D fires are controlled by scooping sand or dry powder media onto the fire. It is important that the buckets of sand stored in the laboratory for this purpose be checked periodically for the presence of cigarettes or other trash that tends to accumulate in them.

Laboratory design & evacuation routes

Rarely does anyone have the luxury of designing a brand-new laboratory. However, there are some key elements that should be included in the floor plan of a new laboratory or even incorporated into a laboratory facility that's been inherited "as is."

The laboratory should have an adequate number of doors, exits, halls, and walkways to accommodate the number of workers who need to exit in an emergency evacuation. Exit signs, evacuation routes, and workspaces should all be well lit, and emergency lighting should turn on automatically in the event of a power failure. This is particularly critical if the laboratory is housed in a basement or an inside space without windows. Areas >1000 square feet must have at least 2 exits, 1 of which opens directly onto an exit corridor. The primary and secondary routes should be as far apart from one another as possible. The farthest point in the laboratory must be <150 feet from a door, and hallways should be at least 96 inches across. Materials contained in exit corridors must not restrict the width of the passage and must not be capable of feeding a fire (eg, paper, coats, trash). Walkways within occupied rooms should be at least 40-48 inches across, while storage rooms

 ISBN 978-089189-6463

may have walkways of 36 inches across. CLSI recommends a clearance of at least 44 inches around instruments, with 60 inches preferred. All exits should be well marked and unblocked at all times. Exit doors ideally swing outward, are never locked in the exit direction, and are self closing. Closets and dead end rooms/corridors that could easily be mistaken for an exit along an evacuation route should be marked "NOT AN EXIT." Fire doors made of fire resistant material are strategically placed along an evacuation route and keep a fire contained for at least 1 hour when the doors are shut. When possible, separate spaces should be built to house and therefore isolate high risk procedures. The purpose of fire resistant doors and other similar materials is to keep a fire from spreading so occupants have time to evacuate. Therefore, fire doors must never be blocked or propped open. This is an important consideration when designing a laboratory, since pathways to eyewashes and safety showers must not be impeded by a closed door. Evacuation routes must be large enough to accommodate evacuees and must lead to exterior areas that allow them to easily get a safe distance from the building. Finally, temporary construction or physical remodeling should never compromise evacuation routes. Temporary alternatives must be developed, clearly marked, and communicated to building occupants.

Storage areas should be spacious and well vented, with all electrical switches and receptacles nonsparking and explosion proof. Fire resistant construction materials are desirable wherever possible, but local fire codes can provide exact requirements based on the laboratory's contents and purpose.

Automatic extinguishing systems such as sprinklers and gas suppression systems must be considered on a case by case basis because of the increased potential of Class B & C fires in laboratories, and the use of chemicals that should not be exposed to water. Unattended operations, especially those that employ flammables, typically require automatic fire suppression systems. Good examples of these are histology tissue processors and solvent recycling systems. Solvent recycling systems can generate hazardous atmospheres during their operation, and it is essential that electrical devices in the vicinity do not create sparks, and that adequate ventilation and fire suppression are present. Automatic systems that are activated only in the location of the fire are superior. In facilities with sprinklers, clearance of 36 inches around and 18 inches below each sprinkler head must be assured. This also means that no objects in the room can be placed higher than 18 inches below the lowest sprinkler head. The local fire marshal has the last word on where and how automatic fire suppression systems must be employed in a facility, so it is important to work with local authorities when making these decisions. Regardless of the presence of sprinklers, adequate fire safety equipment must be easily available throughout the laboratory.

Once the space is designed, a fire plan should be posted strategically around the laboratory. An example of a fire plan for a small laboratory is shown at the end of this Unit. This plan must include a map of the laboratory, all of the nearest exits, the primary (optimal) and secondary evacuation routes from each location, the location of equipment (eg, extinguishers, fire blankets, respirators, alarms, telephones), and the actions that personnel should take (eg, activate alarm, turn off gas). Evacuation routes must not rely on elevators because fires can cause power

 ISBN 978-089189-6463

interruptions. When possible, evacuation should be directed away from high risk areas and toward low risk areas. Good housekeeping, always critical in preventing fire, also extends to keeping evacuation halls and aisles clear; evacuation routes should be checked approximately once per month. Storage of hazardous materials along evacuation routes must be forbidden.

Laboratories should schedule quarterly fire drills so that every employee participates at least once per year. If at each fire drill, 25% of the staff is designated to do a mandatory full evacuation, by year's end every employee will have participated in an evacuation with minimal interruption in the laboratory operation. An outside meeting place should be designated for all evacuees, so staff can verify that everyone has exited the building. Large facilities may wish to designate certain employees to bring with them the current staffing schedule so that they can identify all those who should have been in the building at that time.

Fire safety equipment

All safety equipment must be properly located, easily visible, and well maintained. All fire extinguishers, alarms, smoke detectors, etc should be on a maintenance and inspection schedule. Each piece of equipment should have a tag on it documenting the last inspection. Since the size and layout of every facility is different, the reader should refer to OSHA, NFPA, and local guidelines when determining which devices below should be purchased, how many are needed, and where they should be placed.

Fire alarms

Fire alarms must be scattered throughout a facility and be both auditory (for the visually impaired) and visual (for the hearing impaired, eg, with strobe lights signaling the alarm). Auditory alarms and public address systems must be clearly audible from any part of the laboratory, including storerooms. If a facility has a central alarm system, the activation of any one alarm should be detectable at the central facility and ideally should also alert the local emergency responders. Alarms should be tested at least quarterly, typically in conjunction with fire drills. If any alarm, such as a smoke detector, is battery operated, the batteries should be replaced or tested at regular intervals, typically every 6 months. Using simple triggers, such as scheduling battery changes when, for example, daylight savings time begins and ends, are useful. Staff should receive clear training on how to activate the alarm system and be given a hierarchy of supervisors and first responders to contact in an emergency.

Sand buckets

Sand can be poured on most fires to extinguish the source of oxygen without hazard. It is important to keep these buckets free of debris because they may be the most readily available tools at the onset of a fire. OSHA Standard 1910.157 requires that extinguishing material for Class D fires be kept within 75 feet of Class D materials.

Fire extinguishers

The optimal fire extinguishers vary with the type of fire. Representative extinguishers are shown in **f2.2**. Dry chemical extinguishers are the best all purpose extinguishers and are the extinguishers of choice for most sites. The gas from compressed gas extinguishers, such as CO_2, dissipates rapidly, so they should not be used in enclosed spaces.

f2.2 Types of fire extinguishers and labeling: Class K and pressurized water extinguishers are silver in color while other types of extinguishers are red. The carbon dioxide (CO_2) extinguishers have enlarged discharge horns as compared to the dry chemical extinguishers. The class K extinguisher has no use in the laboratory but is shown to highlight how it must be distinguished from water extinguishers by the labels and symbols. Extinguishers are labeled as to the fire type(s) on which they are appropriate with letters (A, B, C) and/or the relevant pictogram(s). Note in these examples that the A,B and C pictograms are present but that a red line crosses out the type(s) of fires on which the extinguisher must not be used. [www.osha.gov.]

However, gas extinguishers are less damaging to electronics and may be useful in rooms in which computers are housed. Since dry chemical extinguishers leave significant residue, a "clean agent" fire extinguisher is often the best choice to preserve expensive devices. The NFPA publication 2001, "Standard on Clean Agent Fire Extinguishing Systems," defines a clean agent as "electrically nonconducting, volatile, or gaseous fire extinguishant that does not leave a residue upon evaporation." The most common material in this category is Halotron, which is discharged as a liquid, but evaporates rapidly, leaving little residue. Water extinguishers are limited to Class A fires, so they are used less frequently and are not recommended for laboratories.

It is important to note that fire extinguishers are limited in their size, range, and amount of extinguishing agent, and that fire extinguishers have only 2 purposes. They can be used on small fires, such as fires in trash cans, frying pans, or tabletops, or they can be used along an evacuation route to keep the path clear. If a fire has the following characteristics, an extinguisher should not be used and evacuation is the best course:

1. the fire is large and/or partially hidden
2. flammable solvents and gases are involved
3. flames exceed normal adult height
4. the base of the fire cannot be accessed from a normal standing position
5. the area is filled with smoke and/or breathing is difficult
6. the heat of the fire is too intense to approach it at the extinguisher range
7. the fire position could compromise the evacuation route

OSHA Standard 1910.157 requires a fire extinguisher within 75 feet for Class A fire risk and within 50 feet for high Class B fire risk. In general, a fire extinguisher should be placed at every laboratory door,

 ISBN 978-089189-6463

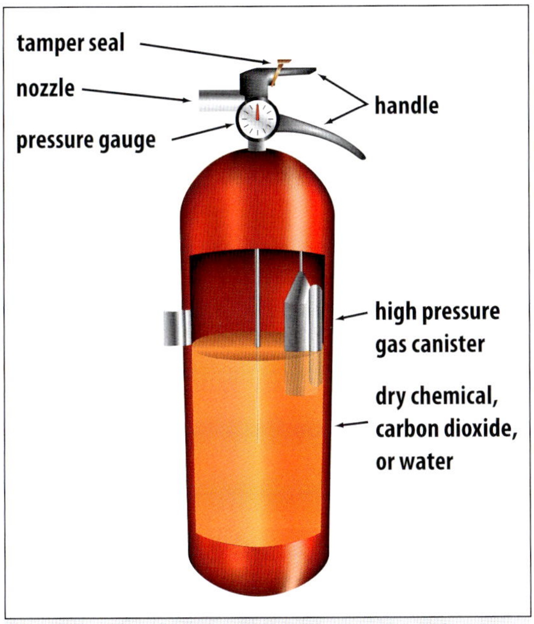

f2.3 schematic of a fire extinguisher [www.osha.gov]

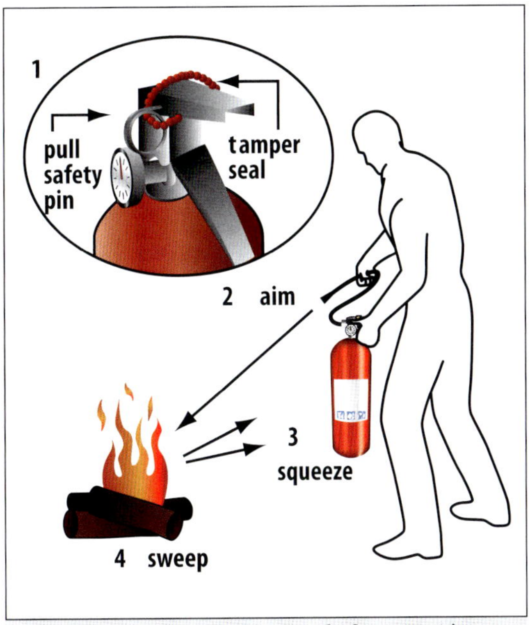

f2.4 illustration of the "PASS" process for fire extinguishers [www.osha.gov]

and a large room should have a second fire extinguisher in the area farthest from the door. Large extinguishers should be wall mounted with the handle ~3 feet from the floor, and small extinguishers can be mounted with the handle between 3 1/2 and 5 feet from the floor.

Fire extinguishers should be visually inspected every month to be sure that the hose is intact, the plastic seal on the pin is undamaged, and the gauge is showing a full charge **f2.3**, **f2.4**. More detailed inspection and/or testing should occur annually. Both monthly and annual inspections must be recorded on the tag affixed to each extinguisher.

"PASS" summarizes the essential features of operating a fire extinguisher. This acronym stands for

Pull out the pin

Aim the extinguisher

Squeeze the double handles

Sweep the base of the fire

Detailed instructions on fire extinguisher operation include the following:

1. Check the "ABC" categorization of the fire extinguisher, and make sure it is proper for the class of fire that you are attempting to extinguish
2. Pull out the locking pin located in the handle
3. Before you approach the fire, squeeze the double handles briefly to make sure that the agent will discharge; it is dangerous to approach a fire with a nonfunctioning extinguisher
4. Make sure you have your back to an unblocked escape route
5. Approach the fire and stand at the approximate maximum reach for the extinguisher you have. You can always move closer if the agent is ineffective at the maximum distance. If you are outside, attempt to stand on the upwind side of the fire. This side is cooler, and smoke should blow away from you. The maximum reach and discharge time vary greatly with each extinguisher type and size. General guidelines are as follows for each type:

- pressurized water extinguishers: maximum reach ~40 feet, maximum discharge time at full release ~60 seconds
- dry chemical extinguishers: maximum reach 12-20 feet (only 8-10 feet for small home extinguishers), maximum discharge time at full release <30 seconds
- gas extinguishers: maximum of ~3-8 feet, maximum discharge time at full release <30 seconds. Because gas dissipates quickly, you must get very close to the fire to use a gas extinguisher; therefore, one should not be used for large fires. In addition, gas extinguishers should not be used in enclosed spaces because the gases will further reduce the percentage of breathable oxygen

It is important to note that the discharge time of all extinguishers is very short.

6. Aim the nozzle of the extinguisher (do not touch the discharge horn if using a CO_2 extinguisher because it can get extremely cold)
7. Squeeze the double handle
8. Direct the discharge at the base of the flame using a side to side sweeping motion
9. Continue to discharge the extinguisher even after the fire is out to prevent reignition
10. Do not turn your back on the fire area, even after the fire is out. Walk backwards away from the fire area, watching for reignition
11. Recharge all fire extinguishers after use, even if some agent is left. Discharging the device opens the valves, and gas or agent leakage may well prevent the extinguisher from working again

An additional class of fire extinguisher, Class K, was developed for kitchens that use high temperature oil fryers. Such fires respond poorly to water, CO_2, and dry chemical extinguishers. Class K extinguishers do not currently have any uses in the laboratory and should not be retrieved from the kitchen to fight a laboratory fire. Since they are silver, they could be mistaken for a water extinguisher, so examining the label is crucial.

Ideally, staff should have the opportunity to train with real fire extinguishers, even without an actual fire, to increase their confidence in the event of a fire. There are laser extinguisher simulations on the market used for training purposes, since the use of real extinguishers can get messy and expensive.

Fire hose (Class A fires only)

A fire hose should be pulled entirely off the rack for use. The hose should be laid as straight as possible to avoid interference with water flow. A fire hose uses high pressure, so the operator should be braced, holding the hose with both hands, while another person turns it on. Again, the base of the flame should be swept using a slow side to side motion, with continued spraying after the fire is out to prevent reignition.

Fire blanket

A fire blanket can be used to put out flames on a person's body, to contain a small fire, or to aid in the evacuation of a fire area. For personal use, one should place the right arm through the rope loop and pull while turning the body and wrapping oneself in the blanket. To exit, one should crawl on the floor below the smoke, with the fire blanket over one's back. If the fire blanket is being used to extinguish a clothing fire, it is essential to use the "stop, drop, and roll" technique. If the victim remains standing when the fire blanket gets used, it will act like a chimney and direct heat and smoke up toward the victim's face. A graphic symbol for a fire

 ISBN 978-089189-6463

blanket is shown in f2.5. Although they are not devices that could fail, fire blankets should still be inspected periodically to verify that they have not been removed from their canisters.

Respirators

Respirators should be worn in the presence of toxic fumes. If respirators are not available, all personnel must be evacuated; instead of respirators, they may use wet cloths to cover the mouth and nose. Respirators must fit snugly onto the face to prevent any air from entering enter the mask. Personnel should be trained in the use of respirators and undergo fit testing to verify that they know how to wear them properly. Additional discussion of personal protective equipment is in Unit 10.

f2.5 symbol for fire blanket

Personnel

The best evacuation plan and state of the art firefighting equipment are absolutely useless if the occupants of the laboratory are unaware of them. All personnel must be adequately trained in a laboratory's fire plan, and their training should be verified with periodic fire drills that include correct activation of alarms, knowledge of emergency phone numbers, ability to locate and correctly use fire equipment, and accurate knowledge of evacuation procedures. To avoid panic in patients and visitors, many hospitals do not announce fire directly and use terms understood by the staff only, such as "code red." It is essential that staff have such institutional codes memorized and that they assist in notifying nonemployee occupants when they need to evacuate the building.

It is easy to overlook recent employees, who are given a great deal of new information during initial training. This is particularly treacherous because new employees are unfamiliar with procedures and equipment, and may be the ones most likely to accidentally cause a fire. After training, it is wise to review fire and other safety plans with a new employee after the first few months.

Some facilities, particularly large ones, may elect to provide special training to certain staff members for service in the "fire brigade." These staff members may act as first responders or experts in facilitating the evacuation of large numbers of people. Personnel also need to be trained to adhere to an institution's smoking policies and to remind visitors to adhere to smoking policies as well. "No Smoking" signs should be posted in areas that are particularly hazardous, and periodic monitoring for compliance may be necessary.

What to do in case of fire

The guide below is a general plan for dealing with a fire using the acronym RACE:

Rescue: be sure all people are out of danger

Alarm: sound the alarm and follow-up with a phone call

Contain: close doors, windows, etc to confine the fire

Extinguish or evacuate: attempt to put out the fire if it is small enough; evacuate if the fire is too big

Individual institutions or buildings often have unique protocols in their fire plans because of particular features of the site. Individuals are responsible for knowledge of the fire plan in every building in which they live, work, or study. Generic recommendations are as follows:

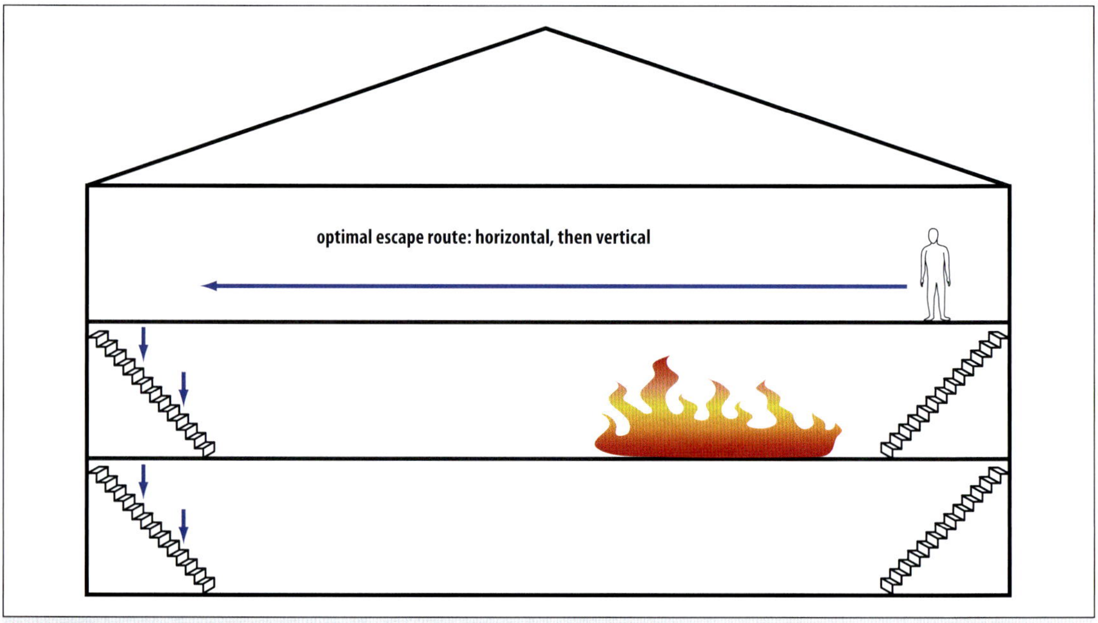

f2.6 horizontal then vertical evacuation

1. People are always more important than property. Evacuation of everyone in immediate danger is the first priority in any fire. Alert all personnel in the immediate area of a fire, and make sure not to overlook restrooms, storerooms, or small, isolated nooks. Injured people, hospital/nursing home patients, children, infants, and the elderly may not be reliable or ambulatory, and therefore may not be able to evacuate themselves. When wheeled beds and wheelchairs are available they can be used during evacuation, but when they aren't available, an excellent alternative is wheeled desk chairs.

   Avoid the elevators because fires can cause power interruptions. Be aware that in multistory buildings, optimal evacuation may be "horizontal first, then vertical," as illustrated in f2.6. Emergency exit lights must be located along all evacuation routes, and must be designed to work when there is an interruption in the central power supply.

f2.7 assembly point symbol

   When possible, it is a good idea to designate a meeting place after evacuation to ensure that everyone has left the building. When the staff members vary from day to day, bringing the employee schedule to the evacuation site helps ensure that everyone expected to be in the facility got out. f2.7 shows the symbol that is to be posted at designated assembly points.

2. The first person to see a fire, no matter how small, must report it immediately. What may be a small fire could grow rapidly out of control, and a delay in reporting could cause precious time to be lost. Reporting might include activating the nearest alarm, calling an emergency number (such as 911), announcing a "code red," or verbally notifying all nearby occupants. Even if a central electronic alarm has been activated, confirm the fire by calling the appropriate emergency number. It is an unfortunate effect of fire plan rehearsals that sometimes emergency responders see an alarm without confirmation and ascribe them to testing or a fire drill. It is important to

 ISBN 978-089189-6463

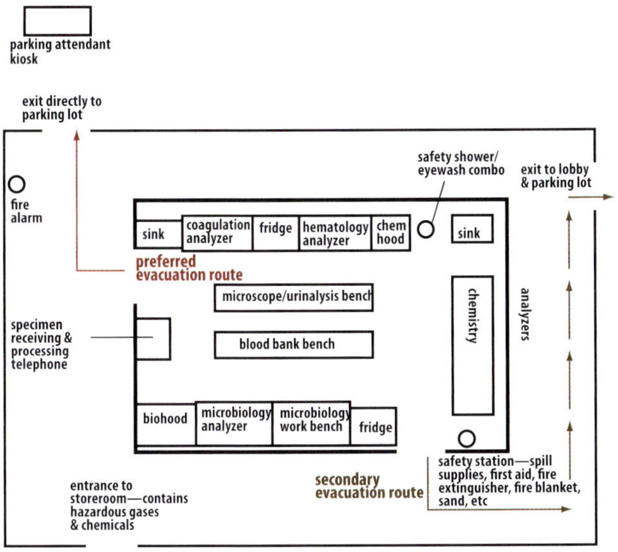

1. **When "Code Red" is announced, make sure all nearby persons are safe and evacuating the premises. Bring the staff schedule. Meet at the parking attendant kiosk and verify all staff on the schedule have evacuated.**
2. **For fire in the lab, as you exit dial "0" and announce, "Code Red, Lab." Pull the fire alarm in the hallway.**
3. **To the extent possible, confine infectious materials and turn off all electricity/ gas sources as you exit. Close the doors behind you as you exit.**
4. **Do not try to use the fire extinguisher to put out the fire unless it is small and you can maintain a clear exit. If in doubt, let it burn.**

f2.8 sample laboratory evacuation plan; note that there are both primary and secondary escape routes and that both escape routes go away from a high risk area

remain calm, and to clearly state the exact location and nature of the fire to the authorities.

3. Close the doors and windows as you leave a fire area. If it can be done safely, turn off the electrical current, especially if it is an electrical fire. When possible, close all gas cylinder valves and the main gas shutoff valve. If the fire is in a hood, close the hood opening and turn off the fans to contain the fire. Finally, open cultures and highly infectious materials should be put into incubators or refrigerators for containment, if there is time.

   If the room is filled with smoke, leave at once by crawling out. The purest air is closest to the floor. If possible, cover your nose and mouth with a damp cloth. As you exit, close the door to contain the smoke.

4. If the fire is not extinguishable with the equipment on hand, shut the door and evacuate. Anything larger than a trash can fire may be too big for a fire extinguisher. For small fires, use the following:
   - Class A fires (ordinary combustibles): something to smother the fire such as sand, wet blankets, water, or the nearest ABC extinguisher.
   - Class B (flammable liquids) and C (electrical) fires: the nearest dry chemical or CO_2 extinguisher, *not water*—the use of water can spread a liquid fire or conduct electricity in an electrical fire

   Fight the fire by keeping yourself between the fire and the exit, and by staying as close to the exit as possible. Never position yourself so that your escape route could be compromised.

5. Administer first aid to anyone suffering from burns or asphyxia.
6. Answer concisely any questions asked by fire response personnel. How the professionals deal with the fire greatly depends on information about the situation that you give them. Alert authorities to hazards present such as flammables, gases, and infectious organisms.

f2.8 is a simple example fire plan that illustrates the basic principles. Every site must customize a plan that meets the needs unique to that particular site.

Summary table: fire safety

Topic	Comments
Fire quadrahedron	1. fuel, oxygen/oxidizer, heat, sustainable chain reaction
	2. remove 1 of 4, fire is prevented or extinguished
Fire hazards	1. Class A: ordinary combustibles (eg, wood, plastic, clothing, paper); use good housekeeping
	2. Class B: flammable liquids and gases; low flash points are the most hazardous, vented storage away from heat and electricity, use no water to extinguish
	3. Class C: energized electrical equipment; use no water to extinguish, try to cut power to facility
	4. Class D: combustible/reactive metals; use sand or special powder media only to extinguish; water intensifies the fire
	5. Class K: kitchen hot oil fires; only use Class K extinguishers; do not use Class K extinguishers on Class A, B, C, or D fires
Lab design	1. adequate number and size of doors, windows, and hallways
	2. well marked, unobstructed exits; at least 2 per 1000 square feet
	3. fire resistant construction materials; at least 1 hour fire doors along exit routes
	4. adequate ventilation and storage areas
	5. no storage of materials in exit halls that is combustible or compromises width
	6. lighted escape routes and exit signs, emergency lighting
	7. clearance of 18 inches in any direction (below, sides) of sprinkler heads
	8. clearance around instruments, desks, etc so evacuees may easily exit lab
Evacuation	1. hold fire drills with 25% of staff every 3 months
	2. plan primary and secondary escape routes to designated meeting place
	3. evacuate toward low risk, not high risk, areas
	4. check to see if "horizontal, then vertical" escape is better
	5. bring staff schedule to determine who should have been in building at the time of the drill
	6. forbid use of elevators during evacuation as electricity may be lost during the fire; emergency lights along evacuation route
Fire equipment	1. audible and visual strobe light fire alarms
	2. sand buckets
	3. fire hoses for Class A fires only: straighten hose and brace feet
	4. fire extinguishers only for small fires: limited reach and limited agent
	a. water for Class A fires only; dry chemical and CO_2 for ABC; Halotron to preserve devices
	b. PASS: PULL the pin, AIM the extinguisher, SQUEEZE the double handles, and SWEEP the base of the fire
	c. always recharge after any use, even if there is agent left
	5. fire blankets: always use with "stop, drop, and roll" for clothing fires
	6. respirators: for use in noxious atmospheres; users must be fit tested
Fire plan	1. train personnel on plan; post plan in a conspicuous spot and practice annually
	2. RACE:
	a. RESCUE all people; evacuation of people, not property, is priority; crawl below smoke to exit
	b. set off the fire ALARM and call emergency numbers to verify
	c. CONTAIN the fire: shut doors, windows, etc; turn off gas and electricity; place infectious materials into incubators or refrigerators
	d. EXTINGUISH the fire if small enough; keep yourself between the exit and the fire so that your escape route is not cut off–EVACUATE if fire is too large

 ISBN 978-089189-6463

Self evaluation questions

Matching. Choose 2 correct answers per item.

1 Class A fires b, g

2 Class B fires f, h

3 Class C fires a, e

4 Class D fires c, d

- **a** Electrical
- **b** Only type of fire on which water is used
- **c** Reactive/combustible metals
- **d** Put out with sand or special powder agent only
- **e** Putting out with water may cause electric shock
- **f** Flammable liquids
- **g** Ordinary combustibles
- **h** Putting out with water may spread fire

5 What is the first step in using a fire extinguisher?

- **a** pulling out the safety pin
- **b** verifying that the maintenance is up to date on the attached tag
- **c** instructing all personnel to get out of the way of the extinguisher's line of fire
- **d** verifying that the extinguisher is appropriate for the class of fire on which it is about to be used
- **e** opening the doors and windows so that the extinguishing agent can be ventilated out of the room

6 Care must be taken to avoid touching the discharge horn of this fire extinguisher because they get particularly cold.

- **a** CO_2
- **b** halon
- **c** water
- **d** dry chemical
- **e** Class K

7 If a fire is small enough to be handled by the laboratory personnel, it is not necessary to

- **a** pull the alarm
- **b** notify the fire brigade
- **c** evacuate other personnel
- **d** make a follow-up call verifying the existence of a fire; however, the alarm should still be pulled
- **e** none of the above; a fire is to be taken seriously no matter how small

 ISBN 978-089189-6463

8 If they can be done safely, all of the following are necessary to do when evacuating an area with a fire EXCEPT:

- **a** closing gas cylinder valves
- **b** crawling out below the smoke
- **c** carrying out irreplaceable items
- **d** closing the doors and the windows
- **e** turning off electricity and equipment

9 The FIRST priority in any fire plan is

- **a** dialing 911
- **b** activation of the fire alarm
- **c** locating the closest fire extinguisher
- **d** ensuring the evacuation and safety of all endangered people
- **e** donning respiratory protection before smoke has had a chance to accumulate

10 The safest institutions have

- **a** at least 1 fire extinguisher per room
- **b** detailed personnel training and frequent fire drills
- **c** a window as well as a door in every room to serve as a fire escape
- **d** state of the art alarm systems and automatic sprinkler systems
- **e** fire extinguishers for every class of fire: A, B, C, D, and K

11 Because of the instrumentation and chemicals in many laboratories, which of the following is (are) not used in every laboratory area?

- **a** a sprinkler system
- **b** fire resistant doors
- **c** an adequate number of well marked exits
- **d** spacious and uncluttered aisles and hallways
- **e** ventilation systems that operate at a rate of 6-12 air changes per hour

Match each symbol to its meaning.

12 Class A fire

13 Class B fire

14 Class C fire

15 Class K fire

a

b

c

d

 ISBN 978-089189-6463

16 Which of the following actions in a fire is (are) CORRECT?

a responder uses a CO_2 extinguisher on a potassium fire
b responder unwinds the first 10 feet of a 100 foot long fire hose to put out a nearby fire
c responder wraps a victim in burning clothing in a fire blanket and walks him quickly out of the building
d victim is on 5th floor, east wing; fire is on 4th floor, east wing; victim goes to 5th floor, west wing, and then takes the stairs to exit the building
e all of the above

17 On which fire(s) below can an extinguisher be used?

a computer on fire
b flames coming through a wall
c floor to ceiling bookcase in flames
d all of the above
e none of the above

18 Given the data below, which solvent is the most hazardous?

Solvent	Flash point, °C	Flash point, °F	Upper explosive limit, %	Lower explosive limit, %
a	−20	−4	2	28
b	17	62	3	19
c	50	122	1	47
d	−11	12	1	7
e	4	38	8	15

19 Which of the following is acceptable?

a storing supplies on top of a flammable storage cabinet
b placing flammable storage cabinets along evacuation routes
c creating primary evacuation routes that pass by hazardous storage areas
d using a Class K fire extinguisher on a laboratory fire when nothing else is available
e none of the above

20 Clearance around a sprinkler head should be _____ inches in all directions.

a 12
b 18
c 24
d 30

21 To minimize damage to devices when a fire extinguisher needs to be used, which extinguisher is the best choice of those listed below?

a water
b carbon dioxide
c Halotron
d dry chemical

Answers

1 **b, g**

2 **f, h**

3 **a, e**

4 **c, d**

5 **d**

6 **a**

7 **e**

8 **c**

9 **d**

10 **b**

11 **a**

12 **d**

13 **a**

14 **c**

15 **b**

16 **d**

17 **a**

18 **a**

19 **e**

20 **b**

21 **c**

 ISBN 978-089189-6463

Unit 3

Chemical safety

All laboratory workers, even those who do not work in "chemistry," use chemicals in their work. This Unit reviews the general principles of safe chemical usage, and the topics discussed here should be covered in the chemical hygiene plan of a facility.

Labels

All chemical containers must be clearly labeled. As a result of the United Nations Global Harmonization System (GHS) project, labeling protocols have changed for participating countries, including the US. The goal of the GHS project was to provide consistency in hazard communication and thus improve safety. Full implementation of the GHS rules was required on June 1, 2015, but manufacturers could ship materials with old labeling until December 1, 2015, to deplete inventory (although recent issues suggest that this deadline may be extended). By June 1, 2016, employers must have fully implemented the new GHS labeling system at the user level. Therefore as this book is written, laboratorians will likely have chemical inventory with multiple labeling systems. When initially developed, the Occupational Safety and Health Administration (OSHA) Hazard Communication Standard did not specify a particular labeling system. Labels merely had to communicate the relevant information to the user. In contrast, the GHS has specific hazard classifications, requirements for pictograms f3.1, hazard

health hazard carcinogen mutagenicity reproductive toxicity respiratory sensitizer target organ toxicity 	**flame** flammables pyrophorics self heating emits flammable gas self reactives organic peroxides 	**exclamation mark** irritant (skin & eye) skin sensitizer acute toxicity narcotic effects respiratory tract irritant hazardous to ozone layer 
gas cylinder gases under pressure 	**corrosion** skin corrosion/burns eye damage corrosive to metals 	**exploding bomb** explosives self reactives organic peroxides 
flame over circle oxidizers 	**environment** aquatic toxicity 	**skull & crossbones** acute toxicity (fatal or toxic) 

f3.1 GHS chemical hazard symbols & hazard categories

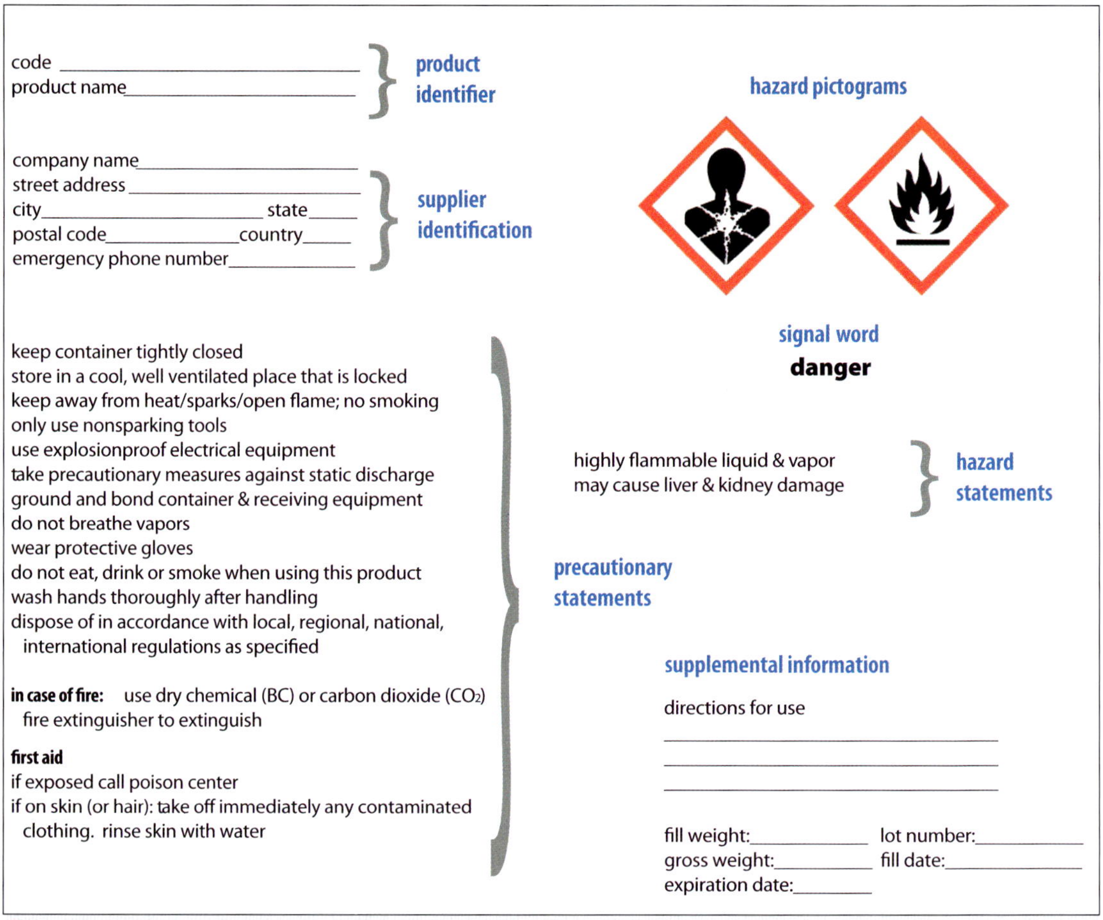
code
product name
product identifier
company name
street address
city state
postal code country
emergency phone number
supplier identification
hazard pictograms
signal word
danger
highly flammable liquid & vapor
may cause liver & kidney damage
hazard statements
keep container tightly closed
store in a cool, well ventilated place that is locked
keep away from heat/sparks/open flame; no smoking
only use nonsparking tools
use explosionproof electrical equipment
take precautionary measures against static discharge
ground and bond container & receiving equipment
do not breathe vapors
wear protective gloves
do not eat, drink or smoke when using this product
wash hands thoroughly after handling
dispose of in accordance with local, regional, national, international regulations as specified
in case of fire: use dry chemical (BC) or carbon dioxide (CO_2) fire extinguisher to extinguish
first aid
if exposed call poison center
if on skin (or hair): take off immediately any contaminated clothing. rinse skin with water
precautionary statements
supplemental information
directions for use
fill weight:
gross weight:
expiration date:
lot number:
fill date:

f3.2 GHS compliant chemical label

statements, and signal words ("DANGER" for high risk hazards and "WARNING" for lower risk hazards). Since the reader is likely to encounter more than 1 system in the near future, the major labeling systems will be described. Manufacturers must comply with GHS labeling requirements, but for secondary containers, other labels can be used as long as the information is complete and does not conflict with GHS. A complete guide to GHS, commonly called "The Purple Book," can be found at www.osha.gov/dsg/hazcom/ghs.html.

GHS symbols shown here are black on a white background with a red border. This format is required in all sectors except transportation. Similar but slightly different GHS symbols are required by the US Department of Transportation (DOT) to better communicate details relevant to that environment. These symbols may be visible when chemicals are shipped to the laboratory but will not be cause for concern.

Label information provided by manufacturers should never be defaced, damaged, or removed as long as the chemical is within the container. It is often necessary to place chemicals in secondary containers or make mixtures of chemicals in other containers for use. Secondary containers must be clearly labeled with at least the information shown below. It is best to label the reagent bottle before it is filled so that there are no unlabeled chemicals in the laboratory even for a moment and mislabeling is minimized. Even containers of innocuous substances such as water should be clearly labeled

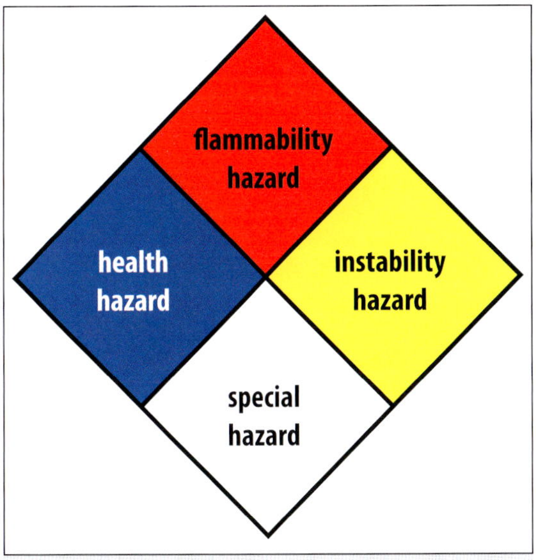

f3.3 generic NFPA diamond

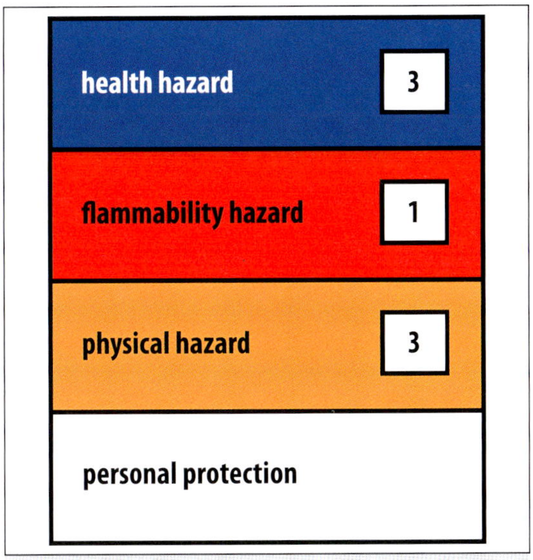

f3.4 generic HMIS rectangle

so that they are not confused with more dangerous compounds.

Permanent chemical labels must ultimately comply with GHS specifications, as shown in the example in f3.2. At a minimum, a reagent label on a secondary container should include the following:

1. identity: recognizable name(s) of the chemical/the concentration present
2. dates: date of receipt, date reagent made, date of expiration (the expiration date of chemicals put into secondary containers should be no farther out than the expiration date on the original containers unless the new storage form introduces instability)
3. personnel: initials of the person who received the chemical or made the preparation
4. hazards: major hazardous characteristics present (eg, health, flammable, physical), route of entry, target organs affected; GHS pictograms/signal words are preferred
5. special: any unique instructions or cautions (eg, refrigerator storage, keep from light)

Per OSHA, labels on secondary chemical containers must be as effective as GHS labels in communicating the hazard(s) present. National Fire Protection Association (NFPA) and American Coatings Association Hazardous Materials Identification System (HMIS) systems (discussed below) are permissible, but their hazard ratings should be consistent with new GHS safety data sheets (SDSs). The only exception to secondary labeling is short term situations, when workers take small amounts of chemical for use during a work shift and keep the container in their possession. Although OSHA may permit this exception, humans are imperfect and best practice is to label everything.

f3.3 & f3.4 illustrate 2 other common chemical labeling systems, the National Fire Protection Association (NFPA) diamond label and the label for the American Coatings Association Hazardous Materials Identification System (HMIS). Numbers placed in the red, blue, and yellow squares of the NFPA label indicate the seriousness of the hazard according to the key in **t3.1**. The health and flammability hazard categories for HMIS labels align with NFPA, but HMIS

 ISBN 978-089189-6463

t3.1 Key for NFPA diamond

Health hazard (blue)	flammability hazard (red)
4. Deadly	4. Flash point below 73°F
3. Extremely dangerous	3. Flash point below 100°F
2. Hazardous	2. Flash point below 200°F
1. Slightly hazardous	1. Flash point above 200°F
0. Not harmful under normal circumstances	0. Will not burn
Special hazard (white)	**Instability hazard (yellow)**
Oxidizer: OX	4. May detonate
Use no water: W	3. Explosive when exposed to heat, shock or water
Simple asphyxiant gas: SA	2. Capable of violent chemical change
	1. Unstable if heated
	0. Stable

uses a physical hazard category rather than an instability category. Also, on an HMIS label, the white bar is used to specify what personal protective equipment should be used, unlike the white area on the NFPA, which identifies "special" hazards; this is because HMIS information is intended for workers using the material, not for emergency responders. Thus, ratings in the 2 systems are similar but not identical. Either one is usually appropriate as long as it does not contradict GHS hazard ratings. This requires careful management and education. For example, because NFPA labels were developed primarily for emergency responders, the classification of health hazards emphasizes chemicals that are acutely hazardous, while GHS and HMIS health classifications also include assessment of chronic toxicity. NFPA and HMIS both use a scale of 0-4 for hazard rating, with higher numbers indicating greater hazard. GHS divides hazards into categories, putting the highest risk substances into category 1; essentially it uses a numbering system that is the reverse of both HMIS and NFPA. Further, there is no "0" category in GHS. It is crucial not to confuse a category rating on a new SDS with the HMIS and NFPA scales.

HMIS and NFPA, as well as many manufacturers, employ icons, nonverbal pictures, or diagrams in conjunction with written instructions to communicate hazards. Visual symbols allow the hazard to be rapidly interpreted, even by nontechnical personnel. It is essential, however, that staff be thoroughly trained in any system if it is to be the exclusive means of chemical hazard communication. OSHA recommends that only 1 system be used at a time to avoid confusion. As a practical matter, before the June 2016 deadline, laboratories should already begin converting all labels to the GHS system and training employees in the new system. This assures compliance with OSHA regulations but also reduces potential confusion from multiple nomenclatures.

Many medical laboratories provide patients with specimen collection containers that contain preservatives. A common example is plastic jugs to collect 24 hour urine specimens. These often contain acid or another chemical to preserve specimen integrity over such

 ISBN 978-089189-6463

a prolonged period. It is important that collection containers be clearly labeled as to the precautions needed for the hazards within. The information must be written at a language level that can easily be understood by people without a laboratory background; therefore, extra detail may be required. It may even be necessary to provide labeling in an alternate language and/or with pictures to insure that the recipients of these specimen containers do not come to harm. The sole use of GHS, NFPA, or HMIS symbols is inappropriate.

Chemical waste must be as meticulously labeled as unused chemical reagents. Many licensed waste handlers will not remove unidentified waste, and laboratories may face trouble and expense identifying unknown chemicals in waste containers.

Storage & inventory

The basic space requirements of chemical storage areas and laboratories are as follows:

1. clear signs on entrance doors/storage units
2. sufficient ventilation (at least 6-12 air changes per hour)
3. good temperature control
4. fire resistant walls, doors, and/or storage units
5. safety equipment for fire/chemical spills
6. sparkproof electrical equipment minimal electrical sources
7. adequate space/signage for the amount of stored chemicals
8. flooring, work surfaces,storage units resistant to chemicals appropriate to their intended purpose
9. security appropriate for amount/types of chemicals present

Airflow in laboratories should be in a "clean to dirty" direction so that surrounding space is not contaminated. Ideally 100% of the "dirty" air should be exhausted to the outside and not recirculated. It is important that negative air pressure be maintained. Negative air pressure means that air is constantly being drawn out, so chemical fumes can be cleared. Positive air pressure would retain/contain substances.

To maintain negative air pressure, the design of many air handling systems require that laboratory doors remain closed when not in use. Self locking or self closing doors that are secured and locked when labs are unoccupied are ideal. Many doors are fire resistant, so keeping them closed reduces fire hazard as well. Heating and cooling systems must be designed to maintain the airflow/pressure necessary without compromising a comfortable temperature. Staff should not need to use separate fans, which will disrupt the airflow design.

Most porous surfaces cannot be adequately decontaminated. Therefore all surfaces possible, particularly countertops and flooring, must be impervious to chemicals/fluids. Carpets are not permitted, and floors should have as few seams as possible, with covered edges. To avoid environmental release, floor drains are not recommended in chemical storage areas. High risk chemicals may require additional protection in storage. The best design, however, cannot replace good housekeeping, inventory management, and organization in the storage area.

Chemicals should be stored in groups of similar reactivities and compatibilities. It may be easier to locate chemicals arranged in alphabetical order, but such an arrangement is potentially hazardous if 2 incompatible chemicals are in close proximity. Important chemical

3

t3.2 Important chemical incompatibilities

Don't mix this	With this
acetic acid	chromic acid, sulfuric acid, nitric acid, hydroxyl containing compounds, ethylene glycol, perchloric acid, peroxide, permanganates, xylene
acetone	concentrated nitric and sulfuric acids, amines, oxidizers
alkali metals: calcium, potassium and sodium, cesium, lithium	water, carbon dioxide, carbon tetrachloride, and other chlorinated hydrocarbons
ammonia (anhydrous)	acids, aldehydes, amides, metals, nitrates, oxidizers, mercury, bleach, calcium hypochlorite, halogens sulfur and hydrogen fluoride
aniline	nitric acid and hydrogen peroxide
azides	acids, heavy metals (especially copper and lead), oxidizers
carbon (activated)	calcium hypochlorite, all oxidizing agents
chromic acid	acetic acid, naphthalene, camphor, alcohol, glycerol, turpentine, and other flammable liquids
cyanide compounds	acids and alkalis
flammable liquids	ammonium nitrate, chromic acid, hydrogen peroxide, nitric acid, sodium peroxide, halogens, or oxidizing agents
formaldehyde	hydrochloric acid, bleach, and other chlorinated compounds
halogens: fluorine, chlorine, bromine, and iodine	ammonium, acetylene, hydrocarbons, finely divided metals
hydrocarbons (benzene, butane, propane, gasoline, turpentine)	halogens, acids, bases, oxidizers
hydrogen peroxide	most metals and their salts, any flammable liquid, aniline, nitromethane
hypochlorites (bleach)	ammonia, acids, activated carbon, formaldehyde, aminoethylcarbazole, diaminobenzidine
iodine	acetylene, acetaldehyde, ammonia
mercury	acetylene, fulminic acid, oxidizers, ammonia
nitric acid	acetic acid, aniline, chromic acid, hydrocyanic acid, carbon, hydrogen sulfide, and substances that are readily nitrated

incompatibilities are listed in **t3.2**. Ideally, incompatible chemicals are housed in different rooms. If this is not possible, they are at least widely separated in the same room, and if inside cabinets, kept behind closed doors. Obviously, water reactive chemicals should not be stored under automatic sprinklers, sinks, or other sources of water.

Chemical stocks should be kept to the minimum amount possible to maintain efficient laboratory operations. Bulk chemical buying may be cost effective, but poses additional safety issues. It may also cause additional waste if chemicals are not used before the expiration date. A separate chemical storeroom, preparation area, and dedicated cabinets and refrigerators are ideal, allowing only the minimum amount of working chemicals to be in the laboratory at any one time. Chemicals sufficient for immediate work can be stored in operating fume hoods or in designated safe storage within the laboratory. Even when chemicals are not water reactive, storage under a sink is a very poor choice.

 ISBN 978-089189-6463

At the very least, a water leak could spread one or more chemicals into the lab. Some accrediting organizations categorically forbid storage under the sink.

All stock chemicals should have a date of receipt on the label. A "first in, first out" system of inventory management should be in place for safety, quality, and expired chemical waste reduction. Rotating the oldest stock to the front and placing new shipments in the back takes extra effort, but all staff must be trained and monitored for compliance with this policy. In many institutions, supply management is the responsibility of nontechnical staff, who may not be aware of best chemical practices, so these employees must be included in relevant training.

Large containers and hazardous corrosives must be stored as close to floor level as possible. Nothing dangerous should be on a high shelf because it could fall on someone. Supplies should not be stacked too high because of a tipping hazard, and a space of ~24 inches between the ceiling and the highest item is recommended. If supplies are stored on high shelves, nonskid ladders or step stools should be provided. Ideally, no chemicals should be stored above shoulder height. Nothing should be stored close to a heat or electrical source. Whenever possible, particularly for large quantities, storage should be in metal safety cans or plastic coated glass. (There may be a surcharge for coated safety glass, but it is usually nominal relative to the extra margin of safety it provides.) It is also advisable that large containers be stored inside a deep pan or other secondary containment whenever possible.

Shelving and storage units of nonreactive materials specially designed for chemicals are necessary. Storage of liquids in trays or on shelves with raised edging prevents bottles from falling off

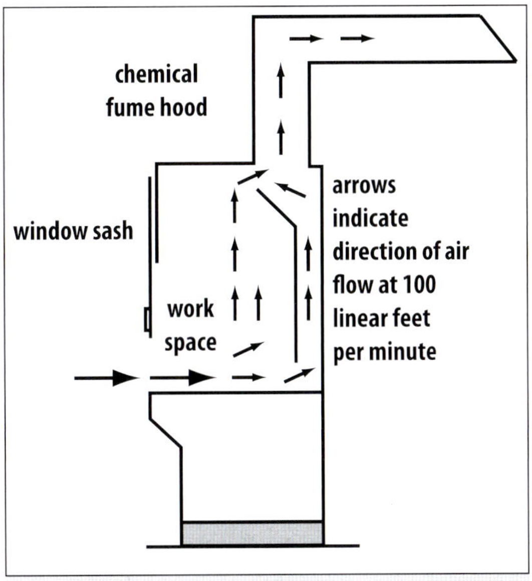

f3.5 chemical fume hood schematic

and helps to contain liquid in the event of a spill. Units made of chemically impervious materials should not be lined with paper towels or other materials. Ideally, shelves should be relatively shallow to prevent of accumulation of chemicals at the back and to allow easy viewing of inventory. Storage should only be 2-3 bottles deep on shelves. Incompatible chemicals should not be stored on the same shelving unit as a spill on a higher shelf could easily leak to a lower shelf. Since spills travel downward and occur most often when chemicals are being moved, the least used chemical should be stored on a higher shelf than the most used chemical. Specially vented safety cabinets are useful for storage of volatile liquids because they disperse dangerous fumes.

In the laboratory, metal safety cans and fume hoods can be used for the storage of working amounts of volatile chemicals. If hoods are used, they must be kept on at all times. As shown in **f3.5**, a chemical fume hood is enclosed on all sides with a window on one side for reaching in and working with the chemicals. It has an exhaust system that draws air up from the

 ISBN 978-089189-6463

work surface and out the top of the hood, away from the worker and the laboratory. Fumes must be exhausted to the exterior of a building at a sufficient distance from the air intake. A special device is required to measure airflow in a chemical fume hood, and ideally a velocity sensor with an alarm is incorporated into a fume hood to provide continuous monitoring. In the absence of a sensor, however, a small, light object such as a piece of yarn can be taped to a spot in the hood, allowing the operator to visually determine whether air is flowing. At least once per year, fume hoods must be maintained and inspected by qualified individuals to verify that airflow is sufficient (face velocity of ~100 linear feet per minute). Hoods should bear a label indicating the last date inspected, and all maintenance/inspection records must be retained onsite.

To provide the correct amount of airflow, manufacturers generally require that staff keep the window shield on the fume hood pulled down as far as possible (~12-18 inches) so that just the hands can enter the hood. Airflow is restricted if the hood is overloaded with objects, so supplies should be kept more than ~8 inches from the hood face. Sometimes laboratory staff paint a bright line inside the hood, so that users know how far back to keep objects. Conversely, objects should never be so far inside the hood that workers must put their heads inside to reach them. Air turbulence (such as foot traffic) at the hood face can reduce efficiency, so hoods should be placed out of high traffic areas and ventilation points to draw air freely. Workers cannot be crowded into too little hood space. With too many people at the opening, the hood draws poorly, and it creates the hazard of workers bumping into one another. ~2.5 linear feet of hood space should be allocated per worker. Finally, since chemical fume hoods and biological safety cabinets are designed differently and cannot be used interchangeably, some laboratories need to have both. Canopy hoods placed over work areas are unacceptable because they are only for use with nonhazardous materials such as water vapor. Backdraft hoods placed above the backsplash of a workbench are acceptable in some applications, particularly with chemicals that are heavier than air, as described later regarding formaldehyde and xylene.

The hazard of storing flammable chemicals in refrigerators was discussed in Unit 2. Because vapors cannot escape a closed refrigerator, any ignition source introduced upon opening the refrigerator (such as a light coming on) can cause a fire. Refrigeration of other chemicals that give off vapors may also be a poor idea, especially if a vapor is toxic. A person opening a refrigerator with toxic vapors inside is likely to inhale a substantial amount. Chemicals that give off vapors should be tightly sealed at all times, but especially in a refrigerator. Screwtop caps are preferred because foil, plastic wrap, corks, and glass stoppers are more likely not to make tight seals. Food and drink should never be stored in a laboratory refrigerator.

Trash or unpacked shipments of laboratory supplies must not be allowed to accumulate in the aisles of a storage area. All persons who use the storage area are responsible for good housekeeping in that area. Nonlaboratory personnel who enter the storeroom need to follow the same restrictions as laboratory personnel with regard to not smoking, eating, drinking, etc.

Only authorized personnel should be allowed in a laboratory, and safety officers should assess the level of security required

 ISBN 978-089189-6463

for their particular facility, including areas of chemical storage. Certain chemicals, drugs, biologicals, and equipment may be targets for theft by terrorists, drug dealers, and other criminals, so it may be necessary to lock storage facilities as well as limit laboratory access. The Department of Homeland Security (DHS) has published "Chemical Facility Anti-Terrorism Standards" and a list titled "Chemicals of Interest." This list specifies the amount of a particular chemical that triggers required compliance with DHS standards. Medical laboratories generally do not house the amounts of chemical that would qualify them to follow these standards; they do, however, use smaller amounts of some of the chemicals listed. Safety officers should verify the status of their particular laboratory at the DHS website: www.dhs.gov/xlibrary/assets/chemsec_appendixa-chemicalofinterestlist.pdf.

Safety data sheets

A key provision of the OSHA Hazard Communication Standard is the use of material safety data sheets (MSDSs), which provide users with comprehensive information about any hazardous chemical they encounter in the workplace. As part of the GHS project, this document was renamed safety data sheet (SDS). Manufacturers are required by OSHA to provide an SDS for every hazardous chemical they produce. Although OSHA did suggest standards for organization and format prior to the GHS, the only mandate was what information to provide. This made it more difficult for users to quickly find the information they sought, as each manufacturer had its own way of organizing SDSs. The GHS format for SDSs has the 16 mandatory categories shown in **t3.3**, which must be presented in the order shown, using GHS nomenclature. This format has emergency information closer to the front of the SDS, and features such as paper size, color, and margins are specified so that SDSs will look more alike

t3.3 Components of a GHS compliant safety data sheet (SDS)

1.	Complete identification of the substance or mixture; complete identification of the supplier, eg, name, address, emergency phone number
2.	Hazards identification: GHS classification, including mandatory label elements and precautionary statements, other hazard information
3.	Composition/information on ingredients: complete composition and all relevant names and identifiers (eg, common names, CAS number)
4.	First aid measures: relevant treatment for exposure for first responders
5.	Firefighting measures: relevant information for user and professional fire suppression, including particular hazards in fire for this chemical
6.	Accidental release measures: actions to be taken during spills
7.	Handling and storage: requirements for temperature, ventilation, etc
8.	Exposure controls/personal protection: engineering controls and relevant protective equipment required
9.	Physical and chemical properties: eg, appearance, odor threshold, pH, flash point, solubility, decomposition temperature
10.	Stability and reactivity: inherent stability, chemicals/conditions to avoid
11.	Toxicological information: eg, thresholds of acute and chronic toxicity, expected signs/symptoms of exposure
12.	Ecological information: eg, harm to the environment, persistence in the environment
13.	Disposal considerations: safe handling and methods
14.	Transport information: eg, transport hazards, UN specifications
15.	Regulatory information: relevant to the country for which the SDS was produced
16.	Other information including information on preparation and revision of the SDS

and be easier to use. Appendix 3.1 contains a sample SDS in a GHS compliant format.

The essential function of the SDS is to communicate the hazards associated with each chemical. A chemical mixture requires an SDS if it contains >1% of a hazardous chemical or >0.1% of a carcinogen. To meet OSHA regulations, only chemicals that are classified as hazardous need to be inventoried and have appropriate SDSs maintained for them. Manufacturers do not have to identify nonhazardous materials on the SDSs, and many will not, especially if they consider it proprietary to their product. Some products composed entirely of chemicals not classified by OSHA as hazardous may not have a published SDS (for example, urine dipsticks). In such cases, it is advisable to ask for a letter from the manufacturer stating why no SDS is available and to keep it with the SDSs to document that the absence of the SDS is deliberate and not due to negligence.

OSHA also requires SDSs for consumer products with which workers come into contact more often than a consumer would. For example, although consumers buy bleach, they probably do not use it as frequently as laboratory employees, who decontaminate countertops with it daily, and so are at risk of a hazard from prolonged exposure. Thus, even though they are in general use by the public, items such as bleach, disinfecting hand soap, correction fluid, and copy machine toner could require SDSs depending on how frequently they are used in the laboratory.

Because laboratory procedures change over time, SDSs for particular chemicals may no longer be necessary. Obsolete SDSs should not be discarded, especially if the chemical was involved in an exposure. Old SDSs should be archived for 30 years in case of chemical exposure claims. However, if the manufacturer updates the SDS, the new SDS should be retained, and the old, outdated SDS discarded, unless the chemical formulation has changed or the chemical has been involved in an incident.

Employees must, at all times, have ready access to SDSs for the chemicals they use. Maintaining an accurate chemical inventory with current SDSs can be a daunting task, especially in large facilities. OSHA does not specify the medium in which the SDSs must be stored, so many sites are replacing bulky paper binders of SDSs with electronic files accessed within a computer network. Also, in the event of an emergency, the laboratory may have to be evacuated, and the SDSs may be difficult for hazardous materials teams to access unless an alternative means has been identified. Increasingly, facilities are relying on electronic SDS databases or phone services that can be accessed by from anywhere. There are many advantages of an electronic system, but there is also the risk of data loss and inaccessibility of information in the event of a power failure or maintenance downtime. A backup plan must be in place, which can be as simple as storing the SDSs on a battery operated laptop computer. Also, SDS phone services provide instant access to the SDS for any chemical by phone 24 hours a day.

Although SDSs are valuable, there are 2 major problems with relying solely on SDSs as the primary tool for communication and policy formulation. Most SDSs contain rather detailed technical information that may be difficult to interpret, particularly by staff with minimal chemical expertise. Secondly, SDSs are often written only for 100% pure chemicals. Thus, the SDS for concentrated hydrochloric acid vastly overstates the hazard of a 5% solution used in the laboratory. While it is essential that SDSs be available to all staff, it is appropriate for safety officers to interpret

 ISBN 978-089189-6463

CHEMICAL HANDLING, STORAGE & DISPOSAL SHEET

SDS: Attached

NAME OF PRODUCT: HDL cholesterol standard

PURPOSE OF USE: Calibrating HDL cholesterol procedures

ACTIVE INGREDIENTS & CONCENTRATIONS WHERE SIGNIFICANT: 50 mg/dL cholesterol in 15% isopropanol, 1% sodium azide

POTENTIAL AS A HAZARDOUS MATERIAL: Alcohol is flammable. Azides cannot come in contact with metals

DISTRIBUTOR: *(manufacturer's name, address)*

METHOD OF STORAGE OR SPECIAL HANDLING WHERE NECESSARY: Keep away from heat and light

METHOD OF DISPOSAL: May be flushed into sanitary sewer as long as pipes are not metal

f3.6 sample MSDS cover sheet

the information in the SDSs to employees for clarity and for applicability to the concentration of chemicals in use. A "cover sheet" for each SDS, as illustrated in **f3.6**, is also useful in summarizing the pertinent data and allowing employees to rapidly access institutional policy for a chemical. Another strategy is to include clear safety instructions in the standard operating procedures for a particular chemical. This may also be addressed in the chemical hygiene plan (CHP). Regardless of the means of communication, every employee must understand what to do with a given chemical.

Hazardous characteristics

Corrosives

Corrosives are chemicals such as acids (low pH, especially <2.5) and bases/alkaline compounds (high pH, especially >12.5). Sometimes they are referred to as caustics. Corrosives cause destruction and/or irreversible alterations in other compounds, especially metals and metallics, as well as human and animal tissues. Fumes emitted by these compounds can also be corrosive. Certain dehydrating agents and oxidizers may also have corrosive properties, even though they are not acids or bases. For example, bleach can be corrosive to metal.

Common laboratory acids include acetic, sulfuric, nitric, trichloroacetic, perchloric, and hydrochloric acids. Common bases are ammonium hydroxide, potassium hydroxide, and sodium hydroxide. These chemicals can cause severe, irreversible injury to eyes and skin as well as the respiratory and gastrointestinal tracts when inhaled or ingested. When a corrosive agent comes into contact with the eye or skin, it must be flushed away immediately and thoroughly for ~15 minutes using a safety shower or eyewash. Water washing can be supplemented with special buffers prepared for this purpose. When acids and bases are mixed together, they generate heat. Therefore, strong neutralizing agents *must not* be used as they produce a heat reaction and worsen any burn. Immediate medical attention is always required.

Acids are compounds that produce hydrogen ions (H+). Skin exposed to acid gets "burned" and scarred. Accidental ingestion is revealed by a sour or bitter taste, similar to vomit (which contains hydrochloric acid). When mixed with water, strong acids generate heat, so treatment of acid burns with water requires massive, uninterrupted amounts of water. Hydrofluoric acid is particularly treacherous because of the delayed appearance of symptoms after an exposure, so it is important to flush even asymptomatic exposures to this acid.

Although all acids have the properties just described, they are not all the same. 2 important categories are oxidizing and nonoxidizing acids. In oxidizing acids, the anion of the acid is a stronger oxidizing agent than H+,;examples include sulfuric, nitric, and perchloric acids. In nonoxidizing

3

acids, the anion of the acid is a weaker oxidizing agent than H+; examples of these include hydrofluoric, hydrochloric, phosphoric, and acetic acids. Most organic acids are nonoxidizing. The distinction is important because some acids react violently with organics (perchloric, nitric), and some acids are incompatible with each other. Acetic acid and acetic anhydride are common examples of nonoxidizing acids that must be stored separately from oxidizing acids.

Bases, or alkaline compounds, produce hydroxyl ions (OH–). Skin exposed to bases gets emulsified and dissolves, becoming a thick, sticky liquid. This is why degreasers and drain openers are often strongly alkaline. Like acids, bases can cause exothermic reactions when mixed with water, so extensive flushing of exposed skin with water is similarly required. A unique hazard of the base ammonium hydroxide is its potential when mixed with other compounds (eg, acids) to give off toxic ammonia.

Because acids and bases react strongly with each other, they must be stored separately. Corrosives, especially oxidizing acids, can also react with flammable compounds so separate storage is required. Acids and bases that are in solid form should not be stored with other dry chemicals but rather should be included in storage with other corrosives.

When working with corrosives, safety glasses must be worn and chemicals manipulated under a hood and near/in a sink. Large amounts of corrosive liquids or solids must not be added rapidly to a mixture because of increased likelihood of splashing and generation of heat. Automatic pipetting devices should be used to add corrosive chemicals to water; *never* add water to the corrosive ("acid to water, not water to acid"). Slow pipetting of the corrosive down the side of the container so it flows into the water minimizes splashing.

Ignitables

Ignitable liquids are characterized by their flash points. As stated in Unit 2, a liquid's flash point is the minimum temperature at which it vaporizes sufficiently to produce an ignitable mixture with air near the surface of the liquid. Chemicals with flash points <100°F (38°C) are called flammables or Class I solvents, and those with flash points between 100°F (38°C) and 140°F (60°C) are called combustibles or Class II solvents. Class I solvents, especially those with flash points below room temperature, are extremely hazardous and must be used under fume hoods and stored under vented conditions away from heat. Class II chemicals are less hazardous unless they are heated. Gases and solids may also be flammable and are handled similarly to flammable liquids. Chemicals labeled as pyrophoric may spontaneously combust when exposed to air at <130°F, and must be stored and handled carefully. Additional details are provided in Unit 2.

Health hazards

There are several categories of chemical health hazards, examples of which are listed in **t3.4**. Health effects can be rapid and acute, as from a large spill. If an SDS lists an STEL (short term exposure limit) or a CV (ceiling value), a chemical can be acutely hazardous. Health effects can also be chronic from exposure to small amounts of chemical on a daily basis over years. The SDS will list other permissible exposure limits (PELs) in this case. OSHA has many requirements for handling chemical health hazards, but specifically targets carcinogens for special precautions. Lists of carcinogens can be found by

 ISBN 978-089189-6463

t3.4 Examples of chemical health hazards

Hazard	Definition
poison	is toxic to humans in some capacity
hepatotoxin	can damage the liver
nephrotoxin	can damage the kidney
neurotoxin	can damage the nervous system
irritant	can be aggravating to body part in contact with chemical, but usually not toxic
sensitizer	induces allergic reaction
mutagen	can induce genetic mutation/DNA changes
teratogen	can cause birth defect in fetus
carcinogen	can cause cancer

consulting lists from OSHA, the National Toxicology Program, and the International Agency for Research on Cancer. Chemicals on any of these 3 lists must be handled as a carcinogen, and many antineoplastic drugs found in clinical settings fall into this category. OSHA also publishes a list of hazardous chemicals that must be part of a medical surveillance program when they are used in the workplace.

All chemical health hazards cannot be covered in this text or in employee training programs. In addition to providing SDSs to workers, OSHA requires that a "target organ poster" be displayed so that workers know what signs and symptoms may be associated with chemical toxicities. A modified sample from the OSHA website is contained in Appendix 3.2. Toxic chemicals likely to be encountered in clinical laboratories are discussed in more detail below.

For certain chemicals, OSHA establishes PELs, ambient levels of chemicals that cannot be exceeded. To perform environmental monitoring, it is necessary for personnel to wear sampling devices during their work with the chemical in question. Devices are usually worn high on the collar, as close to the air being breathed as possible, when high exposures are expected. Measurement should be performed on the busiest workdays and/or during the tasks that generate the most exposure. The amount of exposure is then determined from the sample collected, and it is compared to the PEL established by OSHA. By law, employers must disclose the results of environmental monitoring to employees. Environmental monitoring is required at least once per year or if anything changes, and OSHA requires that records be retained for 30 years. Adequate or specialized ventilation (as in autopsy or histology suites), and respiratory protection and/or containment (such as chemical fume hoods) must be provided to ensure that employee exposure is below an established PEL. At this writing, OSHA is still enforcing PELs that were established decades ago, and it is in the process of updating them. In the interim, OSHA suggests that employers work toward complying with PELs from other sources. These other sources based their PELs on more modern data and can be found at this website: www.osha.gov/dsg/annotated-pels/index.html. It is likely that new OSHA PELs will be in substantial agreement with these other sources, so implementation of additional safety measures now will not be a wasted effort.

Antineoplastic drugs

Chemotherapeutic drugs for cancer treatment are very dangerous, and special protection is required for those administering these drugs. Exposure in the testing laboratory should only be through contact with specimens from these patients. When the drugs are administered, they distribute throughout the patient's body and excreta, and are effectively diluted. As of this writing, the normal protective protocols for biohazards that might be transmitted from human specimens are sufficient to protect laboratorians against the hazards of the small amounts of chemotherapeutic agents expected. These specimens may be disposed of using normal protocols. However, laboratorians involved in the manipulation of these drugs should refer to the Clinical and Laboratory Standards Institute (CLSI) document GP17-A3 for the additional protection required, including gloves known to be impervious to the agents being handled, a hood or other means of respiratory protection, and segregation of these agents from other laboratory areas.

Nitrogen

Nitrogen gas has been called "asphyxiating gas" because when released it reduces oxygen concentration in the air and creates hazardous hypoxic conditions. (This can occur with other gases too.) It is critical, when working with gas cylinders of any type, that the valves are kept tightly closed when not in use and that all fittings be checked periodically for leaks. Since nitrogen is a major component of room air, dangerous levels of nitrogen will not be smelled or detected, so ventilation and evacuation are urgent if a significant leak is suspected.

Mercury

Elemental mercury is extremely poisonous. Mercury can be absorbed through the skin, by inhalation, and by ingestion. Its effects are cumulative, and chronic exposure to mercury can lead to serious toxic effects. Mercury spills of even small amounts are very dangerous because mercury has a very low vapor pressure and can contaminate an entire room if it has poor ventilation. Evacuation or respiratory protection is essential in the event of a significant spill of mercury. Ordinary vacuum cleaners can disperse mercury into fine airborne droplets that are not filtered by vacuum bags, so special cleanup procedures are required. Commercial mercury spill kits contain an absorbent powder that can convert mercury into a harmless amalgam in crevices, and they can be posted in convenient laboratory areas. Enclosed suction devices or vacuums specially adapted for mercury cleaning may also be used. Disposal of mercury and its compounds is highly regulated, and it is *never* acceptable to flush mercury down the drain or place it in ordinary trash.

Because of its low vapor pressure, mercury should only be used under a hood. Clinical laboratories rarely need to manipulate mercury, and the most frequent source is thermometers. The use of organic liquid thermometers circumvents the need for mercury handling procedures, and it is recommended that whenever possible mercury thermometers or other devices with mercury be replaced. Many sites ban mercury use.

An additional source of mercury is fluorescent light bulbs. Not all fluorescent light bulbs contain mercury, but those that do may be regulated as a hazardous waste. The sale of objects containing mercury is now banned in some states, and it is important to consult local waste regulations before disposing of old devices that still contain mercury.

Chemicals in histology/autopsy suites

Paraffin, formaldehyde & xylene

Histology laboratories use various chemicals and instruments to process, fix, and stain tissue. Paraffin is not particularly dangerous, but if cleaning protocols are poor, paraffin can build up on the floor and cause a slipping hazard. Equipment such as tissue processors must have a clearance of at least 5 feet from the paraffin dispenser and from the flammable chemicals in the histology

 ISBN 978-089189-6463

section, unless separated by 1 hour fire resistant construction. Many chemicals in histology are toxic and/or flammable, and should be handled with great care. Proper airflow and air exchanges are especially important to maintain, so workers in these laboratories must be conscientious about keeping doors closed, vents unblocked, hoods on, and so forth, in accordance with how the laboratory was designed. A comprehensive discussion of the hazards of the many chemicals used in histology is beyond the scope of this text, and the reader is referred to other sources. 2 chemicals of particular importance, formaldehyde and xylene, are discussed below.

Formaldehyde

Formaldehyde is a combustible, a carcinogen, a strong chemical irritant, and a sensitizer in some people, causing allergic reactions. The OSHA Formaldehyde Standard (29 CFR 1910.1048) outlines specifications for handling this chemical, including monitoring of environmental levels. According to OSHA, staff who are exposed to environmental levels that exceed 0.1 parts per million (ppm) must have annual training in formaldehyde handling. This will often include people outside the lab who are obtaining specimens and preserving them in containers with formaldehyde (eg, patient care areas, surgical suites, specimen couriers).

To fix and preserve tissue, medical laboratories frequently use formalin, a water/methanol/formaldehyde (37%-50%) mixture, and formalin is covered by this OSHA standard. Formalin is not easily combustible, but formaldehyde gas can ignite when air concentrations exceed 7%, so ventilation is crucial. When mixed with hydrochloric acid, bleach, and other chlorine containing compounds, formaldehyde can form bis-chloromethyl ether, an OSHA regulated carcinogen, so it is important to keep these compounds separate.

There are many work practices that can minimize formaldehyde exposure. How specimens are removed from formalin, containers are kept sealed, fixed specimens are washed in water before cutting, cassette buckets are kept cool, and personal protective equipment is deployed all contribute to minimizing staff exposure. Formaldehyde is a useful fixative; once it has penetrated tissues and is bound to them, it is no longer dangerous, and organisms in the specimen are killed. Work practices that eliminate the chemical from specimens once it has done its job should be given priority. That said, waste formaldehyde is dangerous to all living organisms, and proper removal and disposal is crucial.

OSHA has established PELs for formaldehyde. At least 6-12 air exchanges per hour of 100% fresh air (not recirculated air) are necessary to keep formaldehyde exposure sufficiently low. CLSI suggests that >15 air exchanges per hour must be required in morgues and anatomic pathology areas because of the heavy chemical burden in those areas. Humans smell formaldehyde at ~ 0.8-1.0 ppm. The OSHA 8 hour time weighted average PEL is 0.75 ppm, and the 15 minute STEL is 2 ppm. As a convenient rule of thumb, if staff members can smell the formaldehyde, the area may be over the PEL.

Formaldehyde PELs may readily be exceeded if the chemical is spilled. Evacuation and/or ventilation to avoid breathing the chemical are important. Respirators with formaldehyde cartridges are ideal, of course, but formaldehyde is highly soluble in water. Humidifying the room (if possible), and a wet cloth over the mouth and nose during evacuation

3

minimizes formaldehyde inhalation quickly and simply for everyone. The first sign of excess formaldehyde inhalation is usually a sore throat, so this is an indication to seek care. Accidental ingestion may be lethal, and it is urgent that exposed individuals obtain medical treatment.

Commercial formaldehyde spill kits are available with materials that neutralize formaldehyde as well as respirator cartridges designed to block formaldehyde. In some cases, if formaldehyde is neutralized and its pH brought to neutral, formaldehyde waste can go down the drain, but this practice must be verified against local regulations. It is important that staff members be trained in spill containment and waste disposal for this chemical, and that the appropriate materials and protective equipment are provided anywhere the chemical is used and stored. Again, this may include areas outside of the laboratory, so it is important to provide protection in all areas of a site. CLSI document GP17-A3 has a very useful set of calculations in Appendix E, which permit safety managers to evaluate formaldehyde exposures and ventilation devices in their own particular situations.

Xylene

Xylene has a low flash point and is toxic to various organ systems. Its PEL is 100 ppm, but humans have the ability to smell xylene at 1 ppm. If staff members can consistently smell xylene in the workplace, environmental monitoring may be necessary. In some applications, limonene can be substituted for xylene. Limonene is a skin and respiratory irritant and can be a sensitizer, but it is significantly safer than xylene.

Formalin and xylene are both heavier than air, so it is recommended that ventilation be used at the back of countertops or at floor level. Ceiling vents for air circulation and/or updraft chemical fume hoods may not be as effective in keeping the PELs below acceptable limits. A common activity that causes formaldehyde levels to exceed PELs is pouring off liquid from specimens before their disposal. Appropriate ventilation adjacent to specimen disposal can easily rectify this problem.

Glutaraldehyde

A limited number of high level disinfectants are approved by the Food and Drug Administration (FDA; www.fda.gov/MedicalDevices/DeviceRegulationandGuidance/ReprocessingofReusableMedicalDevices/ucm437347.htm, accessed August 30, 2015). All high level disinfectants that can also function as sterilants contain hydrogen peroxide, peracetic acid, phenol/phenate, isopropanol, and/or glutaraldehyde. These chemicals are important for sterilizing medical equipment that cannot be sterilized in another way (such as with heat). However, because hydrogen peroxide and peracetic acid are somewhat corrosive, glutaraldehyde is far more widely used. In addition, glutaraldehyde can be used as a tissue fixative in histology or for film development in radiology departments. Unfortunately, glutaraldehyde is an acute irritant, a skin sensitizer, and a cause of occupational asthma. Acceptable levels for glutaraldehyde are recommended, and environmental monitoring may be needed to document exposure. A specific OSHA standard and PELs for glutaraldehyde do not exist at the time of publication. Because of the problems associated with glutaraldehyde, many manufacturers make high level disinfection products that contain glutaraldehyde but with another chemical added so that the glutaraldehyde concentration can be lower. The FDA has approved a few glutaraldehyde free

ISBN 978-089189-6463

products for high level disinfection that are listed on its website.

As with its chemical cousin formaldehyde, if a worker can smell glutaraldehyde, the ambient levels could be too high. If glutaraldehyde products are not perfumed, most humans can detect it in the air at concentrations of 0.04 ppm; the lowest recommended exposure levels are 0.05 ppm. Occupational asthma is the most serious problem associated with glutaraldehyde. Therefore, good ventilation and containment are key to reducing exposure. Air exchange rates as high as 10-15 air changes per hour have been recommended if fume hoods are not used. Common sense measures, such as using tight fitting lids on containers for soaking instruments or fixing tissue, should also be used. For prolonged exposure, gloves known to be resistant to glutaraldehyde should be used. Certain glove materials, including latex, may absorb glutaraldehyde and are not recommended. Chemical neutralizers are available for significant glutaraldehyde spills or for disposal of waste glutaraldehyde when required. They should be placed strategically around laboratory areas in which glutaraldehyde is used.

Hydrogen cyanide

Hydrogen cyanide is one of the most powerful poisons known to humanity. With an odor described as "bitter almonds," it is not detectable until the PEL has been exceeded; up to 60% of the population cannot reliably detect the odor. When working with cyanide compounds, it is best to use gloves to prevent absorption through the skin and work under a hood to prevent inhalation. Protocols for disposal of cyanide compounds should be strictly followed. These compounds should *never* be flushed down sinks with plumbing that interconnects throughout the laboratory. The cyanide compounds could mix with acids or other compounds in the pipes and form hydrogen cyanide gas, which could poison the entire laboratory. Therefore, cyanide compounds are classified as both toxic and reactive waste.

Dry ice/carbon dioxide

The frozen form of carbon dioxide (CO_2) is referred to as dry ice because CO_2 gas emanates directly from the solid without an appreciable liquid phase in between. Dry ice is used to ship items that must be kept cold because the freezing point of CO_2 (−78.5°C) is below that of water (0°C), and unlike ice, it does not produce liquid upon melting.

Dry ice, however, is both a physical and chemical hazard. It is so cold that, when in contact with unprotected skin, dry ice can "burn" or, with prolonged exposure, can result in frostbite. Dry ice burns to the eye can cause permanent damage. Dry ice should always be handled with insulated gloves, and if skin is inadvertently exposed, it should be warmed gently with tepid (not hot) water. Dry ice is also hazardous because as the gas is produced, it can

1. "explode" a fully sealed package (the expansion ratio of CO_2 is ~553:1)
2. reduce the oxygen level in an enclosed area to below breathable levels
3. because it is heavier than air, sink and therefore reduce oxygen levels in lower levels of a structure

CO_2 concentrations of ≥11% in a room can lead to unconsciousness in ~1 minute, with only dizziness, headache, shortness of breath, or weakness as possible warning signs. CO_2 has no taste, color, or odor that would serve as a warning. Proper ventilation is essential when using dry ice.

Transport of fragile specimens is a common use of dry ice. Only minimal amounts should be used (2-5 lbs per CLSI), and containers should not be tightly sealed. Given that CO_2 gas will leak from the containers to prevent explosion, it is important that specimen couriers keep car windows slightly opened to allow for sufficient ventilation.

When dry ice is no longer needed, sink disposal is a poor choice because the ice can be cold enough to damage some plumbing systems. Ideally it should be placed in a ventilated area or a hood to sublimate until it completely turns to gas.

Oxidizers

Oxidizers are materials that contain sufficient oxygen or oxidizing species to react with reducing materials and release energy. This group of chemicals includes oxygen, halogens, peroxides, and other compounds. Oxidizers will react with hydrocarbons and must be kept away from hydrocarbon oils and grease. Only the minimum amount required should be stored in the laboratory under sealed, dark, and cold (not freezing) conditions. Peroxides in the solid form are extremely sensitive to heat and mechanical jarring, so great care must be taken to ensure that refrigerator temperatures are maintained above a particular peroxide's freezing point. Oxidizers should be used under a hood with plastic, wooden, or glass materials, *not* metal. Oxidizers can be disposed of by dilution and the addition of a reducing agent (such as potassium iodide) if the amounts are small, but large amounts require special handling. SDSs for each oxidizer have more detail.

Ethers are flammable solvents with low flash points. During storage, they combine with oxygen to form explosive peroxides. Storing ethers in the refrigerator does not prevent peroxide formation and causes flammable vapors to be accumulated in a closed space. Ethers should be stored under a hood or in a vented cabinet in opaque containers. Each ether forms peroxides at a slightly different rate, but the rate is accelerated when the can is opened and the volume of ether declines in relation to the amount of air. Light also accelerates peroxide formation. Therefore, cans of ether should be dated both upon receipt and upon opening so that they may be discarded after the appropriate amount of time. Per CLSI, if there is concern about peroxide formation, mixing 1-2 g of potassium iodide with 10-15 mL of ether will reveal the presence of peroxides if a red-brown color appears. Materials with this reaction must be discarded immediately.

In general, unopened cans of ether must be discarded after 1 year. Inventory management is critical for ethers so that an old ether can does not get pushed to the back of a group of reagents and go unnoticed. In addition, the smallest cans available should be purchased so that only 1 can of ether is opened at a time and emptied quickly to prevent exposure to air. White deposits inside a can of ether or on the cap may indicate the formation of peroxide. The can should not be disturbed and must be discarded by a professional.

Explosives

Explosives are compounds capable of violent reaction under specific conditions of temperature, mechanical shock, or chemical reaction. They often contain both reducer and oxidizer groups that will self react, many times without external oxygen. The thermal decomposition products of these redox reactions are usually simple gases and vapors (CO, CO_2, N_2, O_2, H_2O), and the combination of heat and gas causes up to a 10,000 fold expansion of the gas and a dangerous shockwave. Nitrogen

 ISBN 978-089189-6463

t3.5 Chemical classes that pose physical hazards as classified by the Global Harmonization System

Flammable gases	Flammable aerosols	Flammable solids	Flammable liquids
oxidizing gases	explosives	pyrophoric solids	pyrophoric liquids
gases under pressure	organic peroxides	oxidizing solids	oxidizing liquids
self reactive substances	self heating substances	substances that, in contact with water, emit flammable gases	corrosive to metals

is a common component, and examples include azides (N_3 group), nitroglycerin, and trinitrotoluene (TNT).

Many other materials also have explosive potential. Most flammables will explode if heated sufficiently. Peroxides formed from ethers are explosive as described earlier, and white crystals around container openings indicate this may have occurred. Crystalline picric acid is also an explosive hazard, so picric acid solutions must be protected from drying out. Visible white crystals around the opening of a picric acid bottle indicate possible formation of explosive metallic picrates. When bottles of ether or picric acid have white crystals, they should not be opened, and hazmat professionals should be called. Perchloric acid explodes when mixed with reducing agents or organic materials, and must not be used on wooden workbenches. Dry perchloric acid is extremely explosive, so it is critical that spills be cleaned up properly. Perchloric acid should only be used or heated in fume hoods designed and dedicated for handling this acid. These hoods include a wash down system, or a local scrubbing or trapping system to prevent salt formation. Perchloric acid fumes can form crystals in ordinary chemical fume hoods and turn a valuable piece of safety equipment into a bomb.

Physical hazards

The GHS classification of physical hazards contains chemicals from every group discussed earlier. However, not every chemical in each class is a potential physical hazard. Some chemicals are natively physical hazards, but others are only physical hazards if handled improperly. **t3.5** presents the GHS chemical classes that pose physical hazards.

Incompatible mixtures

t3.2 (ppage 52) lists common laboratory chemicals that when mixed with other chemicals form hazardous mixtures. The SDS for each chemical should list any major incompatibilities for that chemical and should be consulted before it is stored and used.

Handling & usage

Glass containers must be held firmly around and under their bodies with 2 hands, never around less sturdy areas, such as the "neck." Glass containers (especially those containing >500 mL of fluid, per CLSI) should be carried inside a rubber or plastic bucket to prevent breakage and/or contain a spill. When transporting heavy or multiple containers of chemicals from one area to another, a cart with a large rim should be used. Chemical spills in an elevator are very dangerous, so if chemicals must be moved within a building, a freight elevator should be used.

If a freight elevator is not available, only essential personnel should be in a public elevator when chemicals are being moved. Shipping chemicals between locations requires compliance with GHS, US Postal Service, and/or DOT regulations.

Chemicals should never be tasted or purposely smelled. They should never be handled in nonlaboratory areas. Chemical manipulation should be under a hood, over a sink, or on a nonreactive surface, as appropriate. Pipetting, pouring, and transferring chemicals should be done to minimize splashing, fumes, and/or aerosols, even if the operation is in a fume hood. To the extent possible, high risk activities should be isolated and segregated. Reagent caps or stoppers should *never* be placed on a work surface because caps could contain enough chemical to contaminate the surface and cause harm to personnel who later use the area. In addition, the cap could get dirty and contaminate the entire bottle of chemical. Chemicals should only be opened long enough to access the chemical, and then the container should immediately be closed. All work surfaces must be wiped off both before and after use. A surface's contamination may not be visually evident, and it is better not to assume that the previous person to use the area remembered to clean it. Automatic pipetting devices must *always* be used to transfer chemicals, and the appropriate safety apparel should be worn (see Unit 10).

Disposal

In general, there are 4 components of chemical waste management. Additional information on waste disposal is contained in Unit 8.

1. Obey legal requirements. Consult the SDSregulations from the Environmental Protection Agency (EPA),statelocal authorities.
2. Reduce, reuse,recycle. The least amount of chemical possible should be purchased/used. If possible, it should be used repeatedly. Recycling may be difficult, but it is an option that should be investigated. Good organization/planning minimize chemical waste for any cause, including expiration.
3. Segregate, label,date all waste. Hazardous waste should not be mixed with nonhazardous waste because the entire resulting mixture would be classified as hazardous. The contents/age of every waste container should be clear.
4. Store/dispose of waste properly. The EPA has strict quantity limits for stored waste, so regular removal of waste from a site, by a licensed waste handler if necessary, is important.

Sanitary sewer

Laboratory drains may be interconnected, so toxic, malodorous, or irritating chemicals that produce hazardous vapors should not be flushed. Cyanide salts in particular can form toxic hydrogen cyanide gas when mixed with acids. Cyanide compounds have their own unique disposal procedures, or can be collected separately in a safety can until picked up by licensed waste handlers.

Many laboratory reagents contain sodium azide as a preservative. These reagents should not be put in sinks with copper or lead plumbing as the sodium azide will accumulate over time. Copper and lead eventually form explosive azides with sodium azide, so no reagents containing this preservative should be disposed of in sinks with metal pipes. Modern plumbing systems rarely use metal pipes, so this is more likely to be a problem in older facilities. Sodium azide is also toxic and can form toxic gases if mixed with acids. Generally, the amount of sodium azide

 ISBN 978-089189-6463

used in reagents is small, so this hazard is minimal. However, mixing azides with acid should be avoided, and if a pungent, sharp smell is detected, evacuation and ventilation protocols should be initiated.

Proper disposal of strong corrosive types of chemicals is critical because improper disposal can cause damage to plumbing and can cause unexpected reactions in pipes containing incompatible chemicals from other laboratory areas. Before drain disposal, strong acids and bases must be diluted with large volumes of water and/or neutralized until the pH is >3 and <10. Water should be run for several minutes after disposing of the diluted corrosive. However, this does not mean that any chemical can be sufficiently diluted for sanitary sewer disposal. Some types of plastic pipes are damaged by strong chemicals, so evaluate the exact type of plumbing in a facility before disposing of any chemical down the drain.

Chemicals that are candidates for sink disposal are aqueous, neutral solutions that do not contain heavy metals or organic solvents. Buffers and surfactants are good examples. Most chemicals that may be disposed of in the sanitary sewer should be flushed with generous amounts of cold water between chemicals. Chemicals that can be flushed may still need to be segregated because, as stated before, if laboratory plumbing is interconnected, incompatible chemicals may be mixed.

Once the chemicals are disposed of, it may be necessary to triple rinse the empty containers and deface labeling before final disposal, particularly if the chemical was hazardous.

Licensed waste handlers

Chemicals requiring special disposal should be labeled and segregated from ordinary trash for removal by EPA licensed waste handlers. Storage requirements are essentially the same as those for the pure chemical, but if waste is kept in the same place as the unadulterated chemical, it should be clearly marked so that it is not inadvertently used.

Incineration

Incineration should be performed only by professionals in an EPA approved facility.

Landfill or solid waste disposal facilities

Some chemicals may be placed in the "regular" laboratory trash for disposal in a public landfill, while others must be placed in the EPA's hazardous waste sites. Training personnel is essential to prevent hazardous chemicals from reaching a public landfill.

Personal protective equipment

Clothes should be protected by a lab coat or apron. For some chemicals, particularly corrosives and formaldehyde, fluidproof aprons should be worn. Sandals, open footwear, and cloth shoes must *not* be worn. Goggles, face shields, or safety glasses are often necessary, particularly when using concentrated chemicals. Eyeglasses do not suffice in all situations; fluids can still splash under or over the glasses. In some cases goggles must be worn over corrective glasses. Because chemical fumes can penetrate under contact lenses, in some cases they cannot be worn. Toxic and corrosive chemicals demand fluidproof gloves; sometimes double gloving is required. Insulated gloves are prerequisites for handling hot objects, but best practice is to let objects cool first, if that is possible. Gloves should be changed when a leak or tear develops or if they are grossly contaminated. Best practice is to change disposable gloves

 ISBN 978-089189-6463

every 30-60 minutes because chemicals will weaken them over time. Chemicals can dissolve glove materials, so caution must be used when selecting gloves for a particular task as this phenomenon varies with each chemical and glove material (see Unit 10). Hands must be washed when work is complete, whether gloves were worn or not. Finally, if chemicals with toxic fumes are not maneuvered under a hood, respiratory protection may be in order. This is the least preferred method of protection, and it is crucial that device fit testing and training are done.

Chemical spills

There are no accepted criteria for categorizing spill sizes, but CLSI defines a major spill as "a spill that spreads rapidly, presents an inhalation hazard, endangers people or the environment, and/or involves personal injury or rescue and should be handled as an emergency by the department of public safety, fire department, or hazmat team." Spill response will vary as to the size of the spill and the type of chemical(s) involved. Procedures for known chemicals can be very specific, but labs should also have protocols for spills in which the substance is not known.

The Emergency Preparedness and Community Right-to-Know Act, aka Title III of the Superfund Amendments and Reauthorization Act (SARA, Title III), requires certain chemical users to plan for emergencies, report chemical use to adjacent communities, and report accidental release of toxic chemicals. At the time of this publication, research and medical laboratories are exempt under SARA, Title III, regulations. (Note that hospital employees who would be first responders treating victims of chemical spills may have training requirements to meet.) The OSHA regulation pertaining to chemical spill management is 29 CFR 1910.120, Hazardous Waste Operations and Emergency Response, often referred to as "HAZWOPER." This regulation is primarily geared toward large scale industrial spills, but all laboratories should rehearse emergency procedures for chemical spills.

CLEAN is a useful acronym describing the handling of chemical spills:

Contain the spill

Leave the area

Emergency equipment/treatment: eyewash, shower, medical help

Access SDS

Notify a supervisor and/or emergency personnel

A more detailed version of the general sequence of events is as follows:

1. Remove all personnel from harm's way, evacuating/quarantining the spill area as necessary
2. Contain the spill, if possible, during the evacuation
3. Decontaminate with eyewash or shower, as appropriate; give first aid to injured personnel
4. Notify hazardous spill teams, emergency health authorities,relevant supervisors
5. If qualified to do so, put on the necessary personal protective equipment
6. If it can be done safely, stop the spill or leak
7. Turn off heat/electrical sources; if toxic or flammable fumes are present, turn on ventilation
8. Contain the spill by spreading neutral absorbent material around its perimeter
9. Clean the spill
10. Document the incident in detail

 ISBN 978-089189-6463

11. Analyze the incident for appropriate emergency response/development of measures to prevent recurrence

Commercial kits for most types of spills are available and can be posted in strategic sites around the laboratory. For spills of both known and unknown composition, all purpose absorbents such as kitty litter, vermiculite, and sand can easily be stored in buckets and distributed around the laboratory. This allows for immediate spill containment because absorbent material can be placed around the perimeter of large spills to form "dikes" around them. Dikes not only prevent spreading, they reduce the spill surface area and minimize entry of the chemical into breathable air. Small spills could be completely soaked up. Cloths and paper towels frequently are not useful because they may be hazardous if soaked with chemicals, particularly flammables.

If the spill is large (more than ~1 gallon of hazardous material), or it contains a physical or health hazard, personnel should be evacuated immediately. The cleanup should be handled by professionals or by personnel with the appropriate equipment. A good example might be the presence of toxic fumes that require a respirator. When possible, heat and electrical sources should be turned off, especially after a flammable spill. Flammable spills also require that the area be ventilated to prevent fire from accumulated fumes.

If mercury and formaldehyde cannot be eliminated from the lab, specific spill kits should be purchased for these materials. Ordinary baking soda is useful for acid spill cleanup; CO_2 gas bubbles off until the acid is neutralized. Similarly, citric acid can be used for bases. This is *not* to be construed as appropriate treatment for spills on people. These neutralizers are for surface cleaning only. Spill kits often have no expiration dates, but they should be dated when they are put in service and routinely inspected with other safety equipment for continued usability.

If the spill involves a person's clothing and skin, all affected clothing should be removed immediately and the affected areas flushed with water. Do not remove clothing by pulling it over the victim's head; rather, cut it off to prevent contamination of the eyes. A safety shower should be used for body spills, and an eye wash is adequate for eye and face splashes. Chemicals can also get trapped under contact lenses, so if a spill involves a person wearing contact lenses, every effort should be made to remove them while flushing the eyes with water.

It is critical that laboratory workers know the proper procedures for each chemical with which they work. Information on specific chemicals should be obtained from manufacturers, OSHA, the EPA, and local health and environmental departments *before* they are used, and this information should be placed in the standard operating procedures of the CHP.

Chemical safety management

The principle of managing chemicals is simple: use the least hazardous and the least amount practical. This means scrutinizing processes and substituting less hazardous chemicals or changing processes when possible. Once this has been done, robust standard operating procedures must be crafted, and the CHP must reflect the best practices expected at a site. Staff then are trained and monitored for compliance, and incidents are analyzed to discover ways to improve the current protocols.

Summary table: chemical safety

Topic	Comment
Labels	labels on primary chemical containers must be GHS compliant; do not remove or deface 1. product identity and source 2. precautionary and hazard statements 3. red, white, and black GHS hazard pictograms 4. signal words "DANGER" for high risk and "WARNING" for lower risk 5. first aid procedures 6. miscellaneous: eg, lot number, quantity, received/expiration dates
	label secondary containers before filling; includes waste and nonhazardous substances like water 1. name and concentration of chemical 2. dates: receipt, made, expiration 3. initials of person who made or received the chemical 4. hazards present (eg, poison, flammability) and special instructions
	NFPA descriptive labels (scale of 0-4; 0: no hazard, 4: maximum hazard) red: fire hazard — blue: health hazard yellow: instability hazard — white: special hazard
	HMIS descriptive labels (scale of 0-4; 0: no hazard, 4: maximum hazard) red: fire hazard — blue: health hazard orange: physical — white: personal protective equipment
	pictograms: during transition period to GHS implementation, learn all 1. general symbols, eg, personal protective equipment, hazard classes, no smoking 2. DOT hazard symbols 3. GHS hazard symbols (see text)
Storage	1. follow government and NFPA regulations; use SDS & CHP 2. inventory: minimize stocks, organize chemicals to separate incompatibles, "first in, first out" 3. keep storage area labeled, clean, temperature controlled, ventilated, and secured 4. store dangerous material on low, "lipped" shelves; no crowding; only 2-3 bottles deep 5. use refrigerator storage cautiously. Fumes can accumulate inside 6. ensure fume hoods maintain 100 linear feet per minute air flow; hoods in low traffic areas, annual maintenance, sash only ~12" high when working, 2.5' per person, objects away from air intake. May need backdraft hood for vapors heavier than air 7. avoid under sink storage; no water reactives under water source
SDS	1. comprehensive source of information on a chemical 2. required for all hazardous chemicals; document nonhazardous status of others 3. MAINTAIN obsolete SDSs for 30 years especially if involved in incident
Chemical classes	1. corrosives: strong acids and bases 2. flammables (flash point <100°F) and combustibles (flash point >100°F) 3. health hazards: toxins, carcinogens, teratogens, mutagens, irritants, sensitizers 4. oxidizers 5. explosives 6. physical hazards

 ISBN 978-089189-6463

Summary table: chemical safety (continued)

Topic	Comment
Incompatible mixtures	1. SDS for each chemical and CHP should list major incompatibilties 2. keep incompatibles separate in storage, usage and disposal 3. review unit for examples of incompatible chemicals
Handling & Usage	1. use cart with raised sides for transporting; avoid public spaces 2. move glass containers >500 mL inside rubber or plastic bucket 3. hold containers firmly around and under their bodies, not by slender "necks" 4. hold bottle cap in hand; do not lay on countertop unless inverted 5. always use automatic pipetting device; add acid to water, not water to acid 6. do not eat, drink, or smell chemicals 7. no splashing or aerosols; use fume hoods when appropriate or if airborne drops/fumes present 8. decontaminate work surfaces before and after work; wash hands when work complete
disposal	1. comply with regulations for storage and disposal (local, state, EPA) 2. reduce, reuse, recycle waste when possible 3. segregate hazardous from nonhazardous waste; label and date 4. licensed waste handlers for hazardous waste 5. caution with sink disposal, many chemicals ineligible; 1 chemical at a time with lots of water in between
Protective equipment	1. goggles, safety glasses, face shields; prescription glasses NOT sufficient 2. fumes penetrate contact lenses so they may be forbidden 3. verify chemicals will not penetrate gloves in use 4. fluid resistant aprons and lab coats
Spills	1. CLEAN: Contain spill. Leave area. Emergency treatment. Access SDS. Notify authorities 2. neutral absorbents for all spills and kits for particular chemicals in use 3. apply absorbent material at spill perimeter first to contain spread 4. eyewashes and showers for body spills: 15 minutes; Get treatment
Management	1. least hazardous chemical in least amount possible 2. clear instructions in Chemical Hygiene Plan and standard operating procedures

 ISBN 978-089189-6463

Self evaluation questions

1 All of the information below is essential when labeling a secondary reagent container EXCEPT:

- **a** date reagent made
- **b** concentration of reagent
- **c** manufacturing source of reagent
- **d** initials of the person who made reagent
- **e** hazards of the reagent and special handling required

2 Which of the following needs to be labeled?

- **a** a beaker of water
- **b** chemical waste that will not be reused
- **c** a reagent that contains nonhazardous chemicals
- **d** a squirt bottle of bleach being used to clean a countertop
- **e** all of the above

3 According to the NFPA diamond below, this chemical's most significant hazard is

- **a** health
- **b** reactivity
- **c** instability
- **d** flammability
- **e** radioactivity

4 Which chemical type below best fits the HMIS label as shown below?

- **a** mutagen
- **b** corrosive
- **c** sensitizer
- **d** teratogen
- **e** pyrophoric

For questions 5-8, match the GHS symbol to its hazard class.

5 corrosive ______

6 explosive ______

7 oxidizer ______

8 severe health hazard ______

a

b

c

d

 ISBN 978-089189-6463

9 Each item below is stored correctly EXCEPT:

- a ether in a fume hood
- b Class I solvent in a vented safety cabinet
- c solid sodium chloride on room temperature shelving
- d concentrated hydrochloric acid in an overhead shelf with a lip for containing spills
- e peroxide in an explosion proof refrigerator at a temperature *above* its freezing point

10 Each of the following is an important principle of chemical inventory management EXCEPT:

- a open doors during working hours to ventilate chemical fumes
- b physical separation of incompatible chemicals
- c least used chemical on higher shelf than most used chemical
- d shelved bottles not >2-3 deep
- e oldest stock used first

11 When fume hoods are in use:

- a airflow should be maintained at ~100 linear feet per minute
- b at least 1.5 feet of linear work space should be allowed per person
- c the sash should be in the fully up position to draw in the maximum amount of air
- d objects should not be pushed to the back of the hood; they should be close to the front for easy reach to prevent users from leaning into the hood space
- e all of the above

12 All of the information below should be found on an SDS EXCEPT:

- a flammability of a chemical
- b emergency spill procedures
- c health hazard of a chemical
- d common name of a chemical
- e closest disposal site for a chemical

13 Each technique below is correct EXCEPT:

- a adding acid to water
- b wearing gloves when using poisons
- c working under a fume hood with flammables
- d carrying bottles of chemicals firmly by the neck
- e using respiratory protection when cleaning a mercury spill

14 If you can smell _______, it is very possible that its environmental concentration is too high for safety.

- a xylene
- b formaldehyde
- c glutaraldehyde
- d hydrogen cyanide
- e all of the above

For questions 15-20, match the phrase to the option below.

15 ___ corrosive
16 ___ health hazards
17 ___ Class I, flammable
18 ___ Class II, combustible
19 ___ oxidizer
20 ___ explosive

- **a** flash point <100°f
- **b** reduces compounds and releases energy
- **c** irritants, poisons, sensitizers and carcinogens
- **d** strong acid or strong base
- **e** flash point >100°f
- **f** detonates under specific stress, for example, high temperature

21 Ethers should be discarded after 1 year because they can form explosive ____________________. If this has occurred, you may see ____________________.

22 In what way(s) is perchloric acid dangerous?

- **a** it is corrosive
- **b** it will react with wood
- **c** the dried acid is explosive
- **d** fumes will crystallize in fume hood ductwork and crystals are explosive
- **e** all of the above

23 The preservative ____________ must never be flushed down a drain with metal plumbing.

24 Oxidizers should never contact ____________.

25 When can chemicals be flushed down the sink?

- **a** when flushed with large amounts of water
- **b** when they are nonhazardous and water soluble
- **c** when they are flushed individually with a water wash in between
- **d** all of the above

26 Bleach should not be mixed with

- **a** ammonia
- **b** formaldehyde
- **c** acids
- **d** all of the above

27 Which chemical below requires disposal by a licensed waste handler?

- **a** reagent preserved with 0.5% sodium azide
- **b** dry ice
- **c** mercury
- **d** 5% hydrochloric acid
- **e** all of the above

 ISBN 978-089189-6463

28 A chemical that is known to harm an unborn baby is a

- **a** mutagen
- **b** corrosive
- **c** sensitizer
- **d** teratogen
- **e** pyrophoric

29 A chemical that is known to induce allergies in some people is a

- **a** mutagen
- **b** corrosive
- **c** sensitizer
- **d** teratogen
- **e** pyrophoric

30 Dry ice can be hazardous because

- **a** its cold temperature is a physical hazard
- **b** it may reduce the breathable level of oxygen to below safe limits
- **c** it is colorless, odorless, and tasteless, so high levels are difficult to detect
- **d** its large expansion ratio may cause confined dry ice to explode its container
- **e** all of the above

31 In the event of a chemical spill, which action below should generally be done first?

- **a** ventilate the area
- **b** quarantine the area
- **c** create a "dike" around the spill to contain it
- **d** put on personal protective equipment
- **e** ensure that all personnel are safe, evacuating them as needed

32 The chemicals _______________ and _______________ are heavier than air and may be best controlled using a backdraft hood.

33 A chemical particularly harmful to the liver would be classified as a

____

34 When should a work surface be cleaned?

- **a** as soon as work is completed
- **b** before work begins
- **c** a & b
- **d** when the surface is dirty/contaminated

35 What is the maximum time that disposable gloves can be worn when working with chemicals?

- **a** 15 minutes
- **b** 60 minutes
- **c** 2 hours
- **d** 8 hours

Answers

1 c

2 e

3 a

4 e

5 b

6 d

7 a

8 c

9 d

10 a

11 a

12 e

13 d

14 e

15 d

16 c

17 a

18 e

19 b

20 f

21 peroxides; white crystals in the bottle or around the opening

22 e

23 sodium azide

24 metals or reducing agents such as oils and hydrocarbons

25 d

26 d

27 c

28 d

29 c

30 e

31 e

32 xylene; formaldehyde

33 hepatotoxin

34 c

35 b

 ISBN 978-089189-6463

Appendix 3.1: sample SDS compliant with the UN Global Harmonized System from www.osha.gov

1. Identification

Product name: chemical stuff
Synonyms: methyltoxy solution
CAS number: 000-00-0
Product use: organic synthesis
Manufacturer/supplier: My Company
Address: My Street, My Town, TX 00000

General information: 713-000-0000
Transportation emergency number: CHEMTREC: 800-424-9300

2. Hazards identification

GHS classification

Health	Environmental	Physical
Acute toxicity: category 2 (inhalation), category 3 (oral/dermal)	Aquatic toxicity: acute 2	Flammable liquid - category 2
Eye corrosion: category 1		
Skin corrosion: category 1		
Skin sensitization: category 1		
Mutagenicity: category 2		
Carcinogenicity: category 1b		
Reproductive/developmental: category 2		
Target organ toxicity (repeated): category 2		

GHS label

Hazard statements	Precautionary statements
DANGER! Highly flammable liquid and vapor Fatal if inhaled Causes severe skin burns and eye damage May cause allergic skin reaction Toxic if swallowed and in contact with skin May cause cancer	Do not eat, drink or use tobacco when using this product Do not breathe mist/vapors Keep container tightly closed Keep away from heat/sparks/open flame No smoking Wear respiratory protection, protective gloves and eye/face protection Use only in a well ventilated area
Hazard statements	**Precautionary statements**
Suspected of damaging the unborn child Suspected of causing genetic defects May cause damage to cardiovascular, respiratory, nervous, and gastrointestinal systems and liver and blood through prolonged or repeated exposure Toxic to aquatic life	Take precautionary measures against static discharge Use only nonsparking tools. Store container tightly closed in cool/well ventilated place Wash thoroughly after handling

3. Composition/information on ingredients

Component CAS number weight %
Methyltoxy 000-00-0 80
(See section 8 for exposure limits)

4. First aid measures

Eye: Eye irritation. Flush immediately with large amounts of water for at least 15 minutes. Eyelids should be held away from the eyeball to ensure thorough rinsing. Get immediate medical attention.

Skin: Itching or burning of the skin. Immediately flush the skin with plenty of water while removing contaminated clothing and shoes. Get immediate medical attention. Wash contaminated clothing before reuse.

Inhalation: Nasal irritation, headache, dizziness, nausea, vomiting, heart palpitations, breathing difficulty, cyanosis, **tremors, weakness, red flushing of face, irritability. Remove exposed person from source of exposure to fresh air. If not breathing, clear airway and start cardiopulmonary resuscitation (CPR). Avoid mouth-to-mouth resuscitation.**

Ingestion: Get immediate medical attention. Do not induce vomiting unless directed by medical personnel.

5. Fire fighting measures

Suitable Extinguishing Media: Use dry chemical, foam, or carbon dioxide to extinguish fire. Water may be ineffective but should be used to cool fire-exposed containers, structures and to protect personnel. Use water to dilute spills and to flush them away from sources of ignition.

Fire Fighting Procedures: Do not flush down sewers or other drainage systems. Exposed firefighters must wear NIOSH-approved positive pressure self-contained breathing apparatus with full-face mask and full protective clothing.

Unusual Fire and Explosion Hazards: Dangerous when exposed to heat or flame. Will form flammable or explosive mixtures with air at room temperature. Vapor or gas may spread to distant ignition sources and flash back. Vapors or gas may accumulate in low areas. Runoff to sewer may cause fire or explosion hazard. Containers may explode in heat of fire. Vapors may concentrate in confined areas. Liquid will float and may reignite on the surface of water.

Combustion Products: Irritating or toxic substances may be emitted upon thermal decomposition. Thermal decomposition products may include oxides of carbon and nitrogen.

6. Accidental release measures

Keep unnecessary people away; isolate hazard area and deny entry. Stay upwind; keep out of low areas. (Also see Section 8).
Vapor protective clothing should be worn for spills and leaks. Shut off ignition sources; no flares, smoking or flames in hazard area. Small spills: Take up with sand or other noncombustible absorbent material and place into containers for later disposal. Large spills: Dike far ahead of liquid spill for later disposal.
Do not flush to sewer or waterways. Prevent release to the environment if possible. Refer to Section 15 for spill/release reporting information.

7. Handling & storage

Handling

Do not get in eyes, on skin or on clothing. Do not breathe vapors or mists. Keep container closed. Use only with adequate ventilation. Use good personal hygiene practices. Wash hands before eating, drinking, smoking. Remove contaminated clothing and clean before re-use. Destroy contaminated belts and shoes and other items that cannot be decontaminated. Keep away from heat and flame. Keep operating temperatures below ignition temperatures at all times. Use non-sparking tools.

Storage

Store in tightly closed containers in cool, dry, well-ventilated area away from heat, sources of ignition and incompatibles. Ground lines and equipment used during transfer to reduce the possibility of static spark-initiated fire or explosion. Store at ambient or lower temperature. Store out of direct sunlight. Keep containers tightly closed and upright when not in use. Protect against physical damage.
Empty containers may contain toxic, flammable and explosive residue or vapors. Do not cut, grind, drill, or weld on or near containers unless precautions are taken against these hazards.

 ISBN 978-089189-6463

8. Exposure controls/personal protection

Exposure limits: component, methyltoxy - TWA: 3 ppm (skin) - STEL: C 15 ppm (15 min.)

Engineering controls: local exhaust ventilation may be necessary to control air contaminants to their exposure limits. The use of local ventilation is recommended to control emissions near the source. Provide mechanical ventilation for confined spaces. Use explosion proof ventilation equipment.

Personal protective equipment (PPE)

Eye protection: wear chemical safety goggles and face shield. Have eye-wash stations available where eye contact can occur.

Skin protection: avoid skin contact. Wear gloves impervious to conditions of use. Additional protection may be necessary to prevent skin contact including use of apron, face shield, boots or full body protection. A safety shower should be located in the work area. Recommended protective materials include: Butyl rubber and for limited contact Teflon.

Respiratory protection: if exposure limits are exceeded, NIOSH approved respiratory protection should be worn. A NIOSH approved respirator for organic vapors is generally acceptable for concentrations up to 10 times the PEL. For higher concentrations, unknown concentrations and for oxygen deficient atmospheres, use a NIOSH approved air-supplied respirator. Engineering controls are the preferred means for controlling chemical exposures. Respiratory protection may be needed for non-routine or emergency situations. Respiratory protection must be provided in accordance with OSHA 29 CFR 1910.134.

9. Physical & chemical properties

Flashpoint: 2°C (35°F)
Autoignition temperature: 480°C (896°F)
Boiling point: 77°C (170.6°F) @ 760 mm Hg
Melting point: -82°C
Vapor pressure: 100.0 mm Hg @ 23°C
Lower flammability Limit: >3.00%
Upper flammability Limit: <15.00%
Evaporation rate (water=1): 5(Butyl Acetate =1)
Octanol/water partition coefficient: log Kow: 0.5
Molecular weight: mixture
Vapor density (air=1): 1.7; air = 1
% Solubility in water: 10 @ 20°C
Pour point: NA
Molecular formula: mixture
Odor/appearance: clear, colorless liquid with mild, pungent odor
Specific gravity: 0.82g/mL @ 20°C
% Volatile: 100
Viscosity: 0.3 cP @ 25°C
pH: 7, 8% aqueous solution

10. Stability & reactivity

Stability/incompatibility: incompatible with ammonia, amines, bromine, strong bases and strong acids

Hazardous reactions/decomposition products: thermal decomposition products may include oxides of carbon and nitrogen.

11. Toxicological information

Signs & symptoms of overexposure: eye and nasal irritation, headache, dizziness, nausea, vomiting, heart palpitations, difficulty breathing, cyanosis, tremors, weakness, itching or burning of the skin

Acute effects

Eye contact: may cause severe conjunctival irritation and corneal damage

Skin contact: may cause reddening, blistering or burns with permanent damage. Harmful if absorbed through the skin. May cause allergic skin reaction.

Inhalation: may cause severe irritation with possible lung damage (pulmonary edema).

Ingestion: may cause severe gastrointestinal burns.

Target organ effects: may cause gastrointestinal (oral), respiratory tract, nervous system and blood effects based on experimental animal data. May cause cardiovascular system and liver effects.

Chronic effects: based on experimental animal data, may cause changes to genetic material; adverse effects on the developing fetus or on reproduction at doses that were toxic to the mother. Methyltoxy is classified by IARC as group 2B and by NTP as reasonably anticipated to be a human carcinogen. OSHA regulates methyltoxy as a potential carcinogen.

Medical conditions aggravated by exposure: preexisting diseases of the respiratory tract, nervous system, cardiovascular system, liver or gastrointestinal tract.

Acute toxicity values
Oral LD50 (rat) = 100 mg/kg
Dermal LD50 (rabbit) = 225-300 mg/kg
Inhalation LC50 (rat) = 200 ppm/4 hr, 1100 ppm vapor/1 hr

 ISBN 978-089189-6463

12. Ecological information

LC50 (Fathead Minnows) = 9 mg/L/96 hr.
EC50 (Daphnia) = 8.6 mg/L/48 hr.
Bioaccumulation is not expected to be significant. This product is readily biodegradable.

13. Disposal considerations

As sold, this product, when discarded or disposed of, is a hazardous waste according to Federal regulations (40 CFR 261). It is listed as Hazardous Waste Number Z000, listed due to its toxicity. The transportation, storage, treatment and disposal of this waste material must be conducted in compliance with 40 CFR 262, 263, 264, 268 and 270. Disposal can occur only in properly permitted facilities. Refer to state and local requirements for any additional requirements, as these may be different from Federal laws and regulations. Chemical additions, processing or otherwise altering this material may make waste management information presented in the SDS incomplete, inaccurate or otherwise inappropriate.

14. Transport information

U.S. Department of Transportation (DOT)
Proper shipping name: methyltoxy **Hazard class:** 3, 6.1
UN/NA number: UN0000 **Packing group:** PG 2 **Labels required:** flammable liquid & toxic
International Maritime Organization (IMDG)
Proper shipping name: methyltoxy **Hazard class:** 3 Subsidiary 6.1
UN/NA number: UN0000 **Packing group:** PG 2 **Labels required:** flammable liquid and toxic

15. Regulatory information

US federal regulations

Comprehensive Environmental Response & Liability Act of 1980 (CERCLA):
The reportable quantity (RQ) for this material is 1000 pounds. If appropriate, immediately report to the National Response Center (800/424-8802) as required by U.S. Federal Law. Also contact appropriate state and local regulatory agencies.
Toxic Substances Control Act (TSCA): All components of this product are included on the TSCA inventory.
Clean Water Act (CWA): Methyltoxy is a hazardous substance under the Clean Water Act. Consult federal, state and local regulations for specific requirements.
Clean Air Act (CAA): Methyltoxy is a hazardous substance under the Clean Air Act. Consult federal, state and local regulations for specific requirements.
Superfund Amendments and Reauthorization Act (SARA) Title III Information: SARA Section 311/312 (40 CFR 370) Hazard Categories:
Immediate Hazard: X
Delayed Hazard: X
Fire Hazard: X
Pressure Hazard:
Reactivity Hazard:
This product contains the following toxic chemical(s) subject to reporting requirements of SARA Section 313 (40 CFR 372)

Component	CAS Number	Maximum %
Methyltoxy	000-00-0	80

State Regulations

California: This product contains the following chemicals(s) known to the state of California to cause cancer, birth defects or reproductive harm:

Component	CAS Number	Maximum %
Methyltoxy	000-00-0	80

 ISBN 978-089189-6463

International regulations

Canadian Environmental Protection Act: All of the components of this product are included on the Canadian Domestic Substances List (DSL)

Canadian Workplace Hazardous Materials Information System (WHMIS):

Class B-2 Flammable Liquid   Class D-1-B Toxic   Class D-2-A Carcinogen
Class D-2-B Chronic Toxin   Class E Corrosive

This product has been classified in accordance with the hazard criteria of the Controlled Products Regulations and the SDS contains all the information required by the Controlled Products Regulations.

European Inventory of Existing Chemicals (EINECS): All of the components of this product are included on EINECS.

EU classification: F Highly Flammable; T Toxic; N Dangerous to the Environment

EU risk (R) & safety (S) Phrases:

R11: Highly flammable
R23/24/25: Toxic by inhalation, in contact with skin and if swallowed
R37/38: Irritating to respiratory system and skin
R41: Risk of serious damage to eyes
R43: May cause sensitization by skin contact
R45: May cause cancer
R51/53: Toxic to aquatic organisms, may cause long term adverse effects in the aquatic environment
S53: Avoid exposure—obtain special instructions before use
S16: Keep away from sources of ignition; no smoking
S45: In case of accident or if you feel unwell, seek medical advice immediately (show the label where possible)
S9: Keep container in a well ventilated place
S36/37: Wear suitable protective clothing and gloves
S57: Use appropriate container to avoid environmental contamination

16. Other information

National Fire Protection Association (NFPA) ratings: this information is intended solely for the use of individuals trained in the NFPA system

Health: 3
Flammability: 3
Reactivity: 0
Revision indicator: new SDS

Disclaimer: The information contained herein is accurate to the best of our knowledge. My Company makes no warranty of any kind, express or implied, concerning the safe use of this material in your process or in combination with other substance.

Appendix 3.2: target organ poster

Source: modified from OSHA CFR 1910.1200, Appendix A

This table categorizes chemical hazards by the organ specific effects that may occur, including examples of. It is useful to associate the typical signs/symptoms and chemicals that have been found to cause such listed with known chemical exposures and to educate chemical users regarding effects expected from handling these chemical categories. These examples are presented to illustrate the range and diversity of effects and hazards found in the workplace, and the broad scope employers must consider in this area, but are not intended to be all inclusive; they are a good beginning for conceiving safety training and management protocols.

Hepatotoxins: chemicals that produce liver damage	
Signs & symptoms	Jaundice (yellowing of the skin), liver enlargement, dark orange urine
Chemicals	Carbon tetrachloride, nitrosamines
Nephrotoxins: chemicals that produce kidney damage	
Signs & symptoms:	Edema (swelling), protein in the urine
Chemicals	Halogenated hydrocarbons, uranium
Neurotoxins: chemicals that produce their primary toxic effects on the nervous system	
Signs & symptoms	Narcosis (sleepiness), behavioral changes, impaired motor functions
Chemicals	Mercury, xylene, carbon disulfide
Agents that act on the blood or hematopoietic system: decrease (which produces blood cells), decreasing hemoglobin function & depriving the body tissues of oxygen	
Signs & symptoms	Cyanosis (blue skin), loss of consciousness
Chemicals	Carbon monoxide, cyanides
Agents that damage the lung, irritating or damaging pulmonary tissue	
Signs & symptoms	Cough, tightness in chest, shortness of breath
Chemicals	Silica, asbestos
Reproductive toxins: chemicals that affect the reproductive capabilities, causing chromosomal damage (mutations) and effects on fetuses (teratogenesis)	
Signs & symptoms	Birth defects, sterility
Chemicals	Lead, DBCP (1,2-dibromo-3-chloropropane)
Cutaneous hazards: chemicals that affect the dermal layer of the body	
Signs & symptoms	Defatting of the skin, rashes, irritation
Chemicals	Formaldehyde, corrosives, chlorinated compounds
Eye hazards: chemicals that affect the eye or visual capacity	
Signs & symptoms	Conjunctivitis (inflamed, red eyes), corneal damage
Chemicals	Organic solvents, acids

 ISBN 978-089189-6463

Unit 4

Equipment & electrical safety

The proper procedure for using laboratory equipment—from the simplest piece of glassware to the most complex analytical instrument—is essential to preventing accidents. This unit will cover general principles of the proper use of electrical equipment, identification of electrical hazards, glassware safety, centrifuge safety, safety with sharps (such as needles and scalpels), and steam sterilizer/autoclave safety.

4

Nature of electricity

In simplest terms, electricity is the movement of electrons. Materials through which electrons can easily flow are called conductors. Those that resist electrical flow are called insulators. In general, metals that loosely hold their valence electrons, water that contains ions, and human beings make excellent conductors. (Very pure water is not a very good conductor.) Materials that tightly hold their valence electrons, such as dry wood, plastic, glass, rubber, and ceramics, are good insulators. Water enhances the conductivity of any material, and some substances that are innately poor conductors, such as wood, can readily conduct electricity if wet.

Electricity travels best in closed circuits. A circuit is composed of an electron source (eg, a battery) and a conductor, which connects the electron source to an electron "sink" for a continuous, circular flow of electrons. The number of electrons flowing down a path is expressed in amperes ("amps"), and the force of those electrons is expressed as voltage ("volts"). Humans experience electrical shock when they become part of the electrical circuit, and current enters one part of the body and leaves via another. The severity of the shock is related to the amount of electrical current, its path through the body, and its duration. Even a current of low voltage can cause injury if it is of sufficient duration and affects essential organs, such as the heart. Therefore, *no amount of electricity is considered "safe."*

As shown in **t4.1**, reaction to an electrical shock can range from a faint tingling sensation to cardiac arrest, burns, and death, at levels well below those required for a typical fuse or circuit breaker to open a circuit. Electricity can cause muscular contractions that "freeze" the victim to the electrical circuit or cause involuntary movements, which can lead to injury, especially when the victim is "thrown" from the point of contact by the force of the electricity. Burns are the most common manifestation of electrical shock; therefore, electrical injury should always be considered when an unconscious victim presents with burns. In addition, the heat generated by an electrical malfunction creates a fire hazard for the entire facility.

t4.1. Effects of electricity on the human body
(modified from www.osha.gov/Publications/osha3075.pdf)

Current level (milliamperes)	Probable effect on human body
< 1 mA	Usually not perceptible
1 mA	Perception level. Slight tingling sensation. More dangerous if skin wet.
5 mA	Slight shock felt; not painful but disturbing. Average individual can let go. Strong involuntary reactions (muscle contractions) can cause injury.
6-30 mA	Painful shock, muscular control is lost. This is called the freezing current or "let go" range. Victim can get thrown clear if extensor muscles are excited by the shock.
50-150 mA	Extreme pain, respiratory arrest, severe muscle contractions. Individual cannot let go. Death is possible.
1000-4300 mA	Ventricular fibrillation (the rhythmic pumping action of the heart ceases). Muscular contraction and nerve damage occur. Death likely.
10,000 mA	Cardiac arrest, severe burns and probable death.
15,000 mA	Lowest current at which typical fuse or circuit breaker opens a circuit

General management of electrical hazards

The Occupational Safety and Health Administration (OSHA) recommends 5 means to manage electrical hazards:

Insulation

Electrical wires should never be exposed, and frayed insulation should be replaced before any equipment is used. Insulation on devices should never be intentionally compromised. Personal protective equipment made of insulating materials, specifically designed for protection against electricity, is also available.

Guarding

High voltage electrical equipment should be placed in areas of restricted access so that only qualified personnel may work in proximity to it. Sufficient space and security are needed around live electrical work, even if the situation is temporary, and unqualified laboratory personnel should not be allowed to operate or maintain laboratory instruments and equipment. The universal electrical hazard symbol is shown in **f4.1**.

Grounding

By virtue of its size, the earth can absorb large amounts of charge and remain electrically neutral. When an instrument is connected to the ground by a wire that easily conducts electricity, any excess electricity it accumulates will preferentially pass into the earth and not on to a laboratory worker. All electrical equipment must be grounded, usually with 3 prong plugs. (The only exception is certain double insulated, sealed, plastic devices that have polarized plugs to orient

f4.1 electrical hazard symbol

ISBN 978-089189-6463

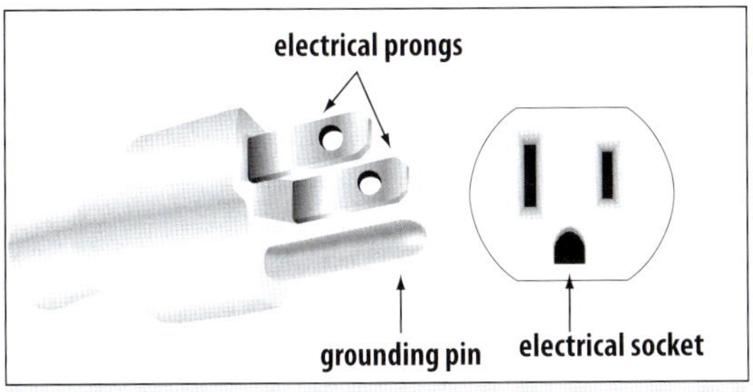

f4.2 schematic of an electrical plug

the plug's prongs to the socket correctly.) A schematic of a 3 prong plug is shown in **f4.2**. The 2 flat prongs carry current, and the large round pin connects the instrument to the grounding system of the building. The grounding pin of an electrical plug should never be cut off or bypassed. Even if an instrument is grounded and coated with an insulator, however, never assume that it is incapable of causing shock. Grounding and polarity on all electrical outlets should be checked at least annually, and all electrical equipment should be on a maintenance and inspection schedule, including checks for grounding and current leaks.

Circuit protection devices

Fuses, circuit breakers, surge protectors, ground fault circuit interrupters (GFCIs), and arc fault circuit breakers are devices that automatically shut off electrical flow in the event that excess current has been detected. They provide an extra layer of protection in addition to grounding. These devices should be clearly labeled and easily accessed. Circuit breakers, GFCIs, arc faults, and some top model surge protectors shut off current mechanically and can be reused. Fuses and less expensive surge protectors contain wires that melt when too much current travels through them and therefore must be replaced after this happens. Fresh fuses have zero resistance to electrical flow, and as long as excess current does not enter an instrument, power should flow freely. A schematic

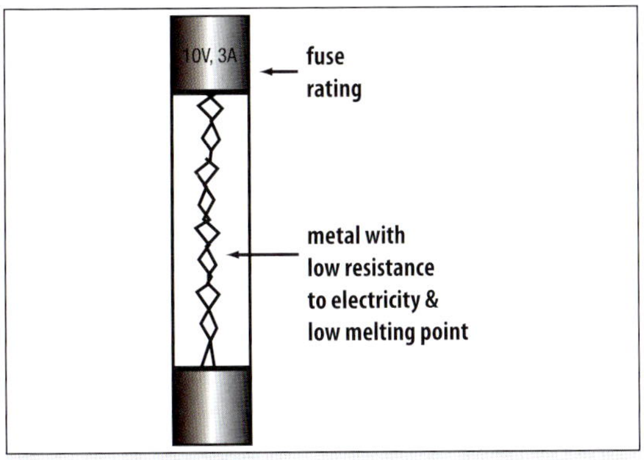

f4.3 schematic of a fuse

of a fuse is shown in **f4.3**. As indicated, fuses are rated for amperage and voltage. The amperage rating is the upper limit of current that can pass through the fuse without melting it. The voltage rating is the lower limit of voltage that is strong enough to "jump" the gap in the fuse created by the melted wire. Manufacturers select fuses that are appropriate to the amount of electricity an instrument can tolerate, so ideally, a spent fuse should be replaced with another fuse of identical rating. If an identical fuse is unavailable, it may be acceptable to use a fuse with voltage or amperage, but never the other way around. If any of these devices shut off current, the cause of the current surge should be determined before reactivating the circuit or replacing the fuse. Repeatedly tripped circuits and blown fuses indicate a problem that should be corrected.

Safe work practices

The major safe work practices recommended by OSHA are "de-energizing electric equipment before inspecting or making repairs, using electric tools that are in good repair, using good judgement when working near energized lines, and using appropriate protective equipment." These issues will be discussed below as they relate to the laboratory.

 ISBN 978-089189-6463

Lockout/tagout

The OSHA Standard "Control of Hazardous Energy," CFR 1910.147, is also known as the "lockout/tagout" standard. Provisions apply during installation, maintenance, testing, or repair of equipment as well as renovation/construction. All of these tasks can be hazardous if electrical current is accidently turned back on or if equipment is activated. During maintenance, staff members must often turn off the electricity to a particular area, and access to electrical panels or instruments should be denied to prevent inadvertent reactivations with a physical locking device (a "lockout") or with prominent signs/tags (a "tagout"). Lockouts are preferred. Only the person who initiated the lockout/tagout is permitted to remove the lock/tag when the work is safely completed. Therefore, each lockout/tagout must bear the name of the person who initiated it. All staff members who initiate lockout/tagout protocols or who would encounter lockouts/tagouts must be thoroughly trained in the protocols used at their institution. During electrical maintenance in an area, additional protocols may be needed, such as foregoing the use of metal stepstools and ladders while wires are exposed.

Although OSHA does not call it "lockout/tagout," similar protocols should be used when an instrument is malfunctioning and is removed from service. It can also apply to other hazard sources such as gas, water, waste, and radiation, which are turned off on occasion and could harm workers if they were reactivated without notice. In many cases, workers from multiple departments or outside contractors are involved, so instructions for lockout/tagout must be clear.

Emergency generators

During power supply interruptions, many homes and public facilities have the capability of restoring power with emergency electric generators. In hospitals and similar institutions, continuous uninterrupted power is required for patient safety. For example, in a hospital, an emergency generator must be in operation within 10 seconds of power loss, so each facility is required to have at least 2 emergency generators, which allows for servicing of 1 without violating the 10 second requirement in case of a power loss. The institution itself must set priorities for emergency power supply, such as lighting along evacuation routes, so during an incident, laboratories may have limited access to power and must plan carefully for continuous operations.

Generators should only be installed by qualified electricians who can then train users on proper protocols for the device. In most cases, the main circuit breaker should be locked in the "off" position so that the generator is not connected to the normal power lines while it is operating. When a generator is connected to the normal power lines, a serious problem called backfeed can arise, in which the generator unexpectedly sends power to the normal power lines and which can fatally electrocute repair workers in the vicinity. To prevent backfeed, generators should never be plugged directly into a wall outlet unless special power transfer switches have been installed.

Laboratories wired to emergency generators often have particular outlets that should be used as sources of electricity when the generator is on. Sometimes these outlets are color coded, oftentimes red. Many generators cannot provide power to every outlet, so placement of laboratory instruments should be strategic. Even when

 ISBN 978-089189-6463

generators are activated during an outage, there are typically a few seconds during which electricity is unavailable. This can be extremely disruptive to some instruments, so the installation of uninterruptible power supplies (UPSes) with battery power may be necessary. In addition, UPS devices with long battery lives can enable medical laboratories to continue services during emergency events or natural disasters until patients can be evacuated. Staff should be familiar with how the laboratory should be powered in the event of an outage to prevent any loss of valuable equipment or electrical injury caused by nonstandard situations.

Use of electrical equipment

Because some instruments and heating devices use large amounts of electric current, laboratory electrical requirements may be significantly different from domestic power needs. Laboratories, however, can be located in facilities originally intended for other purposes. Electrical wiring and power sources in a laboratory facility should be inspected by experts to ensure that dangerous inadequacies do not exist. GFCIs are the circuit protection devices of choice in high risk areas, especially for outlets near sources of water, and should be considered if the lab cannot guarantee separation of electrical supply and fluids.

Surge protectors should be used to protect delicate equipment, but all surge protectors are not created equal. The higher priced models generally provide increased protection against excess current. Using a surge protector rated for the highest amperage and voltage possible is often a wise investment.

OSHA requires that some electrical devices conform to stipulations set forth by private professional bodies. OSHA's Nationally Recognized Testing Laboratory (NRTL) program identifies organizations, such as Underwriters Laboratories, Inc (Northbrook, IL), that meet its standards, meaning devices are in compliance as long as they have been approved by these bodies. Many laboratory instruments are specially designed to operate sparkfree in hazardous environments (eg, in the presence of flammables), so substituting these instruments with domestic consumer goods, eg, ovens, blenders, hotplates, is not appropriate if hazardous atmospheres are likely.

The user's manual is a valuable resource for an instrument's proper operating procedure. Users must consult it if they have never, or have not recently, used a particular device. Directions for cleaning and servicing the instrument are particularly important. The wrong cleaning agent, such as acetone or water, can decrease the effectiveness of insulating materials and cause shock. In addition, insufficient maintenance of an instrument may not only cause interference in the instrument's function but also expose the user to hazardous conditions. Instruments should be on a maintenance schedule and be checked annually for proper grounding, current leakage, frayed wires, or compromised housing. The user's manual should be readily accessible to every individual who operates a particular instrument. Even when being operated properly, many devices generate heat, especially refrigerators and freezers. All devices should be surrounded by space sufficient to prevent heat buildup.

When an instrument is not in use, it should be turned off, unplugged, and dried thoroughly unless the manufacturer explicitly directs otherwise. Any electrical equipment, especially a motor that is operated in an area with flammable vapors, must be nonsparking and explosion proof. This includes temporary equipment brought into the area, such as floor polishers or vacuum cleaners.

Water, unless specially treated, contains ions that make it highly conductive of electricity. Uncontained liquid around electrical equipment is to be avoided at all costs. *Never* plug wet wires into a circuit. *Never* handle electrical equipment with wet hands, or while standing in or near water. *Never* operate wet motors and instruments. For devices needed in emergency situations, portable GFCIs should be used when water cannot be reliably eliminated.

Permanent extension cords are not permitted, and circuits should never be overloaded with too many pieces of equipment (ie, no "octopus" outlets). Wires and extension cords are rated for the amount of electricity that they can support, so users must check that they are rated sufficiently for the equipment. In addition, only a single extension cord of the proper length should be used; multiple cords and multiplugged adaptors should not be joined together in chains. Just as surge suppressors do, extension cords generally increase in cost as they increase in their ability to sustain higher currents. Extension cords conducting current in excess of their capacity get extremely overheated and are a serious fire hazard. In addition, if an extension cord is generating heat and portions of it are tightly coiled together, the heat is retained and the hazard is compounded. Best practice if an extension cord must be used, is to use only the length of cord necessary and extend it fully.

Identification of electrical hazards

Signs of potential electrical malfunction include the following:

1. repeatedly tripped circuit breakers/ GFCIs or blown fuses
2. malfunctioning equipment or equipment that produces a "tingle" when touched
3. instruments, wires, cords, connections, outlets, or junction boxes that are too warm
4. worn/frayed insulation or exposed wires

Instruments associated with the above should be unplugged and marked "OUT OF SERVICE" or "DO NOT USE." Arrangements for service should be made promptly. Only persons qualified for instrument repair should work on malfunctioning equipment. Instruments that have been turned off sometimes have high voltage points remaining from the previous "on" cycle, which can cause shock. Certain types of malfunctions in medical equipment must be reported to the Food and Drug Administration (FDA), so staff should be aware of these guidelines when there are instrument problems.

User's manuals often include troubleshooting guides. These directions must be followed explicitly when any repair is attempted. Repairs that are not listed in the user's manual should *not* be attempted. Some general rules to follow when troubleshooting instruments are as follows:

1. Turn off the power, and unplug the instrument whenever possible. Per OSHA, "Each circuit encountered will be considered live until proven otherwise." Occasionally troubleshooting may require that the instrument remain on (eg, to

 ISBN 978-089189-6463

check whether a light bulb is burned out), making it especially important to follow directions for correct troubleshooting. Never perform a repair with the instrument energized, if the repair can be performed while the instrument is off.

2. Most jewelry conducts electricity, so during a repair, remove all jewelry and follow the "1 hand in pocket" rule. Both hands on an instrument can complete a circuit that passes through the chest, if electricity is present. This rule is still advisable even if the instrument is off, in case any high voltage points remain.
3. Keep a clear path to the master switch or breaker box in case power needs to be interrupted quickly. A minimum clearance of 3 feet in front of electrical panels is required at all times by OSHA; nothing should be on the floor. In the event of shock, try to cut power to the entire laboratory. must be used to separate an electrical shock victim from an electrical source, the hands. Burns must be treated with cool water, but only far away from the source of the electrical problem. Because electrical current can cause severe internal injuries that may not be visible, victims should always be examined by medical personnel.

Exposed instrument parts, frayed wires, and electrical cords can cause shock or fire. All worn cords or plugs should be replaced immediately by those qualified to do so and with the correct replacement part. An important source of strained and frayed wires is repeated unplugging of an instrument by grabbing the cord rather than the plug itself. All users should unplug equipment by pulling on the plug and should be particularly alert for exposed wires adjacent to the plug.

Electrical fires *cannot* be extinguished with water and should only be controlled with the use of a carbon dioxide or dry chemical extinguisher. (See Unit 2 for more details.)

Glassware safety

Glassware, even when handled correctly, is a frequent laboratory hazard because it is easily breakable. Any glassware that is cracked, chipped, flawed, or otherwise showing signs of stress should be discarded. Metal stir rods and old cleaning brushes with metal exposed should be avoided as they will scratch and weaken the glass. Similarly, metal clamps should not be overtightened. Glassware should be stored in such a way as to avoid stress. For example, it is improper to put glassware into drawers where it can bump around every time the drawer is opened. Glass should also not be placed where it can easily be struck. For instance, pipettes sticking out of the tops of flasks or glassware left close to the edge of a work surface can easily be knocked over. If glass or sharp objects, eg, needles or scalpels, begin to fall, the rule of thumb is to let them fall. Catching these objects can result in hand injury.

When glass connections or stoppers become stuck, they should never be forced. It is important to keep glass connections lubricated and to use appropriate hand protection for making glass connections. Certain chemicals, such as oxidizers, should not come into contact with oil based lubricants. Water, soapy water, or glycerin can work well to lubricate glass.

Mouth pipetting is not acceptable, so appropriate devices must be attached to the tops of slender glass pipettes. When inserting a glass pipette into an automatic pipetting device, the pipette must be grasped very close to the end that is being inserted into the device. Holding the

4

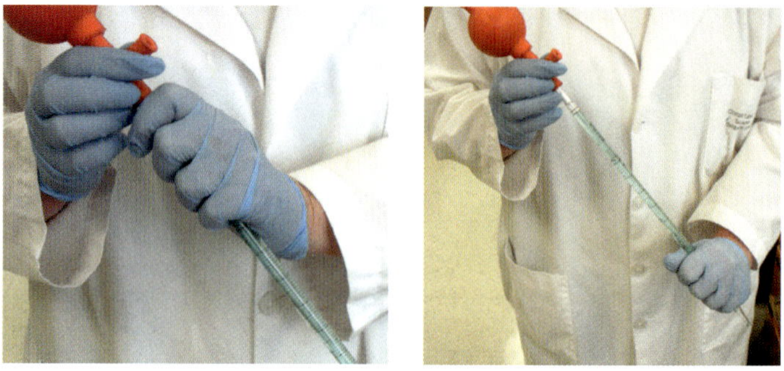

correctly inserting pipette into bulb

incorrectly inserting pipette into bulb

f4.4 inserting pipetting device

f4.5 holding volumetric flask safely

pipette at the opposite end could cause the pipette to break in the middle **f4.4**.

Glassware should be made of borosilicate glass such as Pyrex or some other heat resistant material, and heated glass should only be handled with gloves or tongs. Rapid heating and cooling should be avoided because this stresses virtually all glass. As shown in **f4.5**, glassware should be handled with 2 hands, 1 of which is on the bottom or around the body. The "neck" of most glassware is fragile, and any container in excess of ~500 mL must never be carried or suspended by clamps solely by its neck. This particularly applies to vessels such as volumetric flasks, which have long, delicate necks that are easily broken.

The type of glass and protection must be chosen based on the procedure. For example, glass under pressure (such as that used with suctioning pumps) should be housed in wire jackets. Glassware must also be compatible with the chemical reaction(s) it will house. For example, glass cannot be used with hydrofluoric acid, hot phosphoric acid, and strong hot alkalis.

Glassware must be cleaned thoroughly, as soon as possible after each use. Allowing glass to soak in water or soapy water is good practice to prevent debris from becoming solidified when thorough cleaning must be delayed. If a chemical or biological hazard has been contained in a particular piece of glass, it is important to neutralize the effects of the hazard *before* cleaning. If an automatic washer is not in use, best practice is to wear rubber gloves and to line the sink with a rubber mat. The gloves will improve one's grip on the soapy, wet glass and protect hands in the event of breakage, and the mat will "soften" the sink surface. It is safer to let glass air dry than to towel dry it because wet glass is slippery and more easily dropped. If a final distilled water rinse "beads up" on the glass, it has not been sufficiently cleaned, so the process should be repeated.

In the event of breakage, broken pieces of glass should never be removed by hand. Mechanical means, such as sweeping, tongs, or forceps, must be employed. If the glass breaks during washing, the water in the sink should be completely drained before cleanup. Broken glass must be disposed of in a specially designated puncture resistant container. Placing broken glassware in ordinary trash can puncture garbage bags and harm unsuspecting individuals. Also, because

 ISBN 978-089189-6463

broken glass creates a hazard in vacuum cleaner bags, using a vacuum cleaner is unacceptable.

Thin, delicate glass objects should be treated as sharps and disposed of accordingly. This includes Pasteur pipettes with long slender ends, capillary tubes, and the like. For disposal purposes, glass slides are also considered sharp. Slides easily get broken, so they must be discarded into puncture resistant containers. Glass slides that were used to examine fresh, unstained fluids and tissues are also considered infectious. Slides that have been processed through staining and fixation, and then permanently coverslipped are not considered infectious.

When possible, glass should be replaced with plastic. Available are plastic tubes for blood collection as well as plastic capillary tubes for tests in which the plastic does not interfere. Plastic beakers and flasks are also readily available, but because some chemicals can dissolve plastics, it is important to check any container before it is used with a particular chemical. 1 important caveat: plastic is breakable and can create sharp edges. Safety officers should evaluate the risk of disposal for each plastic object and create policy accordingly. Plastic at high risk for breakage, especially if contaminated, should be disposed of in puncture resistant containers just as glass is.

Safety with sharps

Sharp objects such as needles, syringes, scalpels, blades, and bone saws may be used in a laboratory or autopsy setting. They require cleaning or disposal, and each presents not just a physical hazard but also an infectious hazard if contaminated with biological material. The first and preferred strategy is to eliminate the use of sharps whenever possible.

The second strategy is to provide the safest devices and the best operating procedures possible. Sharps must be covered when not in use and stored securely. Security is important not just for inadvertent injury but also because needles and syringes can be targets for theft by drug users. Workers should plan their activities carefully and minimize the amount of time that a sharp is exposed. There is often pressure, especially in medical environments, to work quickly, but work with sharps must be deliberate and as slow as necessary to be safe. Further, other workers in the vicinity should be aware that one is performing a task with sharps. Sharps must be positioned away from users and others in the area, and work areas should have good task lighting and be isolated from traffic.

Safety engineered devices generally cost more than conventional devices. Because both disposable sharps and their protection devices typically must be discarded as a unit, the volume of hazardous waste and disposal costs are also increased. In situations with potential exposure to human materials, however, there is no choice. The Needlestick Safety and Prevention Act went into effect in November 2000. As part of the Bloodborne Pathogens Exposure Control Plan, safety managers in healthcare settings must incorporate as many safety devices as is reasonable to minimize employee exposure to injury from sharps. In many ways, these practices are cost justified. In 2005, the Centers for Disease Control and Prevention (CDC) estimated that accidental needlesticks contaminated with

 ISBN 978-089189-6463

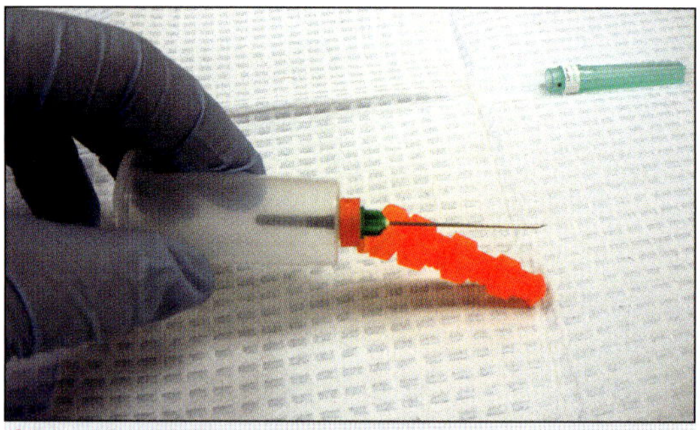

f4.6 safety needle cover; cover can be flipped onto needle with one hand behind the sharp at all times

human materials cost between $900 and $5,000 to treat. If an employee acquires a disease from an accidental exposure, treatment costs could be thousands of dollars with many lost work days. Because costs vary widely with devices and the types of exposures, the CDC now offers an online calculator (www.cdc.gov/sharpssafety/appendixE.html) to assist safety managers in calculating the cost of exposures in various scenarios, helping to justify costs associated with safety improvements.

The OSHA enforcement procedures document CPL 2-2.69 (www.osha.gov) states that sharps devices in medical settings must meet the following criteria:

1. a fixed safety feature provides a barrier between the hands and the sharp after use
2. the safety feature allows or requires the user's hands to remain behind the sharp at all times
3. the safety feature is an integral part of the device and not an accessory that is added by the user
4. the safety feature is in effect before disassembly, and remains in effect after disposal for environmental safety and to protect users and trash handlers
5. the safety feature is as simple as possible and requires little or no training to use effectively

Once activated, the covering on the sharp should be very difficult to disengage, and when the entire device is discarded, casual accidental contact should not cause injury. Good examples are skin puncturing sharps that automatically retract after use, and "self sheathing" needles and scalpels. f4.6 illustrates a device meeting the OSHA criteria outlined above. Employees who try to circumvent safety devices should be disciplined or terminated.

Microtomes are devices that thinly slice paraffin embedded tissues. Other than specimens with infectious prion disease (see Unit 5), infectious organisms are killed by the chemicals used in processing paraffin blocks, so the primary hazard is from the blade. Cryostats are similar devices except that a refrigeration unit is incorporated to permit thin slicing of frozen, unprocessed tissue. Freezing does NOT kill all infectious microbes, so used cryostat blades must be considered infectious.

Cryostats and microtomes have exceptionally sharp blades. Extreme care must be taken when cleaning or changing the blades of these devices. Direct contact with the blades should be avoided during manipulations by using blade guards and tools such as forceps, brushes, and probes fitted with magnetic tips. Portions of the devices that move, such as the cryostat wheel, should be secured during these procedures. Puncture resistant gloves made of stainless steel mesh are ideal to wear when using these, but single or double gloving with nitrile or latex gloves also reduces risk significantly. Exposed blades or motorized microtomes must not be left unattended. Blades must be secured or discarded when work is complete.

In addition to using safety devices, work practices should be scrutinized to minimize hazards. Sharps should never

 ISBN 978-089189-6463

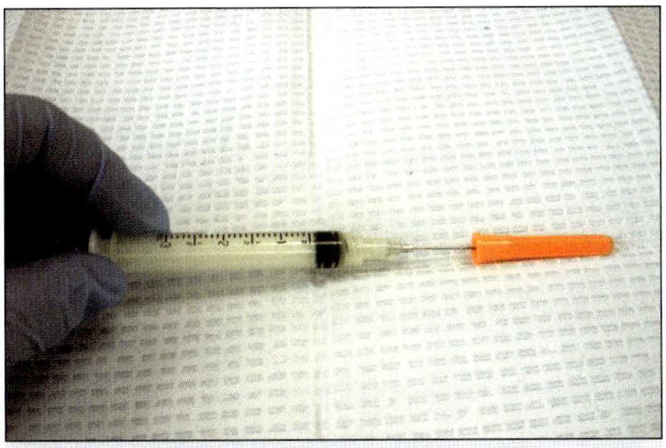

f4.7 the "one handed scoop" to cover needles safely

be used when a nonsharp or alternate method can be substituted. For example, in some autopsy situations, blunt tip suture needles can be substituted for sharp tip needles. Syringes have some of the highest accident rates of all sharp devices. Drawing blood directly into evacuated tubes rather than using syringes is preferable. When possible, plastic capillary tubes and plastic test tubes can be substituted for glass.

Safety engineered devices are usually designed to be single use and discarded as a unit because removal of the sharp and reuse of the device is an OSHA violation. Reuse of devices can also cause biohazard contamination. The use of a safety device does not change the sharp device into a nonsharp one in terms of handling and waste disposal. Sealed sharps must still be discarded in puncture resistant containers and handled as sharp waste.

Disposable needles, scalpels, and microtome blades are readily available, so it is prudent to avoid the use of reusable sharps when possible. Recapping or shearing needles by hand is extremely dangerous and is forbidden unless employers can prove medical necessity, which is extremely rare. More commonly, needles and other sharps are used once and discarded into a puncture resistant container. It is dangerous to overfill sharps disposal boxes. Once they are ~3/4 full, they should be sealed and removed for final disposal.

Disposal boxes for sharps should be within arm's reach of the site of use. Mounted disposal boxes must be at an ergonomically acceptable height for manipulation and clear viewing of the disposal portals. Per CLSI, this means 36-42 inches above the floor for seated workstations and 52-56 inches high for standing workstations. This can generally be achieved in laboratories; however, OSHA recognizes that this may be a problem in some patient care situations. Patients in pediatric units, psychiatric units, and correctional facilities, for example, should not have access to sharps, and staff may need to lock down sharps disposal boxes or keep them in another room. Because it is dangerous to walk any distance with an exposed sharp, devices that automatically resheathe must be used in these situations. If the use of a traditional needle and cap type device is unavoidable or if the safety device fails, the user should use the "1 handed scoop," shown in **f4.7**, to cover the sharp. Using only the hand with the needle, the cap is "scooped up" and covered until it can be disposed of. Again, this should be an extremely rare occurrence.

Used sharps devices must be autoclaved or incinerated to remove infectious hazard. Most states have regulations for medical waste, and in some cases sharps must not only be disinfected but also destroyed beyond recognition to prevent reuse and/or accidental puncture. After use, reusable scalpels may be placed in a pan of disinfectant until they can be autoclaved. The entire pan should be autoclaved to prevent removal of dirty scalpels by hand. If disposable scalpels are used, the entire scalpel should be put into a puncture resistant disposal box.

 ISBN 978-089189-6463

f4.8 centrifuge with closed cover

f4.9 open centrifuge showing properly balanced load

4

Facilities are required to keep "sharps injury logs" to keep track of employee incidents. Such logs are used to evaluate the safety devices in use and work practices that led to injuries. The type and brand of the device must be recorded as well as the procedure being performed and the location in the healthcare facility. As with any accident, a detailed explanation of how it occurred is required for analysis to prevent it from happening again.

OSHA requires that employees who are users of safety engineered devices give input into purchase decisions to prevent managers from buying cheaper devices that may not be the safest or easiest to use. On an annual basis, employees must be surveyed as to their satisfaction with the performance of current devices, and these surveys should be considered along with injuries incurred to determine whether new devices might be superior to currently used devices. At least 1 study has shown that devices that activate automatically without user intervention are associated with fewer accidents, so, for example, these may be preferable. Manufacturers often introduce new products or change prices, so annual surveys of new technologies are also required. Appendix B of OSHA document CPL 2-2.69 contains useful sample evaluation forms for sharps devices and disposal containers.

Any worker who sustains a puncture wound, especially when the sharp is contaminated, should allow the wound to bleed freely rather than put pressure on it. This assists in preventing entrance of infectious organisms into the body. Immediate washing with soap and water should be followed by medical assessment. This includes identifying the risk associated with the infectious source, prophylactic treatment as necessary, victim blood collection for baseline evaluation, testing of source patients, etc. Supervisory personnel must be notified and accident documentation completed.

Centrifuge safety

Centrifuges are common laboratory instruments that spin rapidly to achieve separation of materials based on density. Centrifuges must only be operated on firm, level surfaces with their covers closed, as shown in f4.8. Hair, clothing, or jewelry

 ISBN 978-089189-6463

f4.10 open centrifuge showing improperly balanced load

could become entangled in the moving parts, so centrifuges without safety covers should be avoided. In addition, liquids under centrifugation can produce hazardous aerosols. Therefore, not only should the cover be closed, but all objects in the centrifuge should be capped or covered. This especially applies to flammables, whose fumes can accumulate in a centrifuge if they are not contained. That said, many centrifuges are *not* acceptable for flammables, so centrifugation should either be avoided or a centrifuge constructed for flammables employed. Containers should not be overfilled, especially when using centrifuges with fixed angle rotors, because the spinning can drive the fluid up and out of the side of the container, contributing to aerosols.

Containers with known or suspected infectious organisms, especially respiratory pathogens, should be tightly sealed and spun in centrifuge carriers that are themselves sealed to prevent exposure if the primary container gets broken. This type of double sealing is common, for example, when centrifuging *Mycobacterium tuberculosis* specimen preparations.

Items in a centrifuge must be balanced evenly with regard to position and weight, as shown in **f4.9**. This also means that carriers in the centrifuge need to be balanced. Carriers frequently vary because of manufacturer imprecision, so small adjustments to the weight may be needed. Unbalanced centrifuges, as shown in **f4.10**, shorten the life of the motor, and when they are spinning, they can also "walk" off the edge of a countertop. Once the user starts a centrifuge, he or she should wait until it has achieved full speed and is behaving normally before leaving it alone. Centrifuges vibrate and are not quiet, but unusual noise or motion frequently indicates that a centrifuge is out of balance, so the device should be turned off immediately. If no brake is available, centrifuges should be allowed to stop on their own; *never* stop its spinning by hand.

Visibly cracked or flawed containers must not be spun. If the sound of breakage is heard, the centrifuge should not be opened until it has come to a complete stop. If the centrifuge contains biohazards or respiratory chemical hazards, opening should be delayed for 30 minutes to allow aerosols to settle. If flammables are present, immediate opening of the centrifuge may be necessary to prevent the buildup of fumes. Breakage should be cleaned carefully, using the same techniques specified for regular maintenance and cleaning. Even if no breakage occurs, all centrifuges should be on a maintenance schedule to keep the carriers smooth and rubber cushions clean. This will lengthen the life of the centrifuge and minimize the chance that spinning items will break.

Centrifuges should be inspected at least annually for accurate rate of speed/timing, functionality of safety locking mechanisms, and sound mechanics (eg, motor, lubricant, brushes, electrical

4

f4.11 steam sterilizer/autoclave

wiring, gaskets). Log books must be kept for rotors with limited life span, such as in ultracentrifuges, so that they can be replaced at the appropriate time.

Steam sterilizer/autoclave safety

Steam sterilizers/autoclaves sterilize material using a combination of high pressure and high temperature in the form of steam. A representative unit is shown in f4.11. They can sterilize metal, plastic, glass, wood, and organic material. Caps and covers on all items to be autoclaved should be loose to prevent them from bursting. When liquids are being sterilized, containers should be no more than ~2/3 full, and a longer cycle that uses lower temperatures should be employed to keep liquid from boiling over. Most autoclaves' holders are metal, and when glass is being sterilized, it benefits from being put in heat resistant plastic in order to avoid direct contact with the metal.

Solvents, bleach, and strong oxidizers should not be autoclaved because oxidizers and bleach could combine with organic and cause an explosion. Flammable solvents should never be subjected to high temperatures, so they can never be autoclaved. Because autoclaves discharge steam to the environment at the end of each cycle, breathable hazards such as toxic chemicals and radioactive isotopes also should not be autoclaved. Deactivation of infectious prions requires special treatment, including autoclaving materials in sodium hydroxide (NaOH). NaOH solutions are corrosive to autoclave interiors, and NaOH gas from the heat of the autoclave is hazardous. Special autoclave containers that limit vapor escape should be used, and it is especially important to let the NaOH cool to room temperature before handling. Corrosives can be damaging to some autoclaves, so these procedures must not be undertaken until the device specifications have been checked.

An autoclave should always be operated according to the manufacturer's instructions. In general, water is placed into the appropriate reservoir(s) to provide steam, and heat sensitive tape is affixed to objects to verify that the instrument reaches the proper temperature. To properly sterilize a device, steam must completely penetrate objects. Therefore, autoclaves should not be overstuffed, and objects with heavy layers of caked on material may need to be cleaned before sterilization. Once the autoclave is loaded properly, the door is latched firmly shut, and the autoclave is turned on for the appropriate amount of time.

During the sterilization cycle, the autoclave reaches high temperature and pressure. An open or leaky door will result in forceful blasts of steam. It is important not to operate autoclaves with doors that have damaged or cracked sealing gaskets. An autoclave must be sealed and locked prior to the start of a cycle, and *never* be opened until the pressure has returned to ambient. Even when pressure and temperature gauges indicate that it is safe to open the autoclave, the worker should stand to the side of the instrument and turn his or her face away in case a gauge is malfunctioning.

Ideally, all items from the autoclave should be allowed to return to room temperature before they are handled. Glass is more fragile when heated, so it especially benefits from slow cooling. If objects must be removed immediately, however, heat resistant gloves must be worn, but sometimes even these can be penetrated by steam. If the heat sensitive tape does not show the appropriate color change, the autoclave should be checked and the entire process should be repeated because sterility cannot be guaranteed. When the tape turns the correct color, it is evidence only that correct temperature was achieved, not sterility. Spores from heat resistant quality control organisms such as *Bacillus stearothermophilus* must be autoclaved every week to ensure that all organisms are being killed. It is a good idea to place the quality control organisms deep in the middle of the autoclave load to verify that the steam is penetrating.

Only general principles of equipment most frequently used in medical laboratories have been discussed in this Unit. Almost without exception, every user must be familiar with the user's manual and user specifications for each piece of equipment. Equipment can also be contaminated with other hazards such as chemicals, radiation, and biohazards, so the reader should refer to those units for additional handling protocols.

Summary table: equipment & electrical safety

Topic	Comments
Use of electrical equipment	1. follow OSHA requirements: insulation, grounding, guarding, circuit protection, safe work practices 2. ground all equipment with 3 pronged plugs; use surge protectors for delicate equipment 3. avoid extension cords; if absolutely necessary, make sure they are rated for the amount of current being used, and use only one cord fully stretched out (not coiled) 4. no "octopus," or overloaded, outlets 5. investigate blown fuses, tripped breakers, very warm cords/devices; use GFCIs in high risk, wet areas 6. consult user's manual for new or unfamiliar equipment 7. follow manual for cleaning, servicing and repair; remove jewelry; when possible, unplug and use only 1 hand 8. use nonsparking and explosion proof equipment certified by OSHA approved agency in its Nationally Recognized Testing Laboratory (NRTL) program 9. never plug a wet wire into a circuit; never operate a wet motor; never handle electrical equipment with wet hands or while standing in or near water 10. unplug and tag "DO NOT USE" any instrument that produces a tingle, is malfunctioning, or has frayed wires/ insulation 11. follow lockout/tagout procedures when electrical systems and equipment are being serviced; only the person who placed lock/tag on a device can remove it 12. do not use frayed or broken wires; do not compromise insulation; unplug equipment by pulling out plug, not by pulling on the cord 13. always maintain clear path to and a 3 foot clearance around electric breaker box; maintain clearance around instruments to dissipate heat 14. replace fuses with those of identical ratings; if unavailable, use fuses with higher voltage or lower amperage 15. follow emergency power protocols; prevent generator backfeed 16. suspect electrical shock if victim frozen to equipment or burns present; separate victim with a nonconductor, not the hands
Glassware safety	1. discard cracked, chipped, or flawed glassware 2. lubricate glassware connections; never force glass 3. handle glass with 2 hands from the bottom, not the "neck" 4. use gloves or tongs to handle heated glass; cool and heat slowly to reduce stress 5. store and clean glassware to reduce scratches/stress; avoid metal rods and cleaning brushes with metal 6. insert glass pipettes into devices by holding the insertion end 7. clean glassware after each use; neutralize chemical or biological hazard before cleaning; wear gloves to wash glass, and line the sink with a rubber mat; let glassware air dry 8. sweep up broken glass with a broom and discard in puncture resistant container; no vacuum cleaner, no picking up by hand 9. slender glass (eg, slides, Pasteur pipettes, capillary tubes) disposed of as sharps even if unbroken 10. substitute plastic for glass whenever possible
Safety with sharps	1. working with sharps: good task lighting, slow and careful, hands behind sharp and pointed away from others 2. discard sharps, even if covered, into a puncture resistant container, don't overfill; seal and discard when 3/4 full 3. never recap or shear a needle by hand 4. safety engineered sharps required by Needlestick Safety and Prevention Act; if unavailable, cover with the "1 handed scoop" technique 5. Lock wheel and use blade guards or safety tools on cryostats/microtomes when changing specimens or blades 6. incinerate or autoclave sharps to neutralize biohazards; consult medical waste regulations to see if total destruction is necessary. 7. keep sharps injuries log; consult with employees on purchase of sharps devices; review injuries, devices, employee opinions, and new technologies annually

 ISBN 978-089189-6463

Summary table: equipment & electrical safety

Topic	Comments
Centrifuge safety	1. operate with covers closed and items/carriers balanced with regard to weight 2. cover or cap items in centrifuge to prevent aerosols; do not spin flawed objects 3. allow centrifuge to stop on its own or use brake; do not stop by hand 4. when an item breaks in a centrifuge, allow the aerosols to settle before opening; flammables are the exception 5. clean centrifuges after breakage and on a routine basis following manufacturer's instructions
Steam sterilizer/autoclave safety	1. never open an autoclave until pressure is back to ambient 2. open autoclaves while standing to 1 side with your face turned away 3. allow items to cool to room temperature or handle with gloves 4. loosely cap or cover all items in an autoclave to prevent explosion; containers containing liquids should be no more than 2/3 full 5. do not autoclave oxidizers, bleach, flammables, radioactive materials, toxins, or chemicals with hazardous heat decomposition products 6. follow manufacturer's directions for each autoclave; ensure there is sufficient water for each cycle 7. put heat sensitive tape on items to be sure sterilization temperatures were reached 8. autoclave heat resistant organisms once per week to verify that sterilization is complete

Self evaluation questions

1 Instruments are "grounded" to
- **a** disperse electrical charge
- **b** prevent them from moving while in use
- **c** calibrate them when no samples are present
- **d** calibrate them to sea level specifications of air pressure
- **e** balance them with respect to weight, size, and components

2 Which of the following situations is acceptable?
- **a** operating a wet motor
- **b** using an instrument marked "broken" as long as the control specimens are within range
- **c** operating an instrument with a slight "tingle" as long as the control specimens are within range
- **d** washing an instrument with soap and water as directed by the manufacturer while the instrument is turned off but still plugged in
- **e** none of the above

3 All of the following statements are true EXCEPT:
- **a** use the "1 hand in pocket" rule when servicing equipment
- **b** use nonsparking equipment, especially in areas containing flammables
- **c** frayed electrical cords should not be used at any time, so instruments should be inspected periodically for damaged cords
- **d** when you observe a victim "frozen" to an electrical source, you must pull him/her off forcefully with both hands without delay to prevent permanent tissue damage
- **e** temporary extension cords are not acceptable unless they have been checked by an electrician to ensure they are large enough to carry the current used

4 Give one example of a conductor: ______________________________

Give one example of an insulator: ______________________________

5 Electrical shock occurs when a person accidentally becomes part of a ______________________________.

6 You must replace a 10 amp, 20 millivolt fuse.

What is your first choice? ______________________________

What is your second choice? ______________________________

7 Unplug an instrument by pulling on the ___________ not the ___________.

 ISBN 978-089189-6463

8 A circuit breaker has been turned off and a tagged with a sign that reads, "Maintenance in progress. Do not turn circuit breaker back on." Who has authority to remove this tag and turn the circuit breaker back on?

- **a** a qualified electrician
- **b** the worker's supervisor
- **c** the worker who placed the tag there
- **d** the worker's partner on the maintenance project
- **e** all of the above

9 Each condition below is likely to result from electrical shock EXCEPT:

- **a** cardiac arrhythmias
- **b** involuntary muscle movements
- **c** nausea and vomiting
- **d** inability to move muscles
- **e** burns

10 Choose the CORRECT statement(s):

- **a** an uninterruptible power supply is designed to prevent electrical shock
- **b** outlets in a wet area should be GFCIs
- **c** minimum clearance in front of a breaker box is 18 inches
- **d** since electricity generates heat, it is normal for instruments, power cords and outlets to be very warm
- **e** all of the above

11 Choose the CORRECT statement.

- **a** flasks with minor chips that do not cause leaks are acceptable to use
- **b** the safest way to wash glassware is to use gloves and let the glass air dry
- **c** if broken glass is not contaminated with any hazard, it can be discarded with the regular trash
- **d** a flask that held acquired immunodeficiency virus (AIDS) virus can be washed along the with other glassware as long as it is bleached afterward
- **e** all of the above are correct

12 Which of the following is (are) considered "sharp" with respect to disposal?

- **a** glass microscope slides
- **b** fragile plastic
- **c** glass Pasteur pipettes with slender ends
- **d** slender glass capillary tubes
- **e** all of the above

 ISBN 978-089189-6463

13 All of the following are true regarding safety engineered sharps devices EXCEPT:

- a once the safety device covers the sharp, the entire unit can be discarded into ordinary trash
- b they must not be purchased until employee input is given on the options available
- c they are required at all times unless an employer can prove medical necessity for a traditional device
- d they are acceptable to OSHA if they can easily be operated by 1 hand and the user's hand always remains behind the sharp
- e they should never have the safety features bypassed

14 You just finished drawing blood from a patient, and he started having a seizure. You put the uncovered needle down to get help. As you return, you see the needle and blood start to roll off the table. You should

- a run to catch it
- b try to catch it only if you have gloves on
- c let it fall

15 Sharps disposal boxes are (choose multiple answers)

- a at the correct height if the disposal portal is visible while using the sharp
- b not at the correct height if the disposal portal is visible while using the sharp
- c filled until ~95% capacity to reduce hazardous waste
- d filled until ~75% capacity to prevent overflow
- e within arm's reach of the user without exception
- f within arm's reach of the user unless the area is unsecure

16 Choose the CORRECT statement.

- a specimens should be changed on a microtome with the blade cover off
- b glass pipettes must be held closely to the end that is being inserted into a pipetting device
- c safety engineered sharps devices that are activated by the user are preferred to automatic devices since they could misfire
- d metal stir rods and cleaning brushes with exposed metal do not scratch heat resistant glass
- e all of the above are correct

17 A centrifuge cover prevents all of the following EXCEPT:

- a electrical shock to the users
- b dispersal of aerosols from spinning liquids
- c jewelry, hair, and clothing getting tangled in the centrifuge
- d broken glass flying out when a tube in the centrifuge breaks
- e injury from interaction with centrifuge parts moving at high speed

 ISBN 978-089189-6463

18 A centrifuge is making an odd noise and is moving across the tabletop. You should

- a unplug it immediately
- b open the cover to see what is making the noise
- c turn it off, wait for it to stop, and check it for balance
- d turn it off, unplug it, and mark it "OUT OF SERVICE"
- e let it continue to run, but monitor its movement to be sure it is stable

19 You hear breakage in a centrifuge that is spinning flammables. What do you do?

- a pull the plug immediately
- b turn the centrifuge off immediately
- c let the cycle finish but wait 30 minutes to open the centrifuge
- d cut power to that section of the lab

20 Which of the following materials can be autoclaved?

- a ether
- b perchloric acid solution
- c needles soaking in a pan of bleach
- d glass beaker in a nonmelting plastic tray
- e all of the above

21 All of the following are correct practices for autoclaving materials EXCEPT:

- a objects in the autoclave are loosely covered or capped
- b heat resistant bacterial spores should be autoclaved weekly to demonstrate that the autoclave is sterilizing properly
- c heat sensitive tape should be placed on the objects to ensure that the autoclave reached the correct temperature
- d when the timer indicates that the sterilization cycle is done, the autoclave should be immediately opened to give the contents time to cool
- e the following should be excluded from autoclaving: flammable chemicals, radioactive materials, toxic chemicals, chemicals with hazardous heat decomposition products

22 Sterilization may fail in an autoclave if

- a the objects are caked with debris
- b the autoclave is overfilled
- c there is insufficient water in the unit
- d all of the above

23 What can you autoclave?

- a tube of liquid media 50% full
- b tube of liquid media loosely capped
- c tube of liquid media in longer cycle of lower temperature
- d all of the above

Answers

1 a

2 e

3 d

4 conductors include the following: metals, impure water, living beings; insulators include the following: dry wood, plastic, glass, rubber, and ceramics

5 circuit

6 first choice: identical fuse with 10 amp, 20 millivolt; second choice: fuse lower than 10 amp and or higher than 20 millivolt

7 unplug an instrument by pulling on the plug not the cord

8 c

9 c

10 b

11 b

12 e

13 a

14 c

15 a, e, f

16 b

17 a

18 c

19 b

20 d

21 d

22 d

23 d

 ISBN 978-089189-6463

Unit 5

Biological hazards

A biological hazard is one in which infectious organisms or products from organisms can cause illness or harm. The 5 categories of infectious organisms are bacteria, viruses, fungi, protozoa, and prions. Recombinant DNA/RNA sources are also typically categorized as biohazards.

5

The biohazard symbol

The symbol for biological hazard is shown in **f5.1**. The biohazard symbol must be dark on a bright orange background to comply with the requirements of the Occupational Health and Safety Administration (OSHA). It must be displayed wherever any biohazards are present. This includes:

- **doors**: laboratory entrances, refrigerator doors, and incubator doors are examples
- **trash cans and waste receptacles**: OSHA does allow substitution of red bags for biohazard labeled bags if staff members are fully informed and trained
- **laundry bags**: all laundry placed in a biohazard bag must be processed using procedures designed to protect the worker and decontaminate the laundry
- **specimens and/or specimen containers**: OSHA does not require the symbol on every specimen if all specimens are in a single receptacle that displays the symbol or if a laboratory treats all specimens as biohazards

f5.1 biohazard symbol

- **equipment**: eg, centrifuges, analytical devices, and surgical equipment; if all equipment in a laboratory may be considered biohazardous, the symbol on the door to the laboratory is sufficient, but if a piece of this equipment is sent out for repair, it must be either decontaminated or labeled with the biohazard symbol

Microbial sources & routes of infection

The following are sources of micro-organisms with which a laboratory worker may come in contact. Most of these potential sources are rarely, if ever, sterile and usually must be handled as biohazards. Because the outside of specimen containers can be contaminated, even closed containers are considered biohazardous and handled with gloves.

- blood (whole blood, serum, plasma, or blood clot)
- culture specimens (swabs of various body sites)
- body fluids (eg, cerebrospinal, peritoneal, amniotic, and pleural)
- unfixed microscopic smears
- fecal specimens
- urine
- body tissue and cadavers
- laboratory animals
- plants (sources of contamination and insects, not permissible in a laboratory unless part of the work)
- insects and rodents
- people

The primary vectors of laboratory infection are listed below. In addition to protecting laboratory workers from these exposures, any other person who enters the laboratory must be similarly protected. This includes maintenance workers, instrument service people, vendors, and janitorial staff, many of whom may not be familiar with biosafety protocols and the consequences of exposures.

- inhalation of airborne aerosols or organisms
- mouth contact (eg, eating, drinking, chewing pens) resulting in accidental ingestion or mucous membrane penetration
- direct inoculation by contaminated objects (needle punctures, scratches, abrasions)

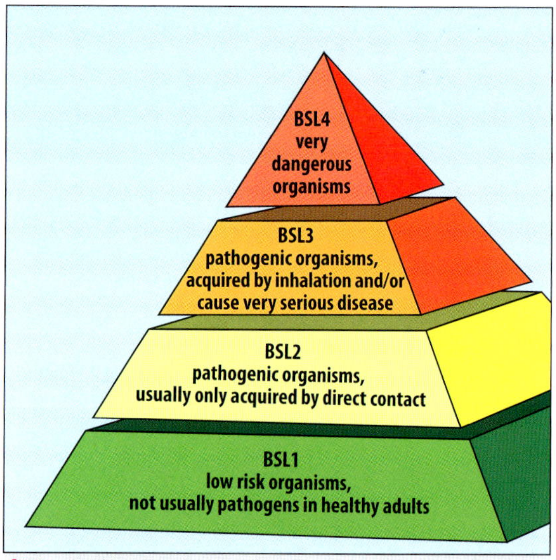

f5.2 CDC biosafety levels

- splashes to damaged skin, eyes, or mucous membranes
- insect transferal of micro-organisms

Source handling will depend on the pathogen most likely to be present and its infectious dose. Additional considerations are the stability of the micro-organism in the environment and whether the matrix housing the organism is likely to be protective. OSHA does not have biosafety levels (BSLs) for microbes, so biosafety programs most frequently rely on the 4 BSLs recommended by the National Institutes of Health/Centers for Disease Control and Prevention (NIH/CDC). These levels specify physical containment and handling for microbes, specimens, and animals based on hazard level. The 4 BSLs are summarized in **f5.2** and **t5.1**. The CDC/NIH publication *Biosafety in Microbiological and Biomedical Laboratories* lists specific organisms and the handling of each. It is an excellent resource for establishing procedures for each of the main components of a biosafety program **f5.3**. NIH categories for recombinant DNA and an alternative risk classification system used by the World Health Organization (WHO) are

 ISBN 978-089189-6463

t5.1 CDC biosafety levels for infectious agents

BSL	Agents	Practices	Safety equipment	Facilities (secondary barriers)
1	not known to consistently cause disease in healthy adults	standard microbiological practices (see **t5.4**)	none required; personal protective equipment: lab coats, gloves, face protection, as needed	easily cleaned, open bench top; sink required
2	associated with human disease with direct contact: percutaneous injury, ingestion, mucous membrane exposure	BSL1 practices plus limited access, biohazard warning signs, "sharps" precautions, biosafety manual defining waste decontamination or medical surveillance policies	primary barriers: class 1 or 2 biological safety cabinets or containment devices for manipulations that cause splashes or aerosols of infectious material; personal protective equipment: lab coats, gloves, face protection, as needed	BSL1 plus autoclave available
3	indigenous or exotic agents with potential for aerosol/ inhalation transmission; disease may have serious or lethal consequences	BSL2 plus controlled access, decontamination of all waste, decontamination of lab clothing before laundering	primary barriers: class 1 or 2 biological safety cabinets or containment devices for all open manipulations of agents; personal protective equipment: protective lab clothing, gloves, eye, face and respiratory protection, as needed	BSL2 practices plus: physical separation from access corridors, self closing double door access, doors closed, exhausted air (not recirculated), negative airflow into lab (lab at lower pressure), entry through airlock or anteroom, handwashing sink near exit
4	dangerous/exotic agents that pose high risk of life threatening disease, aerosol transmitted lab infections, or related agents with unknown risk of transmission	BSL3 plus clothing change before entering, shower on exit, all material decontaminated on exit from facility	primary barriers: all procedures conducted in class 3 biological safety cabinet or class 1 or 2 cabinets *in combination with* full body, air supplied, positive pressure personnel suit	BSL3 practices plus separate building or isolated zone, dedicated supply and exhaust, vacuum and decontamination systems, others in CDC text

Modified from [2009] Biosafety in Microbiological and Biomedical Laboratories, 5e. Government Printing Office, accessed October 16, 2015 at www.cdc.gov

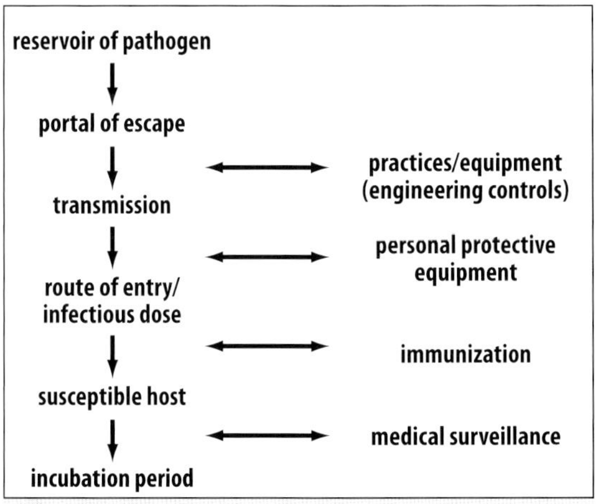

f5.3 Biosafety levels

presented in **t5.2**. Also on the CDC website (www.cdc.gov) is another useful publication, *Guidelines for Safe Work Practices in Human and Animal Medical Diagnostic Laboratories. Recommendations of a CDC Convened, Biosafety Blue Ribbon Panel* at (www.cdc.gov/mmwr/preview/mmwrhtml/su6101a1.htm, accessed December 23, 2015). Protocols developed using these 2 resources are likely to meet the needs of most laboratories. For molecular laboratories, additional protocols in *NIH Guidelines for Research*

t5.2 NIH & WHO biohazard risk classification

	National Institutes of Health (NIH) Guidelines for Research Involving Recombinant DNA Molecules (2013)	World Health Organization (WHO) Laboratory Biosafety Manual (2004)	
	NIH description	**WHO description**	**Types of agents**
Risk group 1	agents not associated with disease in healthy humans	low individual and community risk	unlikely to cause human or animal disease
Risk group 2	agents associated with human disease that is rarely serious and for which preventative and therapeutic interventions are often available	moderate individual risk and low community risk	cause human or animal disease that is preventable/treatable; transmission to laboratory workers, community, livestock, and the environment unlikely
Risk group 3	agents associated with serious or lethal human disease and for which preventative and therapeutic interventions may be available	high individual risk and low community risk	cause serious disease in humans or animals but infection usually self limiting and treatment/ prevention available
Risk group 4	agents that are likely to cause serious or lethal human disease and for which preventative and therapeutic interventions are not usually available	high individual and community risk	cause serious disease in humans or animals that is transmitted between individuals, and treatment/ prevention not usually available

Modified from [2009] Biosafety in Microbiological and Biomedical Laboratories, 5e. Government Printing Office, accessed October 16, 2015 at www.cdc.gov

Involving Recombinant or Synthetic Nucleic Acid Molecules (osp.od.nih.gov/sites/default/files/NIH_Guidelines_0.pdf accessed December 23, 2015) may also be required.

Organisms that are almost always nonpathogens in normal human hosts should be handled using BSL level 1 (BSL1) precautions. Academic laboratories using nonpathogens for teaching, water testing laboratories, and food quality control laboratories are examples of laboratories that might be classified as BSL1. BSL2 controls are designed for organisms that cause disease in humans but are usually acquired through direct contact, ingestion, or injection. Most clinical laboratories are classified as BSL2 because they work with specimens suspected to harbor pathogens transmitted by these routes. Specimens or cultures that may harbor organisms that are acquired by inhalation, such as *Mycobacterium tuberculosis* and many of the pathogenic spore producing molds, should be handled with BSL3 protocols. Also, at a very practical level, the CDC has published "cough etiquette" protocols, which should be observed by workers and patients alike at all times (www.cdc.gov/flu/professionals/infectioncontrol/resphygiene.htm, accessed December 23, 2015). Those working with other high risk organisms with low infectious doses may also need to employ BSL3. BSL4 cautions are typically reserved for highly pathogenic, extraordinary organisms (eg, Ebola virus, smallpox virus) and large amounts of pathogens, as on an industrial scale. In uncertain situations, specimens should be handled conservatively using the most rigorous safety level applicable.

Risk assessment for biohazards requires evaluation of 2 main features, the inherent hazard of the biologic agent and the laboratory procedure(s) involved. The CDC/NIH BSLs are determined based on adult humans with normal immune systems. Staff members with special situations (eg, pregnancy) or compromised immune systems may require additional protection. Laboratory workers should also be familiar with the signs of the diseases caused by the microbes they encounter

ISBN 978-089189-6463

t5.3 Signs and symptoms of HIV, hepatitis, TB, prion disease*

HIV	Hepatitis	TB	Prion disease
initial infection: flulike symptoms, lymphadenopathy later infection: lymphadenopathy, fever, diarrhea, skin/mouth lesions, yeast and viral infections	jaundice, dark urine, malaise, nausea, abdominal pain, fever, fatigue	cough with thick, cloudy and/or bloody mucus lasting >2 weeks, shortness of breath fatigue, muscle weakness loss of appetite, unexplained weight loss, fever, chills, night sweats	Confusion/dementia, difficulty walking/speaking/swallowing, hallucinations, muscle stiffness, fatigue

**Examples only; workers should be alert for signs of infection, eg, unusual fevers, rashes, swollen lymph nodes, diarrhea, that correlate with microbes encountered during the course of work*

in their work (examples shown in **t5.3**). If someone acquires an infection, it can then be treated immediately. Because the exact contents of clinical specimens are unknown, it's recommended that initial handling and processing take place in a biological safety cabinet (BSC). Unless unusual pathogens are known or suspected, most workups on medical specimens can be handled using BSL2 conditions. Because BSL2 procedures are so important for virtually all medical specimens, the CDC recommendations for BSL2 are reproduced here in their entirety **t5.4**. All of the practices listed should be a routine part of biohazard manipulation.

5

t5.4 Biosafety level 2 laboratory practices

Biosafety level (BSL) 2 builds upon BSL1. BSL2 is suitable for work involving agents that pose moderate hazards to personnel and the environment. It differs from BSL1 in that 1) laboratory personnel have specific training in handling pathogenic agents and are supervised by scientists competent in handling infectious agents and associated procedures; 2) access to the laboratory is restricted when work is being conducted; and 3) all procedures in which infectious aerosols or splashes may be created are conducted in biosafety cabinets (BSCs) or other physical containment equipment. The following standard and special practices, safety equipment, and facility requirements apply to BSL2 (modified from www.cdc.gov, *Biosafety in Microbiological and Biomedical Laboratories*, 5e)

A. Standard microbiological practices	1. The laboratory supervisor must enforce the institutional policies that control access to the laboratory 2. Persons must wash their hands after working with potentially hazardous materials and before leaving the laboratory 3. Eating, drinking, smoking, handling contact lenses, applying cosmetics, and storing food for human consumption must not be permitted in laboratory areas. Food must be stored outside the laboratory area in cabinets or refrigerators designated and used for this purpose 4. Mouth pipetting is prohibited; mechanical pipetting devices must be used 5. Policies for the safe handling of sharps, such as needles, scalpels, pipettes, and broken glassware must be developed and implemented. Whenever practical, laboratory supervisors should adopt improved engineering and work practice controls that reduce risk of sharps injuries. Precautions, including those listed below, must always be taken with sharp items. These include the following: a. Careful management of needles and other sharps are of primary importance. Needles must not be bent, sheared, broken, recapped, removed from disposable syringes, or otherwise manipulated by hand before disposal b. Used disposable needles and syringes must be carefully placed in conveniently located puncture resistant containers used for sharps disposal. c. Nondisposable sharps must be placed in a hard walled container for transport to a processing area for decontamination, preferably by autoclaving d. Broken glassware must not be handled directly. Instead, it must be removed using a brush and dustpan, tongs, or forceps. Plasticware should be substituted for glassware whenever possible 6. Perform all procedures to minimize the creation of splashes and/or aerosols 7. Decontaminate work surfaces after completion of work and after any spill or splash of potentially infectious material with appropriate disinfectant

t5.4 Biosafety level 2 laboratory practices (continued)

A. Standard microbiological practices (continued)	8. Decontaminate all cultures, stocks, and other potentially infectious materials before disposal using an effective method. Depending on where the decontamination will be performed, the following methods should be used prior to transport: a. Materials to be decontaminated outside of the immediate laboratory must be placed in a durable, leakproof container and secured for transport b. Materials to be removed from the facility for decontamination must be packed in accordance with applicable local, state, and federal regulations 9. A sign incorporating the universal biohazard symbol must be posted at the entrance to the laboratory when infectious agents are present. Posted information must include the laboratory's BSL and the laboratory BSL criteria: BSL2, the supervisor's name (or the name of other responsible personnel), telephone number, and required procedures for entering and exiting the laboratory. Agent information should be posted in accordance with institutional policy 10. An effective integrated pest management program is required 11. The laboratory supervisor must ensure that laboratory personnel receive appropriate training regarding their duties, the necessary precautions to prevent exposures, and exposure evaluation procedures. Personnel must receive annual updates or additional training when procedural or policy changes occur. Personal health status may impact an individual's susceptibility to infection, or the ability to receive immunizations or prophylactic interventions. Therefore, all laboratory personnel and particularly women of childbearing age should be provided with information regarding immune competence and conditions that may predispose them to infection. Individuals with these conditions should be encouraged to self identify to the institution's healthcare provider for appropriate counseling and guidance
B. Special practices	1. All persons entering the laboratory must be advised of the potential hazards and meet specific entry/exit requirements 2. Laboratory personnel must be provided medical surveillance, as appropriate, and offered available immunizations for agents handled or potentially present in the laboratory 3. Each institution should consider the need for collection and storage of serum samples from at risk personnel 4. A laboratory specific biosafety manual must be prepared and adopted as policy. The biosafety manual must be available and accessible 5. The laboratory supervisor must ensure that laboratory personnel demonstrate proficiency in standard and special microbiological practices before working with BSL2 agents 6. Potentially infectious materials must be placed in a durable, leakproof container during collection, handling, processing, storage, or transport within a facility 7. Laboratory equipment should be decontaminated routinely, as well as after spills, splashes, or other potential contamination a. Spills involving infectious materials must be contained, decontaminated, and cleaned up by staff properly trained and equipped to work with infectious material b. Equipment must be decontaminated before repair, maintenance, or removal from the laboratory 8. Incidents that may result in exposure to infectious materials must be immediately evaluated and treated according to procedures described in the laboratory biosafety manual. All such incidents must be reported to the laboratory supervisor. Medical evaluation, surveillance, and treatment should be provided and appropriate records maintained 9. Animal and plants not associated with the work being performed must not be permitted in the laboratory All procedures involving the manipulation of infectious materials that may generate an aerosol should be conducted within a BSC or another physical containment device

 ISBN 978-089189-6463

t5.4 Biosafety level 2 laboratory practices (continued)

C. Safety equipment (primary barriers and personal protective equipment)	1. Properly maintained BSCs, other appropriate personal protective equipment, or other physical containment devices must be used in the event of the following: a. When procedures with a potential for creating infectious aerosols or splashes are conducted; these may include pipetting, centrifuging, grinding, blending, shaking, mixing, sonicating, opening containers of infectious materials, inoculating animals intranasally, and harvesting infected tissues from animals or eggs b. When high concentrations or large volumes of infectious agents are used; such materials may be centrifuged in the open laboratory using sealed rotor heads or centrifuge safety cups 2. Protective laboratory coats, gowns, smocks, or uniforms designated for laboratory use must be worn while working with hazardous materials. Remove protective clothing before leaving for nonlaboratory areas (eg, cafeteria, library, administrative offices). Dispose of protective clothing appropriately, or deposit it for laundering by the institution. It is recommended that laboratory clothing not be taken home 3. Eye and face protection (eg, goggles, mask, face shield, or other splatter guard) is used for anticipated splashes or sprays of infectious or other hazardous materials when the micro-organisms must be handled outside the BSC or containment device. Eye and face protection must be disposed of with other contaminated laboratory waste or decontaminated before reuse. Persons who wear contact lenses in the laboratory should also wear eye protection 4. Gloves must be worn to protect hands from exposure to hazardous materials. Glove selection should be based on an appropriate risk assessment. Alternatives to latex gloves should be available. Gloves must not be worn outside the laboratory. In addition, BSL2 laboratory workers should: a. change gloves when contaminated, glove integrity is compromised, or when otherwise necessary b. remove gloves and wash hands when work with hazardous materials has been completed and before leaving the laboratory c. not wash or reuse disposable gloves; rather, they should dispose of used gloves with other contaminated laboratory waste; also, hand washing protocols must be rigorously followed 5. Eye, face, and respiratory protection should be used in rooms containing infected animals as determined by the risk assessment
D. Laboratory facilities (secondary barriers)	1. Laboratory doors should be self closing and have locks in accordance with institutional policies 2. Laboratories must have a sink for hand washing, which sink may be operated manually, hands free, or automatically, and should be located near the exit door 3. The laboratory should be designed so that it can be easily cleaned and decontaminated; carpets and rugs are not permitted in laboratories 4. Laboratory furniture must be capable of supporting anticipated loads and uses; spaces between benches, cabinets, and equipment should be accessible for cleaning a. Bench tops must be impervious to water and resistant to heat, organic solvents, acids, alkalis, and other chemicals b. Chairs used in laboratory work must be covered with a nonporous material that can be easily cleaned and decontaminated with appropriate disinfectant. 5. Laboratory windows that open to the exterior are not recommended; existing windows must be fitted with screens if they open to the exterior 6. BSCs must be installed so that fluctuations of the room's air supply and exhaust do not interfere with proper operations; they should be located away from doors, windows that can be opened, heavily traveled laboratory areas, and other possible airflow disruptions 7. Vacuum lines should be protected with liquid disinfectant traps 8. An eyewash station must be readily available 9. There are no specific requirements for ventilation systems; however, those planning new facilities should consider mechanical ventilation systems that provide an inward flow of air without recirculation to spaces outside of the laboratory 10. HEPA filtered exhaust air from a class II BSC can be safely recirculated back into the laboratory environment if the BSC is tested and certified at least annually, and operated according to manufacturer's recommendations. BSCs can also be either connected to the laboratory exhaust system by a thimble (canopy) connection or directly exhausted to the outside via a hard connection. Provisions to assure proper BSC performance and air system operation must be verified 11. A method for decontaminating all laboratory wastes should be available in the facility (eg, autoclave, chemical disinfection, incineration, or other validated decontamination method)

Biosecurity

The CDC uses the term *biosecurity* to mean "protection of microbial agents from loss, theft, diversion, or intentional misuse." Global concern over bioterrorism has been increasing, and the CDC has published several important security protocols for laboratories that handle biologic agents. Many relatively common organisms can be used as biological weapons, so one should not assume that security measures are only appropriate for those laboratories handling exotic organisms that require high BSL procedures. The CDC, the Department of Health and Human Services, and the US Department of Agriculture have developed the "Select Agents and Toxins List," which compiles those organisms/toxins most desirable to terrorists (www.selectagents.gov/SelectAgentsandToxinsList.html, accessed December 23, 2015). Laboratories handling these agents must register with the government and follow certain protocols established for security, including physical containment, inspections, and personnel clearances by the US Department of Justice. Some agents cause human disease and/or could contaminate water supplies, while animal pathogens could disrupt food supplies. There is concern that some of these agents could be genetically altered to be resistant to the usual antibiotics, so if these organisms are encountered in specimens, prompt recognition, antibiotic sensitivity testing, and reporting are vital. Because a laboratory's inventory isn't publicly available, even labs that do not handle these organisms may still be targets for theft. Therefore, all laboratories should perform a risk assessment that includes evaluating the physical security of the premises, employ hiring practices that evaluate the qualifications and risk of potential employees, and maintain strict management of microbe inventories.

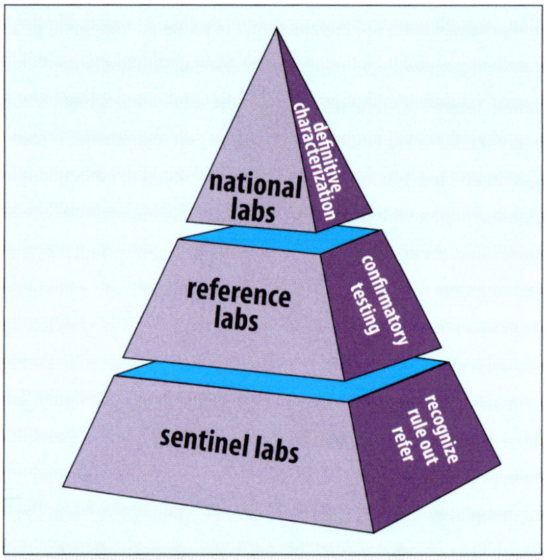

f5.4 Laboratory Response Network

The CDC has established guidelines through the national Laboratory Response Network for microbiology laboratories to follow in the event of the detection of a critical agent or in an actual bioterrorism event (emergency.cdc.gov/lrn/index.asp, accessed December 23, 2015) **f5.4**. Each institution handling micro-organisms should use these guidelines to establish bioterrorism protocols specific to its facility. National laboratories and regional reference laboratories receive referred specimens of suspected agents, and are key in identifying those agents and helping coordinate emergency responses. Such laboratories should meticulously incorporate these CDC guidelines. Most medical laboratories, however, would be classified as sentinel laboratories, which would deal with bioterrorist agents when a victim first seeks medical care. The CDC has transferred responsibility for sentinel laboratory guidelines to the American Society of Microbiology (ASM). Managers of sentinel laboratories should refer to the ASM website (www.asm.org/index.php/guidelines/sentinel-guidelines, accessed December 23, 2015) for help in establishing protocols. Free bioterrorism

 ISBN 978-089189-6463

training for sentinel laboratories can be obtained at emergency.cdc.gov/bioterrorism/training.asp (accessed December 23, 2015).

Some basic elements of a biosecurity plan articulated by the CDC include the following:

- developing and managing a comprehensive program
- devising measures to secure the physical facility and information
- evaluating personnel for security risk and competency in organism identification methods, security protocols, and emergency response
- managing inventory with complete accountability against theft/loss

Controlled access to laboratories is required for BSL2 and BSL3 organism containment. According to the CDC, the security and access control required for BSL2 and BSL3 labs may be adequate for biosecurity at the sentinel lab level, but this should be verified for each individual situation. In addition, a particular concern about bioterrorism is the use of genetically modified agents. Organisms that have been altered to be more dangerous may no longer fit the procedures of the BSL to which they are traditionally assigned. Bioterrorism events may necessitate changes in organism manipulation based on the situation.

Infectious agents of note

Infectious agents usually of greatest concern to medical laboratory workers are *Mycobacterium tuberculosis* and 3 bloodborne pathogens: hepatitis B virus (HBV), hepatitis C virus (HCV), and the human immunodeficiency virus (HIV; the virus that leads to acquired immunodeficiency syndrome [AIDS]). By far, the easiest virus to contract from infected blood is HBV. (Per the CDC, seroconversion after parenteral exposure to HIV is 0.3%, to HCV is 1.8%, and to HBV is 6%-30%.)

The bacillus Calmette-Guérin (BCG) vaccine for tuberculosis is not currently recommended routinely for laboratory workers in the US. However, vaccinations for this and other agents should be evaluated on a case by case basis in laboratories with special situations or with workers who are particularly at risk, such as those who are pregnant, have a chronic disease, or are immunocompromised.

Vaccination against HBV is highly recommended for persons at risk, and this includes most medical laboratory workers. By law, employers must offer the HBV vaccine at no cost to all staff members who may be exposed to blood and body fluids. There is, unfortunately, no similar vaccine for HCV or HIV. HIV infection is not curable, but patients on aggressive antiretroviral therapy currently survive for many years after infection. Recent advances in HCV therapy include drugs that cure a high percentage of infected individuals, but people who harbor the virus chronically have significant liver complications. Therefore, protocols to prevent exposure to blood and body fluids must be followed meticulously (eg, the CDC guidelines described below). Further, workers should be familiar with the signs/symptoms of infection **t5.3**, so that if they are exposed, treatment can begin as soon as possible.

The advent of several antiretroviral drugs to treat AIDS has allowed researchers to establish prophylactic protocols for workers inadvertently exposed to HIV. Many of these protocols are quite effective in preventing HIV infection. Prophylactic protocols for HBV infection are also available. The CDC publishes recommendations for exposure to blood (www.cdc.gov/HAI/pdfs/bbp/Exp_to_Blood.pdf, accessed December 23, 2015), which are updated as new drugs and treatments are developed. Appendix 5.1 lists current recommendations for postexposure programs. Specific drugs used for postexposure prophylaxis are not included because they could change regularly; current CDC recommendations must always be accessed when an exposure occurs.

Healthcare workers who are infected with bloodborne pathogens such as HIV, HCV, and HBV have an ethical obligation to inform their employers so that protocols can be employed to minimize the risk of transmission. Excellent guidelines can be found in the article "SHEA guideline for management of healthcare workers," in the March 2010 issue of *Infection Control and Hospital Epidemiology*.

Tuberculosis (TB) infection has historically been curable with antibiotics, but in recent years, dangerous strains of this bacterium have been isolated that are resistant to multiple antibiotics. Because this organism is almost always spread through inhalation of aerosols, it is far too easily acquired in the laboratory and in the community. TB is more common globally than in the US, so global travel and immigration contribute to the number of cases seen in US hospitals. Therefore, increased attention has been paid to surveillance and prevention of this disease. The success of prevention efforts in the workplace was apparent when OSHA withdrew its own tuberculosis standard in 2003 and began to refer employers to the CDC guidelines instead.

The official CDC recommendations at publication time regarding tuberculosis are entitled, "Guidelines for preventing transmission of *Mycobacterium tuberculosis* in health care settings" (as published in the December 30, 2005, issue of *Morbidity and Mortality Weekly Report*). This document contains the following recommendations regarding tuberculosis: 1) identifying who should be tested for tuberculosis and how often, 2) isolation and management procedures for tuberculosis patients, 3) treatment protocols for workers who test positive, 4) ventilation and safety equipment (including N95 masks), and 5) laboratory and decontamination procedures. Although the multidrug resistant strains of *M tuberculosis* are dangerous, they are transmitted identically to ordinary strains, so special safety precautions are not necessary according to the CDC.

An important development in tuberculosis control is that in addition to the traditional skin test and X rays, blood tests (interferon-γ release assays [IGRAs]) are available. These tests have improved detection of active tuberculosis with less interference from other mycobacteria and BCG vaccination. The CDC indicates that the blood test can be used instead of the traditional skin test in all situations and has published guidelines for its use. Immunocompromised patients, especially those with HIV, have limited immune responses and may show false negative results to either skin or blood tests. People who have received the BCG vaccine, typically those from other countries, would have false positive skin tests but negative IGRA tests if they do not have active TB.

Baseline 2 step skin testing is now recommended for initial evaluation of healthcare workers. 2 tests are administered 1-3 weeks apart; if results of both tests are negative, nothing more is required. Regardless of the result of the first test, if the second test is positive, the worker must be evaluated as a candidate for latent tuberculosis infection therapy. (A positive result following a negative one is believed

 ISBN 978-089189-6463

to be due to immune boosting, allowing better detection of a low level immune response.) Employees still must be monitored annually through skin testing or questionnaires that assess risk of disease. Staff conversions from negative to positive tuberculosis tests must be monitored and followed up. Careful adherence to CDC guidelines is necessary to distinguish conversions from boosted reactions to previous skin tests or vaccination. IGRA blood tests for tuberculosis do not need to be performed in 2 steps and do not suffer from the boosted reaction problem.

Persons who have received the BCG vaccine or who have been successfully treated for tuberculosis will have persistently positive TB skin tests if they have normal immune systems. After a chest X ray has established lack of infection, repeated chest X rays in this population are not necessary unless symptoms of TB develop. IGRA tests are useful in this situation as they are only positive when TB is active.

The infectiousness of the tuberculosis patient/specimen is worse with the following:

- presence of cough, cavitary disease, and/or acid-fast bacilli in the sputum
- involvement of the larynx or pleura in the infection
- inadequate antibiotic treatment
- aerosol generation to include uncovered cough and medical procedures
- small enclosed spaces with air recirculation and/or inadequate ventilation
- inadequate cleaning and disinfection
- improper specimen handling

When suspected tuberculosis cases are admitted, airborne precaution protocols are generally initiated. However, because isolation facilities are limited and extra precautions can be costly, the ability to discontinue the precautions must be prompt. Given that *M tuberculosis* grows slowly in culture, the CDC recommends that 3 negative sputum smears collected at 8-24 hour intervals, 1 from early morning, are sufficient to discontinue airborne precautions. Positive smears and other tuberculosis test results must be reported equally promptly, generally within 24 hours. Because making and staining smears is generally considered nonaerosol producing, smaller laboratories operating at BSL2 can usually provide this service to rapidly categorize patients. All other activities with this organism should be conducted in a BSC and/or using BSL3 protocols.

One class of unusual infectious agents is the prion, a proteinaceous infectious particle. These particles are incompletely characterized and not well understood, but they appear to be composed of only protein. (All other infectious organisms possess either DNA or RNA.) Bovine spongiform encephalitis (BSE; "mad cow disease") and Creutzfeldt-Jakob disease (CJD) are rare, fatal neurodegenerative diseases believed to be transmitted by prions. Chronic wasting disease, which is seen in deer and elk in parts of the central United States, also is caused by a prion, and may be a problem for humans who handle deer and elk carcasses and/or consume the meat.

Current CDC literature states that BSL2 and BSL3 protocols are adequate for handling human prions, depending on the procedure. The blood and body fluids of patients with CJD have a low risk of being infectious. However, contact with neurologic tissue (brain, spinal cord, and adjacent structures) is associated with a very high risk. This category of tissue specimens is relatively rare in most laboratories because it is not lightly removed, but exposure during autopsy can be significant, and special protocols are required. Prions are resistant to many of the standard decontamination procedures, including standard formalin fixation of tissues; additional sterilizing measures are necessary.

In 2014, there was a serious outbreak of Ebola virus infection in West Africa, and several people infected with the virus were treated in the United States. 2 healthcare workers in Texas were infected while treating a patient who had recently been in Africa. The disease is easily transmitted by blood and body fluids from symptomatic infected patients, and specimens from Ebola patients are likely unsafe when handled in typical medical laboratories that operate using biosafety level 2 protocols. At this writing, presumptive testing for Ebola virus is available through the Laboratory Response Network, but all specimens that are preliminarily positive must be confirmed at the CDC. A limited number of sites have been designated as Ebola treatment centers, and ideally all patients with Ebola should be transported there. Reliable decontamination of reusable laboratory instruments and devices is very difficult, so testing at these centers is skewed toward single use, disposable, near patient devices and methods. Instruments used for specimens from suspected or known Ebola patients should be isolated and dedicated to testing these specimens only. It is crucial to avoid and contain aerosols, so manipulations in a BSC are preferred as long as it is not overloaded and airflow is not compromised. Work should be performed behind splash shields if BSCs are not practical. Workers should double glove, and wear fluidproof body coverings and full face shields or masks and goggles. Specimens should only be centrifuged in sealed cups and tested in closed devices that block aerosols. Whole blood testing is preferred when possible to avoid centrifugation. Point of care devices more often use whole blood and are small enough to be placed in BSCs, so these are often the instruments of choice. Specimen exteriors should be disinfected prior to double bagging and sealing in a rigid container for transport, and pneumatic tube systems should not be used. Shipping protocols for these specimens from sentinel labs are outlined by the American Society of Microbiology (www.asm.org/images/pdf/Clinical/pack-ship-7-15-2011.pdf, accessed December 23, 2015), and they should be followed closely. In the event of another outbreak, the CDC is the most reliable source of guidance on handling these patients and their laboratory testing.

Universal, standard & transmission based precautions

In the 1980s, when AIDS was becoming a concern, the CDC issued recommendations that Universal Precautions must be observed when handling blood and other potentially infectious materials. In essence, Universal Precautions dictate that every specimen should be considered infectious, regardless of the source or the known infectious status. Subsequently the CDC replaced the term *Universal Precautions* with *Standard Precautions* and *Transmission Based Precautions* for healthcare workers. The differences between the 3 terms are summarized below.

Universal Precautions applied to blood, any fluid visibly contaminated with blood, semen, vaginal secretions, tissues, cerebrospinal fluid, synovial fluid, vitreous fluid, wound exudates, pleural fluid, peritoneal fluid, pericardial fluid, and amniotic fluid. Generally, sweat, tears,

 ISBN 978-089189-6463

sputum, saliva (in nondental settings), nasal secretions, feces, urine, vomitus, and breast milk do not require Universal Precautions unless they are visibly contaminated with blood. Because the exact source of the many fluids encountered in the medical setting could not be known, many healthcare facilities practiced what was called "body substance isolation" and treated virtually every body substance as if it were infectious. This prevented staff members from having to determine whether blood was "visible" or if an unusual fluid required protection. Since then, the CDC synthesized Universal Precautions and body substance isolation into a recommendation entitled "Standard Precautions." Standard Precautions are taken with all people and specimens, to include blood, as well as all body fluids, secretions, and excretions except sweat (presence of visible blood no longer matters); nonintact skin; and mucous membranes. Transmission Based Precautions are used in addition to Standard Precautions when a specific pathogen that has strong likelihood of person to person transmission has been confirmed. The use of the term *Universal Precautions* in the literature and in infectious disease training materials is so common that the reader can still expect to encounter it. In addition, there are nonhealthcare settings in which Universal Precautions are more appropriate. For example, if Standard Precautions were used in a daycare center with healthy children, merely changing a child's diaper would necessitate personal protective apparel, and this may not be reasonable.

Standard and Universal Precautions for handling human specimens virtually always require gloves. This includes specimen procurement activities such as drawing blood. Gloves should be changed between patients. After gloves are removed, hands must be decontaminated to guard against contamination from any undetected leaks in the gloves. If specimens are apt to splash or spill, fluidproof gowns and face protection are also required. The facility's exposure control plan (ECP) must include a task assessment for every job position. In each assessment, the types of exposure for the particular job are determined and the appropriate protective measures established. Task assessments in the ECP are based on normal, healthy workers, but not every worker falls into this category. Staff members who are immunocompromised, pregnant, sick, or unvaccinated against certain illnesses may be at higher risk from certain agents. For example, work with microbes known to cause fetal damage, such as rubella and cytomegalovirus, may not be advisable for pregnant women. Workers with suppressed immune systems may require additional personal protective equipment. Supervisors should counsel these individuals on a case by case basis and default to providing too much protection rather than not enough.

In the *2007 Guideline for Isolation Precautions: Preventing Transmission of Infectious Agents in Healthcare Settings* (www.cdc.gov/hicpac/2007IP/2007isolationPrecautions.html, accessed December 23, 2015), the CDC added the following 3 recommendations to Standard Precautions to protect patients as well as healthcare workers: Respiratory Hygiene/Cough Etiquette, Safe Injection Practices, and Special Lumbar Procedures. Patients and visitors must be educated to minimize transmission of respiratory pathogens and be provided tissues and/or masks as needed. Single use needles must be used in all cases, even if there is ostensibly no potential for contamination (ie, when injecting medication into an intravenous line), and masks must be worn when accessing the spinal or epidural space.

Guidelines for patient contact vary with the type of illness (Transmission Based Precautions). Contact Precautions are initiated when direct/indirect contact is required for organism transmission; therefore, gloves and gowns may be adequate. Droplet Precautions are initiated when the organism is spread by droplets (>5 μm); however, because droplets travel no farther than 3 feet, protective equipment is generally required when within 6-10 feet of the patient. Organisms in this category include SARS associated coronavirus, *Bordetella pertussis*, *Mycoplasma pneumoniae*, influenzavirus, adenovirus, and rhinovirus. Group A *Streptococcus* and *Neisseria meningitides* require these precautions during the first 24 hours of antibiotic therapy. Organisms in aerosols (<5 μm) that are suspended in the air and remain infectious over long distances require Airborne Precautions, eg, isolation rooms, special ventilation, and masks/N95 respirators for providers and, in some circumstances, patients. Measles virus and *M tuberculosis* fall into this category. Recommendations for other organisms are located in Appendix A of the CDC's *2007 Guideline* described above. Additional measures may be required to prevent the worker from infecting an immunocompromised patient. In general, special precautions should be posted at the patient's door and checked before entering the room. The staff handling a particular patient should be consulted if there are any questions.

Specimen containers should be transported in a plastic bag or other outer container so that they are confined in the event of a leak. Ideally, specimen containers from multiple patients should not be placed in the same outer transport container, because if one leaks, it could affect the others, possibly requiring recollection of all specimens. Paperwork related to the specimen should not be placed in the inner specimen container. Specimens being transported in a pneumatic tube system must not only be sealed in an appropriate outer container but also be sufficiently padded to prevent breakage and contamination of the entire tube system. Outer containers and packing intended for tube systems should be tested thoroughly, using specimen containers filled with water, to verify integrity before actual specimens are sent. Specimen containers that are broken, leaking profusely, or cross contaminated should be rejected according to protocols developed by the laboratory. If specimens are being shipped, they should be placed in unbreakable containers and triple packed with enough absorbent material to contain a leak. The outermost container should display the biohazard symbol or comply with regulations for hazard identification (see Unit 1). To protect the privacy of patient information, containers may also be locked or sealed while being handled by nonlaboratory personnel such as couriers or taxi services. Reuse of disposable materials is not recommended, as any undetected leakage would be transmitted to additional specimens, and with age, the integrity of the materials diminishes.

Histology laboratories & autopsy suites

Many activities in histology laboratories and autopsy suites are inherently high risk. Once tissues are in fixed in alcohol, formalin, or paraffin, they are usually considered a low biohazard risk. However, a considerable amount of cutting and dissection of unfixed bodies, organs, and tissues occurs in these laboratory areas, so staff members are at considerable risk for not only severe cuts

 ISBN 978-089189-6463

but also infection. It is important to note that normal tissue fixation and embalming methods do not inactivate infectious prions; additional protocols may have to be used if these agents are suspected. The reader should refer to the detailed CDC recommendations for histology labs, autopsies, and animal necropsies found in *Guidelines for Safe Work Practices in Human and Animal Medical Diagnostic Laboratories: Recommendations of a CDC Convened, Biosafety Blue Ribbon Panel* (see above).

In general, tissue specimens should be fixed as soon as possible, and manipulation of unfixed specimens with sharps should be minimized. In many cases, double gloves, gowns, aprons, and face protection are required because aerosol formation may be difficult to prevent. To some degree, double gloves reduce the risks associated with puncture because of their extra barrier.

The use of tissue grinders, blenders, and bone saws are particularly problematic. Microtomes that cut tissue embedded in paraffin are less biohazardous than cryostats, which cut unfixed, rapidly frozen tissue (freezing does not inactivate infectious agents). Both have exceptionally sharp blades that must be cleaned, changed, and disposed of with great care. Decontaminating agents for cryostats may vary from 100% alcohol for daily cleaning, to mycobactericidal agents for weekly cleaning, to sodium hydroxide for specimens from patients with undiagnosed encephalopathy and/or known Creutzfeldt-Jakob agent. Bleach, a popular decontaminating chemical elsewhere in the laboratory, must never come in contact with formaldehyde, so its use is generally restricted in histology. In addition, bleach is corrosive and can damage many metal surfaces and instruments, so in autopsy suites, glutaraldehyde is often used it its stead.

Autopsies are very high risk procedures, and the Clinical Laboratory Standards Institute (CLSI) recommends constructing autopsy suites using BSL3 standards (eg, with an anteroom, cleanable surfaces, negative room pressure). Incoming bodies may have embedded sharps (eg, intravenous lines, broken bones) and may leak fluids and waste. Bodies should be transported in plastic bags and removed onto the prosection table with great care to minimize aerosols. When reasonable, staff members should wear heavy duty or metal mesh gloves to protect against sharps injury. Autopsy tables must have a source of running water and adequate drainage, as well as ventilation units positioned to minimize exposure to fumes.

Important protocols that minimize autopsy risk include

- allowing only 1 prosector with 1 sharp at a time to work on a body
- subtituting scissors with blunt tips for scalpels when possible
- placing instruments on a tray to be picked up rather than passed hand to hand
- making an uncontaminated assistant (eg, the circulator or the diener) available to answer the phone, take notes, and fetch supplies while the prosector is performing the autopsy
- using foot operated dictation machines
- wetting bone saws before use and attaching them to vacuums to minimize airborne bone dust
- if tuberculosis, rabies, or prion disease is suspected, wearing N95 masks not to be removed until after work has ended *and* a sufficient amount of time has passed for aerosols to settle

5

 ISBN 978-089189-6463

- minimizing aerosols and splashes from procedures such as placing specimens in preservative, fluid aspiration, and flushing the body, and wearing proper facial protection (eg, face shields, goggles)
- electrical devices such as defibrillators should be deactivated and hardware embedded in the body should be removed first to prevent injury
- the head should be covered while the bone saw is in use if infectious prion disease or rabies is suspected
- double gloving (recommended); the outer glove can be removed if it gets too slippery to allow a safe grip; after a maximum of an hour, gloves should be replaced whether doubled or not, as microtears are more likely

Work with laboratory animals

All laboratories should have an insect and rodent control program in effect to prevent the spread of microbes. Good housekeeping, supply inventory management, and trash removal also minimize places that pests can hide. For example, the inner spaces of corrugated cardboard can harbor insects, so unpacking supplies and disposing of packing material promptly are considered best practice.

Some research laboratories use animals for experiments, and these animals are equally capable of spreading disease if handled improperly. Only healthy animals purchased from reputable sources should be brought into the laboratory. The reader should note that Animal Biosafety Levels (ABSLs) 1-4 have been established by the CDC/NIH; very specific requirements for each level are in place regarding facilities, air circulation, and waste management. Another useful publication is cited above, *Guidelines for Safe Work Practices in Human and Animal Medical Diagnostic Laboratories*. Some very basic guidelines for working with animals are as follows:

- autoclave the entire animal cage with its contents left inside before cleaning and disposal
- quarantine new animals and assume all animals, even the controls, are infectious
- check cages daily and at feeding time for dead animals; remove any dead animals in biohazard bags immediately
- wear gloves at all times when handling animals
- inoculate animals using a press cage or sedation; handle sharps with extreme care
- house cages in specially designed rooms that contain aerosols and waste, and allow for hosing down and decontamination
- perform animal necropsies in BSCs; use safe practices similar to those described above for human autopsy
- autoclave or disinfect all dead animals, waste, instruments, gloves, and gowns as you would any other biohazardous waste
- because animal pathogens are associated with additional risk of disease, get vaccinated for any likely zoonoses when possible; medical monitoring for some zoonoses may be necessary

Research using certain animals and/or pathogens may require additional precautions. In all cases, the animals must be treated humanely to reduce pain and distress, and to keep them in the best of health. Best practices for animal research are beyond the scope of this text, and the reader is referred to the American Association for Laboratory Animal Science (AALAS; www.aalas.org) for additional resources.

 ISBN 978-089189-6463

Aerosols & droplets

Aerosols are fine mists of particles up to 5 μm in size forced into the air. They can require up to 1 hour to settle. Splashing can cause airborne droplets to settle faster. Aerosols and droplets might not be seen or smelled, but they might be inhaled and contain suspensions of pathogens. Aerosols or droplets may result from many common laboratory activities, examples of which are listed below:

- **Activity**: "popping" open the stoppers of blood collection containers
  Solution: open stoppers behind a work shield or with a cover over the opening
- **Activity**: pouring liquid samples such as blood and serum from one tube to another
  Solution: use automatic pipetting devices or Pasteur pipettes, and deliver sample gently
- **Activity**: centrifuging uncapped liquids or overfilled liquid containers
  Solution: centrifuge capped liquids no >90% full in covered centrifuge
- **Activity**: placing hot inoculating microbiological loop in broth or on media
  Solution: allow loop to cool first
- **Activity**: forcibly delivering liquids when pipetting and/or blowing out the last drop
  Solution: deliver liquids gently, allowing them to run down the sides of the container; never blow into a pipette to expel the last drop or to mix the fluid; never pipette by mouth
- **Activity**: sterilizing microbiological inoculation loop in open flame **f5.5**
  Solution: use closed heat source to sterilize loops, such as the Bacti-Cinerator

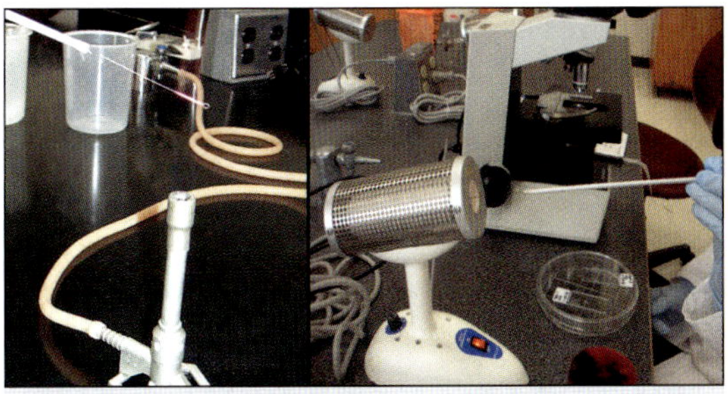

f5.5 left: red hot bacterial loop being sterilized in open flame; aerosols likely; right: bacterial loop being sterilized in closed heat source; aerosols unlikely

- **Activity**: opening centrifuge immediately after breakage of specimen
  Solution: allow the centrifuge to sit for at least 30 minutes before opening
- **Activity**: operating blenders uncovered or vortexing uncovered samples
  Solution: blend, mix, shake, or vortex only covered samples; allow the aerosols inside to settle before opening
- **Activity**: using bone saws in autopsies
  Solution: wet bone saws before cutting and use vacuum attachments
- **Activity**: evisceration in autopsies
  Solution: use single organ (Virchow) evisceration rather than blind evisceration (Rokitansky)
- **Activity**: operating a cryostat to cut tissue without closing window
  Solution: always operate any laboratory instrument with its cover closed

It may be difficult to eliminate aerosols entirely. If aerosols are likely, both organisms requiring BSL2 and BSL3 procedures must be handled in BSCs.

 ISBN 978-089189-6463

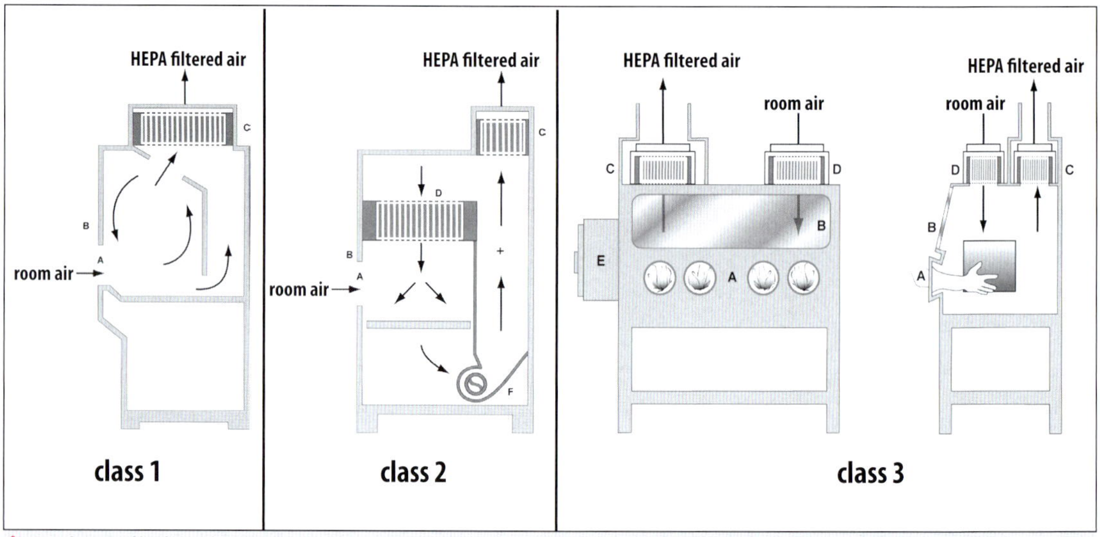

f5.6 classes of biological safety cabinets; A=front opening/glove port, B= sash, C=exhaust HEPA filter, D= supply HEPA filter, E=double-ended autoclave or passthrough box, F=blower (modified from www.cdc.gov/biosafety/publications/bmbl5/BMBL5_appendixA.pdf, accessed December 23, 2015)

5

Biological safety cabinets & air circulation

In a BSC, air flows upward, away from the operator, in order to confine microbes and aerosols to the BSC. The air from the BSC is filtered by a high efficiency particulate air (HEPA) filter, which prevents microbes from being expelled into the exhaust. Schematics of the major types of BSCs are shown in **f5.6**. Class 1 BSCs protect workers, but their contents can be contaminated from the air flowing inward. Class 2 BSCs protect both their contents and the worker; they are suitable for BSL2 and BSL3 work. Among the subcategories of hoods (ie, A, B1, B2, and B3) for class 2 BSCs, B2 is recommended for manipulating tuberculosis cultures because it does not recirculate air and vents 100% to the outside. Class 3 BSCs (sometimes called glove boxes), used for BSL4, are airtight and allow specimens to only be manipulated at the front with gloves. Laminar flow hoods, which are used to make sterile preparations, protect only the product in the hood and not the worker, so they are unacceptable for use with biohazards.

BSCs must be on a regular maintenance schedule to verify that the face velocity of airflow is adequate (usually 75-100 linear feet per minute) and to replace the particulate filters at appropriate intervals. Chemical fume hoods do not have filters and are designed differently, so they cannot be used interchangeably with BSCs. Only very small quantities of chemicals are permitted in BSCs, and if they must be used, air should be exhausted to the outside. Volatile chemicals must be avoided.

Personnel must be trained in the appropriate use of each type of BSC in a laboratory. The checklist for the use of BSCs in Appendix B of CLSI Publication GP17-A3, *Clinical Laboratory Safety*, is useful in establishing protocols. Generally, a BSC must be placed in areas of low air turbulence (low traffic areas) so that airflow is at maximal efficiency, and its sash opening must be in the correct position for proper airflow (usually open no more than 12 inches). A small object, such a piece of

 ISBN 978-089189-6463

yarn, may be taped to the sash, so while using the BSC, staff members can visually determine whether the airflow is adequate. BSCs should not be overloaded or used to store things that are easily kept elsewhere. Supplies should be kept at least 4-8 inches away from the air intake at the face of the BSC, and nothing should be placed inside the BSC that restricts airflow. Air turbulence also disrupts airflow, so devices that will disturb the air, such as blenders and centrifuges, should be placed at the back of the cabinet. Flames also disrupt the air and provide heat that could damage filters, so closed heat sources should be used. Workers should also attempt to reduce the effects of turbulence by 1) turning BSCs on for 4 minutes (per CLSI) before starting work to purge remaining organisms from previous work and establish airflow, 2) gathering all necessary supplies before beginning work to minimize going in and out, (3) inserting supplies and arms gently into the BSC and allowing them to remain still for 1 minute to for organisms to be purged before starting work, 4) moving arms slowly and deliberately while working, to avoid sweeping air into or out of the BSC, and 5) removing arms gently at the completion of work.

To prevent overcrowing, ~2.5 linear feet of workspace for each worker using a BSC is recommended by OSHA, although limiting the BSC to 1 worker at a time is preferred. Workers may need to wear removable sleeve protectors or long sleeves rather than wrist length gloves for better control of contamination after removing their hands from the BSC. Great care must also be taken with objects that are to be removed from the BSC. When possible, objects and trash should be disinfected and sealed *before* removal. This especially applies to BSL3 organisms. Horizontal, not vertical, disinfection pans should be used because they are easier to remove, and the direction of workflow should go from "clean to dirty." Taping signs to the glass sash should be avoided as they can block the view of objects in the hood and interfere with safe manipulation.

Some BSCs are equipped with ultraviolet (UV) lights used to decontaminate the contents inside before they are removed. UV light in the range of 265 nm damages DNA and RNA, and prevents micro-organism replication. It is effective against a wide array of bacteria, viruses, fungi, and even spores (www.cleanhospital.com/pdfs/uv_data_sheet.pdf, accessed December 23, 2015); however, several problems are associated with these devices. First, they must be dusted and cleaned weekly, or the UV light dims, impeding disinfection. Weekly checks of the UV light's intensity are also required. Also, UV light can damage eyes and cause skin cancer, so the devices should not be operated when a room is occupied, unless everyone's skin and eyes are protected and the hood sash is down. Finally, and perhaps worst of all, disinfection with UV light devices is not foolproof. Illuminating all surfaces of an object is difficult, so some portion of it is always in shadow. Also, the further the UV light is from an object, the less effective it is at disinfecting it. In addition, the UV light can only disinfect the surface of an object, so micro-organisms that have penetrated beneath the surface or are situated in a deep layer of organic matter may not be killed. Thus, UV light decontamination of objects inside BSCs is rarely the method of choice. Ethanol (70%) is usually a good disinfectant for the interior of hoods, as long as all sources of heat are first extinguished. Bleach is also very effective, but it is corrosive to metal and requires rinsing.

 ISBN 978-089189-6463

Even if BSCs are in use, the lab must have proper air ventilation (at least 6-12 air changes per hour) and appropriate recirculation. For good containment, the laboratory should be at negative air pressure; airflow should be directed into the laboratory and out via an exhaust so that no air goes from the laboratory into neighboring spaces inside the building. Air can only be recirculated in nonhazardous areas. Most laboratory airflow designs require that doors always remain shut, and some require that BSCs are always switched on. It is important for staff to know if either or both are the case.

Fomites

A fomite is an inanimate object, such as a pencil, test tube, or book, that is not inherently biohazardous but has been exposed to micro-organisms and thus may be capable of transmitting infection. The list of potential fomites in the laboratory is almost endless, and the potential for harm is serious. Fomite transmittal of many infections, such as community acquired methicillin *Staph aureus*, has been documented. To date, no environmental transmission of HIV, HCV, or HBV has been documented; however, HBV in particular has been recovered alive from dried blood spots up to a week old, so fomite transmittal cannot be excluded.

Clothing such as laboratory coats must be considered potentially infectious and should never be worn outside the laboratory, especially not to meals. Contaminated laboratory coats should be hung on separate pegs from clothing intended to be worn outside the laboratory. Grossly soiled clothing used in the laboratory should be removed in leakproof biohazard bags and laundered in hot water and bleach. Ideally, all decontamination should be performed on site. Workers should never take contaminated clothing home; furthermore, OSHA requires that employers provide laundering of contaminated clothing to employees at no cost.

There should be no contact from hand to mouth, nose, or eyes in the laboratory. Eating, drinking, chewing gum, putting on makeup, taking medication, inserting contact lenses, and smoking in the laboratory are strictly forbidden. Foods should *never* be placed in refrigerators or freezers used to store reagents or patient samples.

All laboratory equipment that is exposed in any way to infectious aerosols or handled by gloved personnel should be considered potentially hazardous. This includes items such as telephones, computer keyboards, benchtop supplies, and BSC fronts in addition to pieces of equipment that directly contact microbes such as Petri dishes, pipettes, glassware, and needles. Some laboratories require not only that access to the area be strictly limited, but also that all persons who enter the area wear gloves. Since contamination of items such as telephones and keyboards cannot be detected by sight, it must be clear to staff which items to handle with gloved or bare hands to prevent unnecessary exposure. The use of speakerphones is highly encouraged. Some laboratories label clean areas and equipment to remind staff not to wear gloves, as some electronic devices may be impossible to decontaminate without being destroyed by the disinfecting agents.

Disposable items such as gloves and paper towels should be placed in orange biohazard bags for disposal if their contamination with biohazards is certain. Disposable gloves and lab coats that have been worn for protection but are not visibly contaminated can sometimes be put in ordinary trash, depending on local

 ISBN 978-089189-6463

regulations. Items that are not technically "sharp" but may puncture bags, such as capillary tubes and Pasteur pipettes, should be disposed of in puncture resistant receptacles. Reusable items must be disinfected or autoclaved *before* cleaning. It is essential that these items not be put in ordinary trash or routine cleaning units.

Spills & decontamination

Many porous surfaces cannot be adequately decontaminated, so all surfaces possible, particularly countertops and flooring, must be impervious to chemicals/fluids. Carpets are not permitted, and floors should have as few seams as possible and coved edges.

As stated in Unit 3, CLSI defines a major spill as "a spill that spreads rapidly, presents an inhalation hazard, endangers people or the environment, and/or involves personal injury or rescue, and should be handled as an emergency by the department of public safety, fire department, or hazmat team." Of course the size of the spill matters, but with a biological spill, the organism(s) involved usually matter more. Organisms that require BSL3 or -4 and/or are inhalation hazards are the most dangerous. CLSI specifically lists the following organisms as falling into this category: *Mycobacterium tuberculosis, Neisseria meningitides, Francisella tularensis, Brucella* species, and dimorphic fungi such as *Coccidioides immitis*.

A spill in a BSC is not particularly hazardous as long as the BSC is left on. Gloves must be worn to clean out the inside of the BSC with the appropriate disinfectant. If a spill was large enough to go through the grills, manufacturer's instructions should be followed. In a typical protocol, the drain valve should be closed and disinfectant poured through the grill. After sufficient contact time, usually 20-30 minutes, per CLSI, tubing should be inserted so that the disinfectant can be drained with minimal aerosol generation, and the area can be rinsed. All materials used to clean the spill must be disposed of with a protocol that ensures the organisms are killed.

A spill in the open laboratory is much more dangerous, and during evacuation everyone should hold their breath as long as possible. Universal absorbents such as sand and kitty litter can be used to contain large spills by pouring them around the perimeter of the spill (creating a "dike" around the spill). Aerosols generated by the spill may require immediate evacuation and cessation of air ventilation. Contaminated clothing should immediately be removed, and any residue should be washed off in the shower. Personnel who clean up the spill must wear respiratory protection and gloves at a minimum. They may also need gowns, goggles, respirators, and footwear covers when cleaning a large spill. Whenever possible, re-entry into the area should be delayed for at least 30 minutes to allow the aerosols to settle; BSL3 requires re-entry after spills in the open laboratory be delayed for 60 minutes. A disinfectant of sufficient strength to kill the pathogen(s) spilled must be selected. However, penetration of large spills with high protein content and organism load may be difficult regardless of the disinfectant used. Disinfection should be repeated after the bulk of the spill has been removed.

There are 3 broad classes of decontamination methods. Disinfectants, antiseptics, and sterilants are not the same things. Sterilization procedures by definition kill all micro-organisms, whereas disinfectants and antiseptics may leave some organisms behind. Sterility is

5

 ISBN 978-089189-6463

often not necessary, and the removal of the important human pathogens may be sufficient. In this case, disinfectants are agents used on objects, while antiseptics are used on human skin. The Food and Drug Administration regulates sterilants and antiseptics used for medical purposes, and the Environmental Protection Agency regulates disinfectants used for environmental decontamination.

The choice of the following methods is based on the goal that needs to be achieved:

- heat: boiling, incineration, dry heat (ovens), or autoclaving (steam sterilization under pressure) are effective means to kill microbes, but objects must be heat stable to be used (freezing kills some, but not all, micro-organisms, so cold treatment is not acceptable)
- radiation: UV light or ionizing radiation can sterilize, but these techniques can be cumbersome, especially because exposure of humans to radiation is dangerous; killing efficacy also depends on the light's distance from the source and the exposure time, so it is important to establish protocols unique to each source, including attaining full exposure of 3D objects
- chemical: certain gases can be used to sterilize equipment (ethylene oxide, for example). Numerous low, medium, and high level liquid disinfectants on the market may be suitable for medical laboratories. Low level disinfectants are effective against a wide variety of organisms, but bacterial spores, and many viruses are resistant. Midlevel disinfectants kill most viruses, and fungi, but do not kill spores. High level disinfectants kill virtually everything except high numbers of bacterial spores. Thus, laboratories must know which microbes are likely to be present and select the disinfectant accordingly

Laboratories will most frequently be decontaminated with liquid chemicals, but the efficacy of disinfectants varies widely with the type/quantity of organism and source (eg, culture broth, blood, spinal fluid). Chemicals may penetrate proteinaceous materials poorly, compared to aqueous materials, so chemical concentration and contact time required for killing can vary widely. Gross debridement, such as removal of tissue remnants on autopsy tools, is necessary to ensure adequate exposure to disinfectant. Generally, the disinfectant should be allowed to act for as long as possible (~10-15 minutes) before removal to ensure adequate exposure of the microbes to the agent. Disinfectant may not penetrate large loads adequately, so it may be necessary to disinfect both before and after cleaning.

A simple and easily obtained chemical disinfectant is a fresh 1:10 solution of household bleach (~5.25% sodium hypochlorite). Free chlorine in bleach kills micro-organisms, but it escapes as a gas from squirt bottles. For this reason, the 1:10 bleach solution should be made fresh every 24 hours. Alternately, the Association of Professionals in Infection Control guidelines indicate that a 1:5 bleach solution can be made and used for up to 30 days, and disinfection efficiency will be as good or better than a 1:10 fresh solution. Although this is inexpensive and effective (activity against many spores), bleach is corrosive to some materials and creates toxic products when mixed with ammonia, acid, or formaldehyde. For these reasons, some laboratories may choose to use another midlevel disinfectant such as ethanol, isopropanol, or commercial products. Certain organisms, such as rotavirus, norovirus, and *Clostridium difficile*, may resist some hospital disinfectants, so commercial products should be evaluated carefully. In fact, exposure of *C difficile* to nonchlorine disinfectants has been shown

 ISBN 978-089189-6463

to increase its spore formation as well as its resistance to chemical disinfection.

Work surfaces and the interiors of BSCs should be decontaminated as needed or at least at the completion of work each day. A midlevel disinfectant such as ethanol or 1:10 household bleach will kill almost every organism of concern in a BSL2 laboratory. Most important organisms are deactivated in tissues preserved in 10% formalin in excess of 10 volumes of the tissue.

An important exception to the techniques above is infectious prions. At this writing, additional steps recommended by the CDC if prions are suspected include the following:

- heat resistant instruments: soak in undiluted household bleach or 1 N sodium hydroxide (NaOH) for 1 hour, autoclave for an hour, clean, rinse with water, and then sterilize normally
- surfaces and heat sensitive reusable instruments: flood or soak in 2 NaOH or undiluted household bleach for 1 hour, clean, rinse with water
- waste: incineration is the method of choice for all waste that has contacted prions, including the liquid formaldehyde in which specimens have been preserved

See Unit 4 for additional information on the special containers required for autoclaving NaOH. The CDC has other useful recommendations for prion decontamination, which should be consulted for particular situations.

There may be occasions when high level disinfectants are required to decontaminate environmental surfaces, but this is rare. Many high level disinfectants and sterilants contain glutaraldehyde, a chemical that poses several health hazards. Unit 3 has additional information on the proper handling of glutaraldehyde.

Accidental exposures

Needlesticks, glassware punctures, and injuries involving any piece of equipment that is potentially contaminated should be immediately treated and reported. If a particular patient specimen was involved, that patient may be tested if state law permits. (In some states, patient permission must be obtained before testing can take place.) With their consent, accident victims may also be tested for infectious agents to establish a baseline against which seroconversion may be detected. Treatment protocols depend on the source and organisms suspected, and could include administration of antibiotics, tetanus vaccine, hepatitis B vaccine, hepatitis B immune globulin, and HIV prophylaxis. Procedures should be established at each facility based on current standards of medical care and results of testing. The essential components of infection prevention and exposure protocols are listed in Appendix 5.2. In all cases, OSHA requires that all related medical care at the time of and subsequent to the incident be provided free of charge and be documented in workers' records.

Puncture wounds are not the only means by which pathogens can be acquired. Laboratory coats and gloves should be worn at all times, and goggles should be available if production of large aerosols or splashing is unavoidable. Personnel should always cover any open cuts or wounds and wear gloves in the laboratory.

To reduce the possibility of self inoculation, hands should be kept away from any mucous membranes, such as the eyes, mouth, and nose. Frequent hand washing or decontamination is very important in reducing infections, even when gloves are worn. Unit 10 contains additional information on hand decontamination.

Disposal

All biohazards must be disposed of in a clearly marked container, usually a red or orange biohazard bag. Used biohazard bags are considered contaminated and never reused. Many other types of containers are acceptable as long as the hazardous nature of the waste is clearly indicated. Containers with lids are preferred so that waste can be covered when work has been completed. Sharp objects should never be placed in biohazard bags, and double bagging may sometimes be necessary to prevent leaks. Microbes in the waste must be destroyed before final disposal. This may be accomplished by chemical or radiation disinfection, autoclaving, or incineration. Some states require that biohazard labeling be removed or defaced after trash has been sterilized. Medical facilities must also ensure that specimen labels that personally identify the source patient are also defaced or destroyed. Waste that is infectious must *never* be handled as ordinary trash.

Risk assessment

Taking all of the aforementioned information into consideration, the CDC recommends a 5 step approach to assessing risk and implementing the proper protocols:

1. Identify the biologic agents present or most likely to be present, and assign an initial risk category based on the BSL assigned to that agent. At minimum, BSL2 is assigned to unknown agents in clinical specimens.
2. Review the hazards of the laboratory procedures being conducted. Take steps to minimize dangerous steps in the procedures themselves. For example, potential aerosol generation is unacceptable at any risk category.
3. Make a final determination of the appropriate BSL by considering the inherent hazards of the procedure and those of the agent(s) involved.
4. Train all staff in appropriate safety protocols and document their proficiency. Inspect safety equipment, and document whether all safety equipment is functioning properly.
5. Review the final risk assessment and established protocols with a knowledgeable professional.

The steps above should be reviewed periodically, but especially as information about various organisms evolves. A useful risk assessment worksheet can be found at www.cdc.gov/biosafety/publications/BiologicalRiskAssessmentWorksheet.pdf (accessed December 23, 2015).

 ISBN 978-089189-6463

Summary table: biological hazards

Topic	Comments
Biohazard symbol	1. put on doors, waste, laundry, equipment, and specimens and/or containers 2. must be black on orange background 3. if staff is trained, red bags can substitute for symbol on waste
Microbial sources	1. use Universal/Standard Precautions and CDC/NIH biosafety level guidelines 2. wear gloves and lab coat to handle biohazards; use fluidproof lab coat and face protection and work in a BSC or behind shield if splashing likely 3. assume all lab animals are infectious, and handle them using animal biosafety levels 4. segregate biohazards and decontaminate for disposal 5. evaluate staff with compromised immune systems or special situations 6. follow Transmission Based Precautions for specific diseases 7. conduct risk assessment based on microbes present and tasks; assign BSL2 at a minimum for unknown organisms in clinical specimens, but process all specimens in BSC
Biosecurity	1. consult regulated microbial agents and toxins listed by CDC and USDA 2. consult CDC bioterrorism guidelines for referral labs 3. consult ASM sentinel lab guidelines for most medical labs 4. evaluate restricted access for BSL2/3 labs to see if adequate in sentinel lab 5. develop bioterrorism plan that includes protocols for organism identification and referral, risk assessment of personnel and facility, and information and microbe security
Notable infectious agents	1. know signs/symptoms of diseases caused by microbes in lab; maintain medical surveillance 2. get vaccinated against HBV and test for antibodies when complete 3. consult CDC postexposure protocols, eg, antiretroviral drugs for HIV 4. make sure TB screening is current: single blood test adequate, 2 step skin test for initial screening 5. treat CNS tissue as high risk for prions; use special decontamination procedures for prions 6. isolate Ebola specimens from instruments and main laboratory area; ideal testing is with small point of care devices in BSC
Precautions categories	1. Universal Precautions: use for blood and all body substances likely to have blood or visibly contaminated with blood 2. Standard Precautions: use for all blood and body substances except sweat 3. Transmission Based Precautions: use special protocols based on presence of an infectious agent
Histology and autopsy	1. histology and autopsy high risk from dissection of unfixed tissue; follow protocols; freezing does not inactivate all pathogens 2. minimize risks from sharp objects (eg, bone saws, scalpels, sharps associated with body, cryostat/microtomes) with appropriate handling protocols 3. minimize aerosols (eg, by operating the cryostat closed, using wet bone saws with vacuum, transferring bodies with minimal splashing) 4. decontaminate appropriately with at least a midlevel disinfectant; use special protocols for suspected prions; don't mix bleach and formaldehyde
Aerosols and droplets	to avoid aerosols and droplets: 1. open tubes with cover over the opening, or work behind shield 2. use automatic pipettes and deliver liquids gently, allowing them to run down the side 3. operate all instruments with covers closed 4. cap all liquids to be centrifuged and use covered centrifuges 5. allow aerosols to settle for 30 minutes before opening a centrifuge containing breakage 6. use a closed heat source for sterilizing loops, and allow loops to cool before inoculation 7. only blend, mix, or vortex covered samples; allow aerosols to settle before opening 8. use BSCs if aerosols cannot be avoided; verify airflow and clean HEPA filters on schedule

Summary table: biological hazards (continued)

Topic	Comments
BSCs and air circulation	1. verify airflow (75-100 lfpm) and clean HEPA filters for BSCs on schedule 2. don't use chemical fume or laminar flow hoods for biohazards; avoid strong chemicals in BSCs 3. know BSC capabilities; use at least Class 2 in medical labs a. Class 1: protects worker only b. Class 2: protects worker and experiment, but not sealed c. Class 3: protects worker and experiment, but sealed "glove box" 4. minimize air disturbance around BSCs, and make sure airflow is unrestricted 5. allow 2.5 feet per worker and set up "clean to dirty" workflow in cabinets 6. lab air pressure should be negative with 6-12 air changes per hour, with no air recirculation from hazardous areas 7. decontaminate with alcohol or bleach; use UV light cautiously
Fomites	1. any objects exposed to biohazards are fomites 2. remove lab coats and wash hands when leaving lab 3. do not eat, drink, smoke, apply cosmetics, insert contact lenses, or chew gum in lab 4. separate equipment/computers/telephones that should be used with a gloved hand from those that can be used with bare hands
Spills	1. if inside BSC, leave BSC on, and wear gloves when disinfecting spill 2. If outside BSC, contain spill with absorbent and evacuate if necessary; BSL3 spills require 1 hour to settle 3. disinfection requires gloves and may require respiratory protection, gowns, goggles, and/or footwear covers 4. disinfect with correct level of disinfectant, and let it remain in contact with the spill for 10-15 minutes for maximum effect; disinfection may need to be repeated for large spills, once the bulk is removed
Decontamination	1. major categories: heat/pressure, UV/radiation, chemical 2. chemical disinfectants a. low level (kills some viruses and vegetative bacteria) b. midlevel (kills almost all viruses, fungi and bacteria, including TB): daily 1:10 solution of bleach or every 30 days a 1:5 solution c. high level (kills everything except large numbers of spores): may contain glutaraldehyde (see Unit 3) 3. more rigorous treatment required if prions known or suspected, or if specimen is high risk (brain, spinal cord, or central nervous system tissue
Accidental Exposures	1. report the accident and seek immediate medical attention 2. exposure source and exposed individual may be tested for infections 3. facilities must establish treatment protocols consistent with current medical practice, which may include vaccination, antibiotics, antiretrovirals, tetanus shots, immune globulin, etc 4. encourage vaccination in workers when appropriate; know disease signs and maintain medical surveillance (eg, TB testing)
Disposal	1. use biohazard symbol or red bags; remove symbol when waste decontaminated 2. double bag to prevent leakage; sharps must be placed in puncture resistant container 3. before disposal, waste must be decontaminated by chemical agents, heat, or radiation

 ISBN 978-089189-6463

Self evaluation questions

1 The symbol at the right is for

- **a** biohazards
- **b** electrical hazards
- **c** radiation hazards
- **d** chemical hazards
- **e** compressed gas hazards

2 Specimens from patients who are positive for human immunodeficiency virus (HIV) should be handled at which biosafety level (BSL)?

- **a** 1
- **b** 2
- **c** 3
- **d** 4
- **e** 5

3 For any manipulations EXCEPT making microscopic smears, *Mycobacterium tuberculosis* specimens should be handled at which biosafety level (BSL)?

- **a** 1
- **b** 2
- **c** 3
- **d** 4
- **e** 5

4 What type of containment is necessary for handling nonpathogens in a teaching laboratory?

- **a** class 2 biological safety cabinet
- **b** gowns, goggles, masks, gloves, shoe covers
- **c** standard microbiological procedures: minimal aerosols, hand washing, no mouth pipetting, benchtop decontamination
- **d** all of the above
- **e** none of the above; nonpathogenic organisms require no special handling

5 When you handle human materials that might contain blood or are visibly contaminated with blood, you must assume that each specimen is infectious and treat it accordingly. What is this philosophy is called?

- **a** Universal Precautions
- **b** Standard Precautions
- **c** Transmission Based Precautions
- **d** Ordinary Precautions
- **e** all of the above

 ISBN 978-089189-6463

6 The only body fluid exempt from Standard Precautions is:

a urine
b tears
c sweat
d vitreous humor
e cerebrospinal fluid

7 Which of laboratory worker(s) below might require protection in excess of that recommended by the CDC/NIH biosafety levels?

a pregnant worker
b worker who is HIV+
c worker with a kidney transplant
d cancer patient who recently completed chemotherapy
e all of the above

8 For which of these infections is the recommended baseline the 2 step skin test?

a hepatitis B virus
b hepatitis C virus
c human immunodeficiency virus
d *Mycobacterium tuberculosis*

9 Workers should begin antiretroviral drug therapy prophylactically when exposed to which of the following?

a hepatitis B virus
b hepatitis C virus
c human immunodeficiency virus
d *Mycobacterium tuberculosis*

10 All workers who come into contact with blood and body fluids should get the vaccine for which of the following?

a hepatitis B virus
b hepatitis C virus
c human immunodeficiency virus
d *Mycobacterium tuberculosis*

11 Which of the following has no available vaccine and can result in liver complications if the infection does not respond to medication?

a hepatitis B virus
b hepatitis C virus
c human immunodeficiency virus
d *Mycobacterium tuberculosis*

 ISBN 978-089189-6463

12 Is the following statement true or false: "Only rare, exotic and rapidly lethal organisms can be used as weapons by bioterrorists"?

- **a** true
- **b** false

13 The most dangerous specimens for exposure to infectious prions are

- **a** blood
- **b** liver tissues
- **c** lymph nodes
- **d** central nervous system tissues
- **e** bone marrow biopsies from patients with zoonoses

14 Which protocol is correct for handling laboratory animals?

- **a** monitor for zoonoses when the laboratory animals involved are primates
- **b** only wear gloves when handling animals inoculated with pathogen
- **c** autoclave cages after their contents have been removed and the cages cleaned
- **d** animal autopsies may be conducted on stainless steel counters as long as the work surfaces are decontaminated afterward with 10% bleach
- **e** none of the above

15 All of the following procedures for minimizing aerosols are CORRECT EXCEPT:

- **a** opening a capped tube with a tissue over the cap
- **b** centrifuging capped tubes of blood from an Ebola patient in closed carriers
- **c** keeping a centrifuge closed for 30 minutes after breakage
- **d** delivering liquid to a flask by allowing it to run down the side
- **e** making sure that tubes are completely full of fluid before centrifugation

16 A class 3 biological safety cabinet is most appropriate for biosafety level (BSL)

- **a** 1
- **b** 2
- **c** 3
- **d** 4
- **e** 5

17 Which of the following is NOT a potential fomite?

- **a** pencil
- **b** purse
- **c** telephone
- **d** eyeglasses
- **e** all of the above are potential fomites if exposed to microbiological aerosols

 ISBN 978-089189-6463

18 If a microbiological spill occurs in a biological safety cabinet you should

- **a** leave the fans on, put gloves on, contain the spill with absorbent material
- **b** turn off the fans immediately to prevent microbes from entering the exhaust
- **c** clean the spill with soap and water, apply disinfectant
- **d** a & c
- **e** b & c

19 All of the following can be used to dispose of biohazardous waste EXCEPT:

- **a** autoclaving
- **b** incineration
- **c** landfill burial
- **d** chemical disinfection
- **e** exposure to ionizing radiation

20 Which of the following is a midlevel disinfectant appropriate for decontamination of a BSL2 countertop?

- **a** 1:5 dilution (20%) of household bleach made 15 days ago
- **b** 1:10 dilution (10%) of household bleach made 5 days ago
- **c** commercial product labeled as "tuberculocidal" (kills *Mycobacterium tuberculosis*)
- **d** a & c
- **e** a & b
- **f** all of the above

21 Which of the following inactivates infectious prions?

- **a** formalin fixation
- **b** 10% household bleach
- **c** 30 minutes of autoclaving at 140°C
- **d** all of the above
- **e** none of the above

22 Current bioterrorism guidelines for sentinel laboratories in the United States are published by

- **a** National Institutes of Health
- **b** American Society of Microbiology
- **c** Department of Homeland Security
- **d** Centers for Disease Control and Prevention
- **e** Occupational Safety and Health Administration

 ISBN 978-089189-6463

23 Regulated microbial agents and toxins with potential for bioterrorism are published by

- **a** US Department of Agriculture
- **b** American Society of Microbiology
- **c** Department of Homeland Security
- **d** Centers for Disease Control and Prevention
- **e** a & c
- **f** c & d
- **g** a & d

24 A medical lab scientist is examining a culture from a wound that may be *Franciscella tularensis*. What should be done next?

- **a** notify a supervisor
- **b** verify how to follow-up using the lab's bioterrorism protocols
- **c** move all materials associated with the culture to a biosafety cabinet
- **d** all of the above
- **e** handle the microbe using BSL2 protocols

25 In a BSL2 laboratory

- **a** every instrument must be labeled with the biohazard symbol
- **b** every entrance into the laboratory from public space must be labeled with the biohazard symbol
- **c** waste that is not visibly contaminated with blood can be placed into ordinary trash
- **d** all of the above

26 Which of the following is (are) permitted in laboratories handling microbes?

- **a** plants
- **b** holiday decorations
- **c** pets
- **d** all of the above
- **e** none of the above

27 In the US Laboratory Response Network, community hospital laboratories are considered

- **a** reference labs
- **b** national labs
- **c** sentinel labs
- **d** BSL2 labs
- **e** BSL3 labs

28 A new employee who was born in another country had a negative skin test for TB. 3 weeks later when the skin test was repeated, it was positive. Which is most likely?

- **a** BCG vaccination
- **b** latent tuberculosis
- **c** error in the first test
- **d** error in the second test

29 The Interferon γ Release Assays for TB have which of the following advantage(s)?

- **a** there is no false positive from BCG vaccine
- **b** they only need to be performed once to detect latent TB
- **c** there is no false negative in subjects with compromised immune systems
- **d** all of the above
- **e** a & b
- **f** b & c

30 A person suspected of having Ebola is admitted to your hospital. How should you handle specimens from this person?

- **a** transport them in the pneumatic tube system to minimize exposure in the hospital corridors
- **b** centrifuge them in sealed cups
- **c** test them using normal instrumentation, but be sure aerosol shields are closed and double glove
- **d** use small, near patient devices with disposable testing containers whenever possible
- **e** all of the above
- **f** a, b & d
- **g** b & d

31 Each autopsy protocol below is CORRECT EXCEPT

- **a** minimum of 2 prosectors per body
- **b** at least 1 diener in the room
- **c** priority to remove sharps from body such as IV needles
- **d** BSL3 protocols
- **e** replace scalpels with blunt end scissors when possible

32 Which protocol(s) below is (are) CORRECT regarding the use of biosafety cabinets?

- **a** keep UV light on while manipulating organisms
- **b** load the cabinet with all materials needed for work and then turn it on
- **c** disinfect by spraying with 10% bleach and letting soak overnight
- **d** keep objects at least 4 inches away from the front air intake grate
- **e** all of the above

 ISBN 978-089189-6463

33 Which situation below is most likely to have organisms that are infectious?

- **a** autoclave that completed its full cycle and the indicator tape turned color
- **b** cryostat that contains frozen tissue
- **c** blood spill cleaned with glutaraldehyde
- **d** empty biohazard cabinet exposed all night to UV light

34 You cut yourself on a test tube that contains blood. The cut is about a half inch long and is not bleeding profusely. What should you do FIRST?

- **a** call 911
- **b** call your supervisor
- **c** wash the cut thoroughly with soap and water
- **d** soak the cut in bleach
- **e** fill out an incident report

35 Should laboratories that house micro-organisms be kept locked?

- **a** yes, access should be controlled at all times
- **b** only if they contain something on the "Select Agents and Toxins" list
- **c** only if they are BSL3 or BSL4
- **d** yes, unless they are BSL1

Answers

1 a
2 b
3 c
4 c
5 a
6 c
7 e
8 d
9 c
10 a
11 b
12 b
13 d
14 e
15 e
16 d
17 e
18 a
19 c
20 d
21 e
22 b
23 g
24 d
25 b
26 e
27 c
28 b
29 e
30 g
31 a
32 d
33 b
34 c
35 a

 ISBN 978-089189-6463

Appendix 5.1: HHS and USDA Select Agents & Toxins, 7CFR Part 331, 9 CFR Part 121, and 42 CFR Part 73

HHS select agents & toxins

Abrin
Botulinum neurotoxins*
Botulinum neurotoxin producing species of *Clostridium**
Conotoxins (Short, paralytic alpha conotoxins containing the following amino acid sequence $X_1CCX_2PACGX_3X_4X_5X_6CX_7$)[1]
Coxiella burnetii
Crimean-Congo hemorrhagic fever virus
Diacetoxyscirpenol
Eastern equine encephalitis virus[3]
Ebola virus*
*Francisella tularensis**
Lassa fever virus
Lujo virus
Marburg virus*
Monkeypox virus[3]
Reconstructed replication competent forms of the 1918 pandemic influenza virus containing any portion of the coding regions of all eight gene segments (reconstructed 1918 influenza virus)
Ricin
Rickettsia prowazekii
SARS associated coronavirus (SARS-CoV)
Saxitoxin
South American hemorrhagic fever viruses:
- Chapare
- Guanarito
- Junin
- Machupo
- Sabia

Staphylococcal enterotoxins A,B,C,D,E subtypes
T-2 toxin
Tetrodotoxin
Tick borne encephalitis complex (flavi) viruses:
- Far Eastern subtype
- Siberian subtype
- Kyasanur forest disease virus
- Omsk hemorrhagic fever virus

Variola major virus (smallpox virus)*
Variola minor virus (Alastrim)*
*Yersinia pestis**

Overlap select agents & toxins

*Bacillus anthracis**
Bacillus anthracis Pasteur strain
Brucella abortus
Brucella melitensis
Brucella suis
*Burkholderia mallei**
*Burkholderia pseudomallei**
Hendra virus
Nipah virus
Rift Valley fever virus
Venezuelan equine encephalitis virus[3]

USDA select agents & toxins

African horse sickness virus
African swine fever virus
Avian influenza virus[3]
Classical swine fever virus
Foot & mouth disease virus*
Goat pox virus
Lumpy skin disease virus
Mycoplasma capricolum[3]
Mycoplasma mycoides[3]
Newcastle disease virus[2,3]
Peste des petits ruminants virus
Rinderpest virus*
Sheep pox virus
Swine vesicular disease virus

USDA Plant Protection & Quarantine (PPQ) select agents & toxins

Peronosclerospora philippinensis (*Peronosclerospora sacchari*)
Phoma glycinicola (formerly *Pyrenochaeta glycines*)
Ralstonia solanacearum
Rathayibacter toxicus
Sclerophthora rayssiae
Synchytrium endobioticum
Xanthomonas oryzae

**Denotes Tier 1 Agent*

[1]C = cysteine residues are all present as disulfides, with the 1st and 3rd cysteine, and the 2nd & 4th cysteine forming specific disulfide bridges; The consensus sequence includes known toxins α-MI and α-GI (shown above) as well as α-GIA, Ac1.1a, α-CnIA, α-CnIB; X_1 = any amino acid(s) or DesX; X_2 = asparagine or Histidine; P = Proline; A = alanine; g = Glycine; X_3 = arginine or lysine; X4 = Asparagine, Histidine, Lysine, Arginine, Tyrosine, Phenylalanine or Tryptophan; X_5 = Tyrosine, Phenylalanine, or Tryptophan; X_6 = Serine, Threonine, Glutamate, aspartate, glutamine, or asparagine; X_7 = Any amino acid(s) or DesX and; "DesX" = "an amino acid does not have to be present at this position." For example if a peptide sequence were XCCHPA then the related peptide CCHPA would be designated as DesX.
[2]A virulent Newcastle disease virus (avian paramyxovirus serotype 1) has an intracerebral pathogenicity index in day-old chicks (*Gallus gallus*) of 0.7 or greater or has an amino acid sequence at the fusion (F) protein cleavage site that is consistent with virulent strains of Newcastle disease virus. A failure to detect a cleavage site that is consistent with virulent strains does not confirm the absence of a virulent virus.
[3]Select agents that meet any of the following criteria are excluded from the requirements of this part: Any low pathogenic strains of avian influenza virus, South American genotype of eastern equine encephalitis virus , west African clade of Monkeypox viruses, any strain of Newcastle disease virus which does not meet the criteria for virulent Newcastle disease virus, all subspecies *Mycoplasma capricolum* except subspecies capripneumoniae (contagious caprine pleuropneumonia), all subspecies *Mycoplasma mycoides* except subspecies mycoides small colony (Mmm SC) (contagious bovine pleuropneumonia), and any subtypes of Venezuelan equine encephalitis virus except for subtypes IAB or IC, provided that the individual or entity can verify that the agent is within the exclusion category.

Modified from www.selectagents.gov/SelectAgentsandToxinsList.html, accessed December 23, 2015

Appendix 5.2: infection prevention & post exposure protocols

Practice recommendation	Implementation checklist

Establish an infection control policy	A written plan for infection control consistent with Centers for Disease Control (CDC) and US Public Health Service (PHS) guidelines
	Enforcement of prevention strategies
	  HBV vaccine offered within 10 days of assignment; testing for antibodies 2 month after vaccination complete.
	  Compliance with vaccination and safety protocols (Exposure Control Plan)
	Periodic review of policy to incorporate most recent guidelines

Provide training for exposure protocols	Immediate self treatment
	  Washing needlesticks and cuts with soap and water
	  Flushing splashes to the nose, mouth, or skin with water
	  Irrigating eyes with clean water, saline, or sterile irrigants
	Proper reporting channels and documentation
	Established health care providers to evaluate exposure and provide all necessary treatments per current CDC guidelines 24 hours per day

Evaluation & treatment of exposures	Performed with appropriate counseling and consent
	Evaluation of victim
	  Vaccination history and evidence of immunity (such as hepatitis B antibodies)
	  Type of exposure (ie, percutaneous injury, mucous membrane or nonintact skin exposure, and bites resulting in blood exposure)
	  Immune system competence
	  Collection of blood for evaluation of prior exposure
	  Pregnancy testing, as appropriate
	Evaluation of source
	  Type of fluid (eg, blood, visibly bloody fluid, other potentially infectious fluid), tissue, culture, etc
	  Pathogenicity of microbe involved
	  Patient testing, as permitted by law and patient consent
	    Hepatitis and HIV testing (consider using rapid testing)
	    Other microbes
	  Risk assessment of patient based on history and current status
	Selection of treatment(s) based on current CDC recommendations
	  Infusion of immune globulin, such as hepatitis B immune globulin (HBIG)
	  Medications such as antibiotics and antiretroviral drugs for HIV
	  Vaccination, if appropriate, such as hepatitis B or tetanus vaccine
	Counseling
	  Signs/symptoms that require follow up
	  Possible negative effects of treatment
	  Infectious risk and precautions to take, if any, to avoid infecting others
	  Emotional/psychosocial

Posttreatment follow up	Time intervals per CDC recommendations
	Evaluation of
	  Health status, especially signs/symptoms of infection
	  Effects from treatments
	  Compliance with treatments
	  Compliance with behaviors needed to prevent secondary infection.
	Blood collection to detect change in antibodies from baseline and alter treatment, as appropriate
	Additional evaluation of source patient, as appropriate
	Additional testing, such as liver enzymes, to detect occult infection
	Additional counseling, as needed, for psychosocial effects and compliance with treatment

Analysis of incident & response	Staff adherence to protocols, efficacy of procedures, observed outcomes Protocol changes to prevent accidents, improve response or optimize outcomes

Modified from www.cdc.gov/HAI/pdfs/bbp/Exp_to_Blood.pdf, accessed December 23, 2015

 ISBN 978-089189-6463

Unit 6

Compressed gases

Compressed gases are gases that are confined under great pressure in metal cylinders. They are used for various instruments in the laboratory such as flames, carbon dioxide incubators, and gas chromatographs. Types of gas frequently used are air, nitrogen, oxygen, carbon dioxide, helium, argon and propane. Gases can also be liquefied for applications that require extreme cold.

6

Nature of compressed gas

Gases by their nature completely fill and exert pressure on any container. The higher the temperature, the more pressure the gas exerts (as demonstrated by a balloon popping as it gets too close to a candle). Though high pressure allows more gas to be stored in a smaller container, the gas will readily leak out of even the smallest opening.

The US Department of Transportation defines a compressed gas as exerting a pressure of >40 pounds per square inch (psi) at 70° F. The pressure of the Earth's atmosphere is ~15 psi. A standard 55 inch gas cylinder can have an internal pressure of ~3000 psi, >200× atmospheric pressure. Thus, leaks can readily occur, often with great force. More subtle leaks may be difficult to detect because the gas itself often cannot be heard, seen, or smelled, which is potentially very hazardous. For example, a flammable gas could leak out of its cylinder and create conditions in which a single spark could ignite the entire laboratory. Sudden release of pressure, as seen when a cylinder's valve is sheared off or the contents are heated, can cause the cylinder to become a deadly "torpedo" as the gas is quickly released (just like what happens when a balloon is blown up and let go, without tying it off).

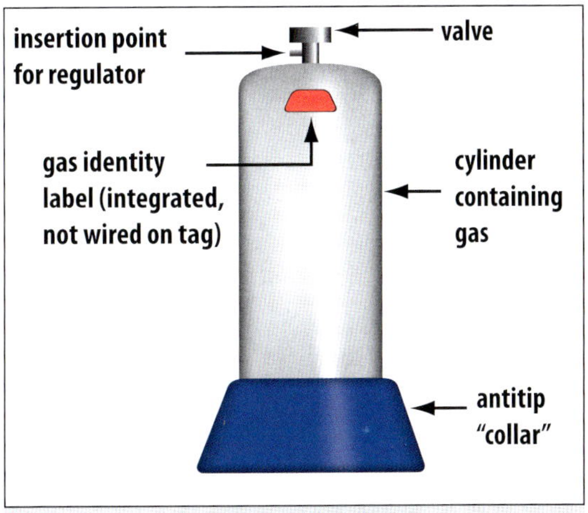

f6.1 schematic of a compressed gas cylinder

Small cans of compressed air are useful for cleaning objects such as computer keyboards. The pressure in these cans is very low compared to compressed gas cylinders. Staff should never use compressed gas cylinders for cleaning purposes.

f6.1 shows a schematic of a typical compressed gas cylinder. A cylinder has a valve at the top, which allows gas to exit when open. A regulator fits into the exit hole and allows users to monitor cylinder pressure and control gas flow from the cylinder. When a regulator is in use, the valve is left open and gas flow is controlled through the regulator. Therefore, the *valve* must be closed to ensure that no gas escapes the cylinder. **f6.2** shows 2 cylinders, one in use with a regulator in place and the other not in use with a valve cover in place.

Compressed gases are also hazardous because of their chemical natures. Oxygen and hydrogen are flammable. Nitrogen has been called "asphyxiating gas" because significant nitrogen leaks reduce the amount of oxygen in the room to fatal levels (anoxic atmosphere). Anoxic atmospheres also occur due to leaks of carbon dioxide and other gases. Gases can also be toxic, corrosive, or reactive. Thus, no amount of leakage from a compressed gas tank is acceptable. Nitrous oxide (aka "laughing gas") is used in medical/dental applications and can cause fatal asphyxiation, but it can also be abused for its intoxicant effects; thus, nitrous oxide cylinders have been targets of theft. Compressed gas cylinders in general could be targeted for theft by terrorists. Cylinders should be locked up and inventory tracked not just for safety, but for protection against unauthorized removal.

Compressed gas cylinders are typically made of steel or some other very sturdy material. Many cylinders are reusable, but laboratories rarely have the capacity to refill them. Typically, the supplier fills empty cylinders, delivers them to the laboratory, and collects the empty cylinders from the laboratory for refilling. Vendors are responsible for testing cylinder integrity every 10 years. The last date of integrity testing should be stamped into the metal. Although users are not responsible for this testing, they are responsible for purchasing from reliable vendors and inspecting cylinders for rust, cracks, worn off labeling, or any other signs of integrity loss. The cylinders in **f6.2** clearly show signs of the wear and tear that can be expected. Since they are routinely moved around, cylinders are constructed to be strong enough to contain the gas while also light enough to transport. It is important not to assume that these containers are too strong to fail.

f6.2 compressed gas cylinders

Safe handling of compressed gas

Compressed gases are hazardous because improper use and storage can result in fire, explosions, asphyxiation, poisoning, or mechanical injury. General guidelines are as follows:

1. **Proper storage.** Gas cylinders have high centers of gravity and tip over easily if unsecured. Cylinders should be used and stored using 2 hour fire construction in secured locations,

 ISBN 978-089189-6463

well away from casual contact by unauthorized persons. Clear signage as to the cylinders' contents and hazards must be displayed on all doors and access points. Cylinders must be stored away from sources of heat and electricity; storage areas should be designated no smoking zones.

Whether in storage or in use, all cylinders must be chained to the wall in a vertical position or secured in a "nontip" or "antitip" base (collar). Valve covers must be on when not in use. Chains should be secure enough to prevent tipping and hold the cylinder in place in the event of a sudden vertical release of pressure. Multiple cylinders packed tightly together can be secured by a single strong chain, but the chain length must be adjustable so that it can fit tightly. Chains must go around the body of the cylinders, not the valve.

To keep incompatible gases apart, ideally cylinders should be stored by type in separate rooms, especially if they are flammable or reactive. Chemicals and waste, especially flammables, salts, and corrosives, should be stored separately from compressed gases.

In all areas where gases are used or stored, it's essential to ensure good ventilation to disperse leaks and lack of moisture to prevent cylinder corrosion. Because some gases are heavier than air, consideration should be given as to where the storage area is located and whether leaking gases could accumulate there. For example, more leaked vapors can accumulate in a basement than an above-ground room with several windows and doors.

Cylinder storage should not be adjacent to building evacuation routes or postevacuation outdoor assembly points. Outdoor storage is permitted, but cylinders must be protected from weather, direct contact with the ground, vehicle/pedestrian traffic, and vandalism. The material most cylinders are made of contains iron, so in hospital settings they must be kept well away from magnetic resonance imagers (MRIs). Gas piped in from a central location must meet similar requirements and building codes.

2. **Safety shutoff valves.** If gas is being pumped in, shutoff valves should be unobstructed and clearly marked. All staff members in the vicinity should know where the valves are and how to use them. Some cylinders have safety release rupture disks that will relieve pressure by melting and expelling gas if the room gets too hot (>125°F). These disks must be inspected and kept in good repair. Any repair or modification of pressure relief valves, piping, or other components must performed only by qualified individuals.
3. **Inventory management.** A "first in, first out" system of inventory management must be used for gas cylinders. Full cylinders should be stored such that the oldest cylinders are used first; those cylinders that have been in storage for too long should be readily identifiable. Only the minimal number of cylinders required for routine laboratory operation should be kept in inventory. A good rule of thumb is to store no more than the cylinder in use, one spare, and/or one week's supply of gas. As stated above, different gas types should be used and stored separately, at least 25 feet apart, to prevent incompatible mixtures. If this distance is not achievable in a single space, specially constructed fire walls may be necessary. Ideally, flammable gases should also be kept separate from nonflammable gases; if the flammable gases burn, they could heat the other cylinders to the point that they explode.

 ISBN 978-089189-6463

Inventory should be tracked to detect unauthorized removal.

4. **Safe transportation.** Valve covers must be on during transportation. Cylinders should be moved using a hand truck f6.3 or other appropriate transportation device, and never rolled, slid, or dropped. Cylinders should never be lifted by their valve caps or valves. Improper transport of gases can result in a sudden, dangerous release of pressure. Large cylinders can be heavy and awkward, so safe transport may require more than one person.

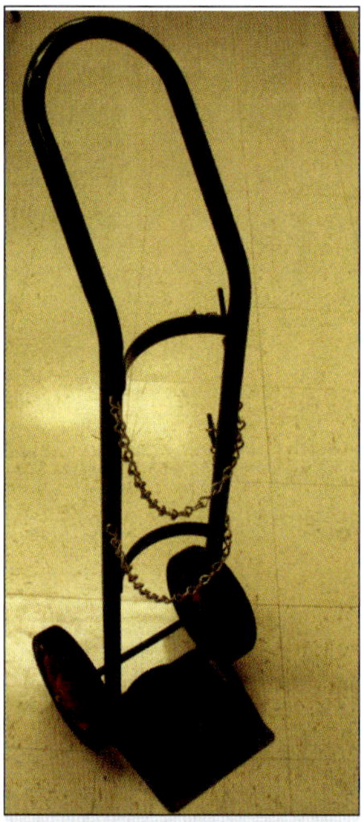

f6.3 hand truck for gas cylinder

5. **Minimal amounts.** The smallest cylinder possible for the job at hand should be used. If gas use is heavy, a large cylinder may help avoid the need to replace the cylinder too often during work. However, larger cylinders present greater hazards, so an optimal balance must be struck.
6. **Labeling.** The contents of each cylinder should be permanently affixed to the outside; wired-on tags are unacceptable. Although each gas type has an assigned cylinder color **t6.1**, never rely on a cylinder's color alone to identify its contents; labels must be checked every time. Department of Transportation and Globally Harmonized System symbols are typically used as well, as shown in f6.4. *Never* use an unmarked cylinder; these cylinders should have be tagged "DO NOT USE" and returned to the manufacturer as soon as possible.

t6.1 Standard colors for medical gases

Gas name	Standard color
medical air	yellow
medical carbon dioxide	gray
medical helium	brown
medical nitrogen	black
medical nitrous oxide	blue
medical oxygen	green
mixture or blend of medical gases	standard colors for each component, eg, 95% oxygen & 5% carbon dioxide = green cylinder with gray band or shoulder

7. **Valve protection.** The weakest area of almost all cylinders is the valve area, so the removable protection caps should be left on until just before use. Valve protection caps should not be removed until the cylinder is secured.
8. **Defect-free cylinders.** "Frozen" or difficult cylinder valves/protection caps must *never* be forced open or inappropriately lubricated. Damage to the valve area can result in rapid pressure release, so hitting the area with tools or inserting tools into a valve is never acceptable. Another cylinder should be used instead, and the defective cylinder returned to the manufacturer. Hydrocarbon based lubricants (eg, oil, grease) can be reactive, especially with oxygen, so only lubricants specifically designed for a compressed gas application should be used.
9. **Proper fittings.** Only those fittings made for the particular gas in use should be connected to a cylinder. Cylinder fittings that connect a cylinder to an instrument are made to fit only specific cylinder types, thus

 ISBN 978-089189-6463

f6.4 top row: DOT gas hazard symbols; bottom row: GHS hazard symbols

preventing incompatible gases from mixing. Therefore, fittings should never be interchanged or lubricated unless according to manufacturer's instructions. To improve the seal of the fittings, many regulators are used with washers, which should be visually inspected before use. It is important to distinguish between sealing washers, which can be reused, and crush gaskets, which are designed for one time use. Serious oxygen fires have resulted from improper sealing of oxygen tanks because crush gaskets have been reused.

10. **Regulators.** Users must learn how to read the regulator's pressure gauges. **f6.5** shows a typical regulator. In general, one gauge measures the amount of pressure within the tank, indicating how full it is. The cylinder valve must be open for the pressure gauge to display this reading; otherwise it reads 0. A second gauge monitors the pressure (flow) of the gas coming out. When the flow regulator is closed, this flow gauge should read 0, even when the tank is full and the pressure gauge shows tank pressure. Proper understanding of how the regulator functions is crucial for safe handling.

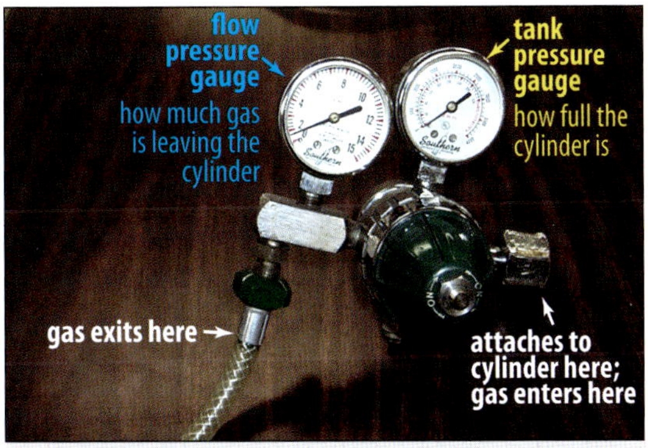

f6.5 regulator

11. **Safe installation.** All users of a gas must understand its hazards and ensure that the cylinder is installed only under safe conditions (eg, no incompatible gases nearby, protective personal equipment worn, connected device in working order, proper operation of cylinder and device). Safety data sheets (SDSs) for every gas should be readily available to users. Users must make sure that the top cylinder valve is closed and that the exit pressure displayed on the regulator is 0 *before* a fitting is removed from a used cylinder. When installing a new cylinder, briefly opening the top valve to allow the gas to expel any dust or debris ("cracking" the tank) may be appropriate, especially with oxygen tanks. When placing a regulator on the new cylinder, all fittings must be secured as tightly as possible to prevent leaks. Because the gauge covers can blow out, users should stand to one side of them with their faces turned away. The top cylinder valve should be opened *slowly* as users listen carefully for hissing sounds, which could indicate leaks. Valves must be quickly closed if a leak is apparent.
12. **No leaks.** Once a cylinder is installed, all connections must be checked for leaks with a soap solution, such as the commercial product Snoop (Ohio Valley Specialty Company, Marietta, Ohio). If there is a leak, bubbles will form in the area directly over the leak. Devices that detect leaks should be installed, if available and/or appropriate.
13. **Daily monitoring and shutoff.** Cylinder pressure should be checked and recorded daily when the gas is in use to detect whether a cylinder is leaking or close to empty. Most users should have a sense of how much gas is used during routine daily operations, so excess gas consumption should be investigated as a possible leak. When the gas is not in use, the cylinder should be closed at the valve to reduce pressure on parts. When the laboratory is unattended, cylinder valves should not be left open unless continuous gas supply is necessary for laboratory operations (eg, carbon dioxide incubators).
14. **Empty tanks.** Empty tanks should be clearly labeled as such and separated from full tanks. If an empty tank is accidentally hooked up to a pressurized system, foreign matter could enter the empty cylinder and cause corrosion, explosion, or fire. Tanks should never be *completely* emptied; this creates negative pressure and suction in the tank. Also, just as one always treats a gun as if it were loaded, one must treat an empty gas cylinder as if it still contains gas. Empty tanks that held incompatible gases should be separated. Empty cylinders must be returned to the vendor for professional refilling. "Transfilling" gas from a full cylinder to an empty one is very dangerous without the proper equipment and skill, and should be forbidden as general laboratory policy.
15. **Emergencies.** Cylinders with slow leaks should be removed from the laboratory immediately and placed in well ventilated area with clear "DO NOT USE" signage. Uncontrolled venting of a gas cylinder is an emergency. If the valve can be shut off easily and safely, this should only be done if not much time has passed; if the gas has been leaking for a significant amount of time, the atmosphere could be toxic, anoxic, or physically dangerous. Thus, spending too much time trying to close a stuck valve or assessing what is wrong with a cylinder could be deadly. Evacuation

 ISBN 978-089189-6463

is the best first choice for all staff adjacent to the leak so that experts can be summoned to evaluate the problem. If possible, ventilation and/or chemical fume hoods should be turned on and doors closed as everyone exits. Victims who have inhaled gas or have passed out must be moved as quickly as possible to fresh air. Unit 2 has more information on establishing evacuation routes.

The Compressed Gas Association (www.cganet.com) has a series of free posters displaying good practice for compressed gas cylinders. Other sample posters are located in Appendix 6.1.

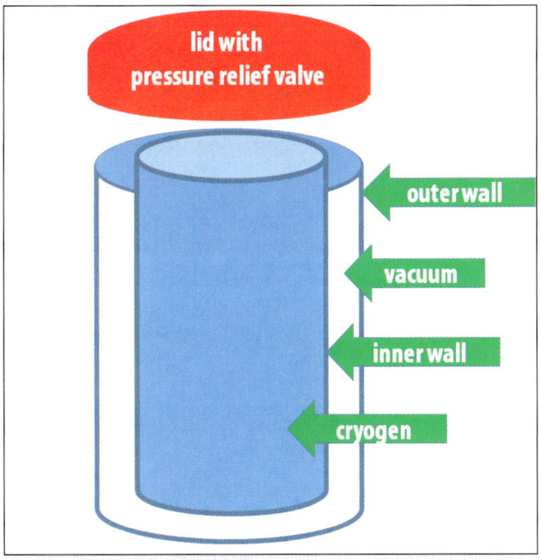

f6.6 schematic of Dewar flask

Cryogens

Gases with boiling points –100°F at 1 atmosphere of pressure are used as cryogens in the laboratory. Many applications in biological research and medical laboratory practice, such as "snap freezing" specimens, use cryogens. Liquid nitrogen and helium are some of the most commonly used cryogens in the laboratory. These gases are explosion hazards if they become warm enough for the liquid form to rapidly turn to gas. Their extremely cold temperature also makes them physical hazards, and in some cases they can be flammable. Since it takes up less space, liquid oxygen is available for medical oxygen supplies, but it causes rapid burning when exposed to fuel. Extremely cold temperatures alone can cause oxygen in the air to condense into ordinary combustibles and increase their inherent flammability; for example, wood saturated with liquid oxygen can be explosive. The materials used for cryogenic applications must be carefully chosen not only to withstand the cold, but also to avoid oxygen condensation. The work area must be kept exceptionally clean and well ventilated, and SDSs for all the cryogens in use should be accessible.

Liquid gases are typically stored in metal tanks, metal cylinders, or Dewar flasks. All should be equipped with pressure relief devices, which must be kept clean and never blocked. Dewar flasks **f6.6** are double walled containers with a vacuum layer in between to provide insulation. Compromise of the insulating layer (loss of vacuum) causes the liquid to warm. If gas starts to form, dangerous pressure could build up. This is likely to trigger a release from the pressure valve, and the gas should be allowed to vent or the container could explode. Cryogens that convert to the gaseous form can also cause anoxic atmospheres, as discussed above, so evacuation and/or aggressive ventilation of the rooms are necessary.

Dewar flasks are often filled from a central supply so that smaller amounts can be used in the laboratory. Gases should only be transferred into flasks that are rated as safe for that gas, and the flask must never be overfilled. A phase separator to remove gas from the liquid should be at the end of the filling line, which is placed directly on the bottom of the flask to prevent splashing. Only metal tubing

 ISBN 978-089189-6463

should be used, as rubber and plastic will crack.

No amount of exposure to cryogen is considered safe. The temperatures required to keep gases liquid can cause catastrophic damage to human skin and the underlying tissues; thus proper personal protection equipment for the eyes, face, and body is essential at all times. Workers should never wear anything that could trap cryogenic fluids against the skin, so watches and jewelry should not be permitted. Should liquid gas spill on clothing, it must be removed quickly so that nothing penetrates to the skin. Gloves, goggles, face shields, laboratory coats, aprons, and other apparel designed to protect against the cold, including specialized "cryogloves," must be worn.

Objects to be frozen must be lowered slowly into the cryogen to avoid splashing, using tongs or a "cryoclaw." Cryogloves are *not* intended to be immersed in cold liquids; they are only for protection against splashes. Test tubes or containers must be rated for used with cryogens. Should the liquid gas splash onto skin, the affected area should be warmed slowly using slightly warm (not hot) water for 15 minutes. Rubbing the affected area could damage skin further, so only water should be used, and medical attention should be sought immediately.

Gases for medical use

The compressed gases used in laboratories usually differ from those gases used in patient care. Medical laboratory staff, however, may have to comply with special procedures to prevent medical gas mix-ups. For example, in 2004, a patient died of nitrous oxide poisoning when an oxygen flow meter was forced into a nitrous oxide outlet (Institute of Safe Medication Practices Medication Safety Alert, 9(24), 2004). A portion of the flow meter designed to prevent insertion into an incompatible gas outlet was broken, and poor lighting in the room made it difficult to distinguish the oxygen tank (green) from the nitrous oxide tank (blue). Medical gas cylinders come in standard colors to identify their contents **t6.1**, but they also must be permanently labeled. As stated above, cylinder labels, not colors, must be checked before use, and fittings should never be forced or interchanged. Cylinders should never be repainted by anyone other than the owner/supplier. Laboratories may use cylinders prepared for industrial use for instrumentation, but these cylinders are *not* acceptable for medical use; facilities that have both industrial and medical use cylinders must label and separate them carefully.

Certain gases or tanks of a particular size may require even more restrictive handling than that outlined herein. As with any equipment or chemical in a laboratory, the rules and regulations that apply to the particular compressed gas should be established before it is used.

ISBN 978-089189-6463

Summary table: compressed gases

Topic	Comments
Hazards	1. contents under pressure; likely to leak, and hazard worse if contents heated; catastrophic pressure release and cylinder becomes "torpedo" 2. contents can be chemical hazards (eg, oxygen and hydrogen are flammable, nitrogen and other gases can reduce room oxygen to fatal levels, some gases toxic, corrosive) 3. valve area is usually weakest point of cylinder; protect it and use valve caps 4. liquid gases used as cryogens: severe physical hazard from cold, explosion hazard if warmed, asphyxiant hazard with gas leaks, and increased flammability of materials due to oxygen condensation 5. potential targets of theft 6. cylinders refilled and reused; important to monitor for damage
Storage	1. chain all cylinders (even "empty" ones) to the wall in a vertical position or use "nontip" base 2. make sure all cylinders are permanently labeled (wired on labels not acceptable); return unlabeled cylinders to manufacturer 3. learn GHS/DOT labels and FDA color codes 4. store away from heat, electricity, chemicals, and waste, in area with good ventilation and lack of moisture 5. segregate empty and full cylinders; clearly label empty tanks 6. separate incompatible gases; separate flammable from nonflammable 7. use "first in, first out" inventory system; maintain smallest inventory possible 8. use good signage; secure against unauthorized or casual access 9. disallow gas use/storage along evacuation routes or near public spaces
Transport	1. use hand truck or other transportation device, especially with heavy cylinders 2. always transport with valve cover on 3. do not pick up by valve or valve cover 4. get help for heavy, large, awkward cylinders
Usage	1. keep shutoff valves to gas lines unobstructed and well marked 2. use the smallest cylinder possible for any job 3. leave valve covers on until just before use 4. never force or lubricate a valve; return to manufacturer 5. never interchange cylinder fittings; do not reuse crush gaskets 6. identify gas by permanent label, not by color or wired on tag 7. close the top cylinder valve (not regulator flow valve) and make sure exit pressure is 0 before removing a fitting 8. learn function of gauges and valves on regulator 9. learn locations of gas shutoff valves and how to use them 10. stand to one side, face turned away, when opening valve for the first time after regulator insertion; close valve quickly if leak heard 10. use soap solution to detect leaks; bubbles will form over leaks 11. close valves if gas not in use, which is safer and reduces pressure on fittings 12. do not completely empty a tank; check pressure daily 13. use proper personal protection and no jewelry/watches for cryogens 14. scrupulously clean work area to avoid oxygen condensation and proper materials for cryogens 15. use tongs or "cryoclaws" to insert objects into cryogen 16. keep cryogens in cool areas and do not block pressure relief valves 17. fill Dewar flasks properly and avoid splashing
Emergencies	1. shut off gas if possible, but only if it can be done safely and very quickly 2. evacuate if leaks uncontrolled; turn on ventilation/hoods and shut doors 3. take inhalation victims quickly to fresh air 4. remove clothing if cryogen spilled onto it 5. treat cryogen burns with mildly warm water; do not rub skin; immediately seek medical attention

6

Self evaluation questions

1 Which of the following is a proper technique in the use of compressed gases?

- **a** store tanks in a horizontal position
- **b** secure only full or partially full tanks
- **c** store all full cylinders in the same area
- **d** keep empty and full cylinders separated
- **e** completely empty tanks to be cost effective

2 A cylinder's tag has fallen off. You should

- **a** open the valve to see what it is
- **b** label it "EMPTY" so no one will use it
- **c** ask around to see if someone knows what it is
- **d** not use that cylinder under any circumstances
- **e** all of the above

3 Cylinder fittings are

- **a** checked for leaks using soapy solutions
- **b** best unfrozen with liberal use of lubricant
- **c** virtually impermeable to gases if installed properly
- **d** interchangeable if cleaned properly before use with different gases
- **e** all of the above

4 All of the following are characteristics of compressed gases that contribute to hazardous nature of gases EXCEPT:

- **a** high pressure
- **b** biohazardous contents
- **c** flammability of certain kinds of gases
- **d** leaks that often cannot be seen or smelled
- **e** dropping a cylinder during transport can shear off the valve

5 Before you remove a fitting on a gas cylinder, the most important thing to do is

- **a** first check the fitting with a soapy solution for leaks
- **b** first lubricate the fitting so that it will come off easily
- **c** make sure the valve protector cap is nearby and ready to be placed back on the valve
- **d** make sure the cylinder valve is turned off and the exit pressure on the regulator is 0

6 The symbol at the right indicates which hazard?

- **a** toxicity
- **b** reactivity
- **c** flammability
- **d** compressed gas
- **e** explosive hazard

6

 ISBN 978-089189-6463

7 How are cryogen liquid gases hazardous?

- **a** pressure from anything in gas phase
- **b** chemical nature of the gas
- **c** physical hazard from extreme cold
- **d** increased flammability of materials from condensed oxygen
- **e** all of the above

8 If the cylinder valve is completely closed on a cylinder, which gauge(s) on the regulator should read 0?

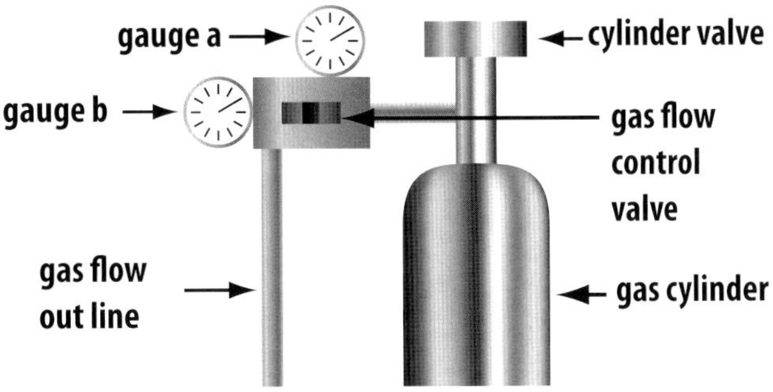

- **a** a
- **b** b
- **c** both a & b
- **d** neither a nor b

9 Specifications for medical gases include all of the following EXCEPT:

- **a** standard colors for each gas type
- **b** prohibition of industrial gas cylinders for medical use
- **c** standard color combinations for gas mixtures
- **d** the ability to bypass regulator fittings to deliver oxygen during a medical emergency

10 A laboratory has an incubator in microbiology that uses ~1 CO_2 carbon dioxide compressed gas cylinder a month. 5 spare CO_2 cylinders are stored in the corner of the room, chained vertically to the wall with their valve covers on. Evaluate this situation.

- **a** this situation illustrates proper storage and inventory of CO_2 compressed gas
- **b** this situation is unacceptable because cylinders should not be stored vertically with chains
- **c** this situation is unacceptable because an excessive number of cylinders are being stored in the laboratory
- **d** this situation is unacceptable because valve covers should be removed when cylinders are placed in the laboratory
- **e** this situation is unacceptable because CO_2 is heavier than air, and these cylinders cannot be stored in the laboratory

11 Propane fuel tanks are stored in the same room as compressed air tanks. Evaluate this situation.

- **a** this is acceptable because these 2 gases are compatible
- **b** this is not acceptable because these 2 gases are incompatible
- **c** this is acceptable if the 2 tanks are at least 10 feet apart
- **d** this is not acceptable because flammable and nonflammable tanks should be separated

6

12 A compressed gas cylinder can be lifted by grabbing its

- **a** valve
- **b** valve cover
- **c** regulator
- **d** all of the above
- **e** none of the above

13 You receive a shipment of compressed gas cylinders, and one cylinder has a couple of dents and pits of rust in many places. The label identifying the gas is peeled off in several places, so you can just barely identify the gas.

- **a** this is to be expected as cylinders age, so the tank is safe to use
- **b** you should check the date on the cylinder to see when its integrity was last checked
- **c** you should connect the gas to your instrument to see if it functions properly before deciding how to proceed
- **d** you should mark the cylinder "DO NOT USE" and return it to the manufacturer

14 You need to "snap freeze" a tissue specimen. What is the best way to do this?

- **a** dispense liquid nitrogen over the specimen in a sink
- **b** insert the specimen into liquid nitrogen while wearing "cryogloves"
- **c** insert the specimen into liquid nitrogen with cold resistant tongs
- **d** any of the above

15 The air conditioning is not working on a hot summer's day, and when you enter the cryogen room, a Dewar flask is venting through its pressure valve. What should you do?

- **a** turn off the pressure valve
- **b** turn up any available ventilation
- **c** notify the supervisor that an evacuation may be necessary
- **d** check the pressure valves on any other liquid gas containers
- **e** a & d
- **f** b & c

16 Wearing unnecessary items such as jewelry and watches when working with cryogens should be avoided because

- **a** they are chemically incompatible with liquid gases
- **b** they interfere with the fit of protective equipment
- **c** they can trap liquid gases against the skin
- **d** they hasten oxygen condensation

 ISBN 978-089189-6463

Answers

1 **d**

2 **d**

3 **a**

4 **b**

5 **d**

6 **d**

7 **e**

8 **c**

9 **d**

10 **c**

11 **d**

12 **e**

13 **d**

14 **c**

15 **f**

16 **c**

 ISBN 978-089189-6463

Appendix 6.1: compressed gas poster

FDA Public
Health Advisory
Medical
Gas
mix-ups
can cause
DEATH
and
serious
injury
FDA
U.S. FOOD AND DRUG ADMINISTRATION
For more information call
1-888-INFO-FDA
or go to
www.fda.gov/cder/dmpq/gases.htm

Gases for medical use are prescription drugs that must be carefully regulated and handled.

Adaptors should never be used and fittings never changed on medical gas containers. If a connection doesn't fit, it isn't supposed to fit. Contact the supplier immediately.

Store medical grade products separately from industrial grade products in well-defined areas.

Educate and train personnel who are directly responsible for handling medical gas to:

- recognize medical gas labels
- examine all labels carefully before hooking containers to the system.

Skilled and knowledgeable personnel should always check the container and connection prior to introducing the gas into the system.

Won't Connect?
Don't Connect!

Patients have been injured – and some have died – because of medical gas mix-ups. This usually occurs when the wrong gas is forcibly connected to the oxygen supply system. Please promote the importance of properly handling medical gases.

- Manufacturers who receive reports of death or serious injury associated with the use of medical gases are required by law to report those incidents to the FDA.
- Hospitals, nursing homes, and other health care facilities should submit reports of such mix-ups (whether or not they resulted in a serious injury) to FDA's voluntary reporting program, MedWatch:

Phone - (800) FDA-1088

Fax - (800) FDA-0178

Mail - MedWatch, Food and Drug Administration
5600 Fishers Lane
Rockville, Maryland 20852-9787

 ISBN 978-089189-6463

Unit 7

Radioactive materials

Radioactive materials are unstable isotopes of elements that change (decay) in some way to form more stable isotopes. In the process of becoming more stable, these materials give off energy and/or particles that can be very harmful to living organisms.

Types of radioactive emissions

Identifying the emission type of a radioactive isotope is important because safe handling methods vary slightly with each type. This Unit discusses 3 types, illustrated in **f7.1**, that could be encountered in the laboratory. A fourth type, neutrons from fission reactions and other atom manipulations, is rarely found in laboratories and will not be discussed.

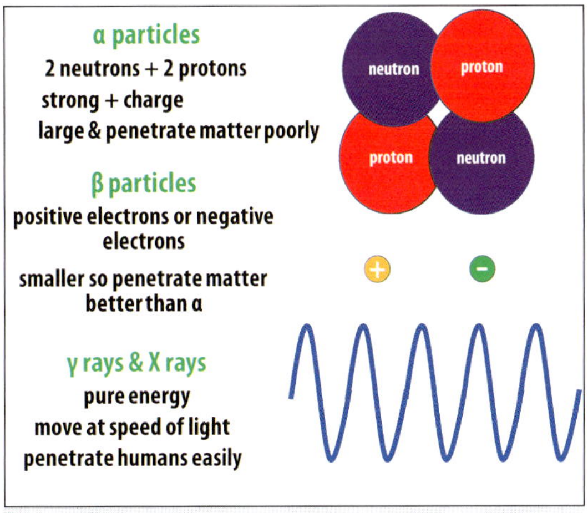
α particles
2 neutrons + 2 protons
strong + charge
large & penetrate matter poorly
neutron
proton
proton
neutron
β particles
positive electrons or negative electrons
smaller so penetrate matter better than α
γ rays & X rays
pure energy
move at speed of light
penetrate humans easily

f7.1 categories of ionizing radiation

α particles

α particles are "naked" helium nuclei, 2 protons and 2 neutrons. Because they are large and particulate in nature, they are not able to travel far or penetrate any substance deeply, including human skin, so while they can cause significant damage to humans because of their strong charge, the damage is only to localized areas. These particles are especially dangerous if they are ingested, injected, or inhaled, as they will cause internal damage. Gloves, plastic, and Lucite easily block α particles; therefore lab coats and gloves made of latex or nitrile provide adequate external protection, and adherence to good laboratory practice (eg, minimal generation of aerosols, no hand to face contact) should provide adequate internal protection.

β particles

β particles can be negatrons (negative electrons) or positrons (positive electrons). Though they are also particles, they are significantly smaller than α particles, so they can travel farther and penetrate more deeply. Similar to α particles, their greatest hazard is internal damage when they are ingested, injected, or inhaled. Lab coats, latex or nitrile gloves, plastic, and Lucite easily block them, and good laboratory

 ISBN 978-089189-6463

practice provides adequate internal protection. Lead shielding should *not* be used for β emitters because the interaction of high energy β particles with lead can *cause* additional radiation (Bremsstrahlung radiation). The most common β emitter to cause this problem is ^{32}P.

γ ray/X rays

γ rays and X rays are forms of electromagnetic radiation made up of only energy and no mass. Because they are pure energy, they penetrate very deeply, and therefore present both internal and external hazards to humans. Several feet of concrete can block them, but as a practical matter laboratories should use lead shields.

Effects of radiation

The radiation under discussion in this Unit is more properly called *ionizing radiation*. Particles and energy from radioactive decay interact with matter to ionize and excite it, which can be very damaging, particularly in biological systems. The damaging and/or destructive effects of ionizing radiation are cumulative and directly proportional to the intensity and length of exposure, as well as the type of radiation. The lens of the eye is vulnerable to cataract formation under radiation exposure. Some cells destroyed by radiation (such as stem cells in the bone marrow) cannot be replaced, and in some cases, damaged cells become malignant cancer cells. Radiation has also been known to cause DNA mutations and damage developing embryos/fetuses. Therefore, it is critical to keep track of the exposure of personnel who work with ionizing radiation, using special monitoring and exposure limits for pregnant workers.

Radiation exposure

Sources of ionizing radiation in laboratories

Radioisotopes have been used to track reactions in many different laboratory procedures. Radioisotopes are attached to molecules of interest, and activity can be tracked though the radioactive emissions. Because there are few obstacles to radioactive decay, radioisotopes can give very reliable signals. Laboratory reagents containing radioisotopes must be clearly labeled as to their content and include the universal radiation symbol (more below).

When radioactive materials are injected into living organisms, they usually disperse throughout the body, becoming more dilute. Therefore, blood, urine, body fluids, and biopsy specimens are usually significantly safer than the injected radioisotope. In medicine, specimens from patients who have been administered radiation typically are not radioactive enough to be harmful to laboratory workers, if they wear the usual protective equipment. An exception to this is brachytherapy, a common treatment for prostate cancer, during which multiple (up to ~100) radioactive "seeds" are inserted into the tumor to deliver high dose radiation for a few months. After treatment, the seeds are no longer significantly radioactive and are not removed unless it becomes necessary to remove the entire tumor. However, when these specimens are received in the laboratory, the seeds must be considered radioactive unless information is provided to the contrary, and as the specimen is being processed, they should be placed in lead containers and disposed of as radioactive waste. The 2 usual isotopes for brachytherapy are ^{103}Palladium and ^{125}Iodine. Both are γ emitters and require lead shielding.

Some laboratory devices contain sources of ionizing radiation. For example, a transfusion service department may

 ISBN 978-089189-6463

irradiate blood units with one such device to kill white blood cells prior to transfusion by placing the units in the device where they are moved into a radioactive core for a specified amount of time. Users may be required to wear dosimeters, which detect radiation exposure and are described further below. Other blood irradiators provide sufficient lead shielding such that radiation leakage is almost nonexistent; users do not have to be monitored for exposure if leak tests are performed periodically. The Clinical and Laboratory Standards Institute (CLSI) recommends testing these devices every 6 months or immediately after they have been moved, undergone maintenance, or sustained damage. Other laboratory devices may have radioactive sources that carry similar requirements. All devices in the laboratory that contain radioactive sources must be clearly labeled and put on strict maintenance schedules to guard against radiation leaks.

Reducing exposure to ionizing radiation in laboratories

The 3 basic means of reducing radiation exposure are time, distance, and shielding. Radioisotope work areas should be separated from nonradioisotope work areas, putting distance between workers in each space and shielding those not working with radioisotopes. Workers should also rotate through the workstations so that no one works for an excessive amount of time in the radioisotope area. The most inexpensive and powerful way to reduce exposure is by maximizing workers' distance from radioisotopes. Exposure reduction is proportional to the distance from the isotopes *squared*. In other words, doubling one's distance from the source reduces exposure to 1/4 of the original level.

reducing exposure to ionizing radiation: the ALARA principle

As Low As Reasonably Achievable

minimize time with radioisotopes
maximize distance from radioisotopes
use proper shielding

f7.2 the ALARA principle

A rem (roentgen equivalent, man) is a unit of exposure that takes into account the biological effects of radiation on the person exposed. (Note: The SI unit for the rem is the Sievert.) Radiation workers should have a means to monitor exposure, such as a thermoluminescent dosimetry device or a film badge. A commonly used isotope is ^{125}I, and because the thyroid gland incorporates iodine to a high degree, workers using this isotope may require thyroid monitoring.

Per the Nuclear Regulatory Commission (NRC), radiation workers must not be exposed to >5 rems annually. However, the NRC also specifies that the "ALARA" principle **f7.2** be followed, meaning exposure must be "as low as reasonably achievable." For example, a shield that reduces exposure by 1 rem must be applied, even in a work setting with an annual exposure of only 4 rems. Exposure must be documented using a film badge or other device unless the total exposure in 1 year is unlikely to exceed 10% of the maximum allowable dose. Because the amount of radiation in many medical laboratories is very low, they are often exempt from monitoring requirements.

As indicated above, it is crucial to avoid inhalation, ingestion, or absorption through the skin and eyes of radioisotopes. Best practices in the laboratory, which are described elsewhere in this text, will also help prevent radiation overexposure. Contact lens wearers may need additional eye protection, not just from splashing but

 ISBN 978-089189-6463

f7.3 symbol for radiation hazard

also from absorption of radioisotopes by some contact lens materials.

t7.1 Common isotopes found in the laboratory

Isotope		Type of decay	Half-life	Shielding required
^{103}Pd	palladium	γ	17 days	lead, if in significant amount
^{125}I	iodine	γ	60 days	
^{57}Co	cobalt	γ	270 days	
^{51}Cr	chromium	γ	27.8 days	
^{3}H	tritium	β– (negatron)	12.3 years	low energy; none typically required
^{14}C	carbon	β– (negatron)	5,730 years	
^{33}P	phosphorus	β– (negatron)	25 days	
^{35}S	sulfur	β– (negatron)	87 days	
^{32}P	phosphorus	β– (negatron)	14 days	Lucite

Safe handling & disposal of radioactive materials

All areas in which radioactive materials are used or stored must display the radioactive hazard symbol **f7.3**, and should be restricted to essential personnel. Any laboratory that uses or stocks radioactive isotopes must be licensed by the NRC and/or by a state agency. Medical laboratories usually are categorized under "General License for Use of Byproduct Material for Certain In Vitro Clinical or Laboratory Testing (10 CFR 31.11)." The NRC recognizes many "agreement states," which have licensing requirements acceptable to the NRC; in these states, the state license and regulations prevail. Each laboratory must determine whether or not it operates in an agreement state and obtain the proper license for its site. Licenses are based on the amount, type, and usage of isotopes within a particular facility, and will specify allowable amounts of radiation, expressed as Curies (US unit, abbreviated as "Ci") or Becquerels (SI unit, abbreviated as "Bq"); in US laboratories, radiation is typically measured in microcuries (μCi). Taking accurate inventory is vital to documenting that allowable limits are not exceeded.

The extent to which an isotope remains radioactive over time is expressed as the isotope's half-life, which is the amount of time that it takes for 50% of an isotope's radioactivity to be lost. Isotopes often encountered in the laboratory are presented in **t7.1**. Isotopes with longer half-lives remain radioactive for more time and are therefore more hazardous. Procedures should be developed to use the smallest amount of isotope possible with the shortest half-life practical.

The safety protocols used by personnel for the handling, storage, and disposal of radioactive materials **t7.2** are very similar to the standard laboratory practices that apply to other kinds of materials. Additional requirements to which staff must strictly adhere may be specified by NRC licensure, state licensure, local regulation, or institutional policy. The following list includes some general guidelines:

Labeling and signage. Any area containing radioactive materials, including refrigerators, devices, work areas, laboratory access points, and storage areas, should prominently display the

ISBN 978-089189-6463

t7.2 Safety protocols for radioactive materials

Topic	Comment
Get appropriate federal, state, and/or local license	follow most stringent applicable regulations for location of lab
Use proper labeling and signage	radiation hazard symbol on all reagents, waste, storage units and lab access points NRC Form 3 posted
Maintain good security and proper storage	radioisotopes locked up against theft; shielded storage if needed restricted access to lab good inventory to identify theft and document compliance with license limits
Wear protective clothing, gloves, face/eye protection	remove before leaving the laboratory discard gloves into radioactive waste stream if contaminated contact lenses absorb radiation; avoid them or use eye protection
For large amounts of radioactivity, work behind shields	Lucite/plastic shields for α and β emitters lead shields for γ emitters only
Avoid hand to face contact	never eat, drink, smoke, insert contact lenses, take medications, or apply cosmetics in the lab wash hands before leaving the laboratory and engaging in hand to face contact activities check hands with Geiger counter after washing, if possible
Do not store food or drink in laboratory refrigerators	use storage designated for food store and administer medication, cough drops, etc outside of the lab
Reduce exposure from handling	never pipette by mouth; use automatic pipetting devices when possible handle isotopes with tongs to increase distance if possible, separate devices used with radioisotopes from other devices use minimum amount of isotope with lowest half-life practical
Use ventilated hood or glove box for airborne materials	minimize aerosols and opportunities to inhale or ingest radioisotopes, ideally making hood use unnecessary
Contain and decontaminate spills	work in trays lined with absorbent material and have spill kits nearby turn off ventilation in the event of a spill, and wash affected areas with decontaminating detergent after cleaning, check spill sites with Geiger counter
Follow federal, state and local regulations for disposal	preferred method is waste "decay in storage": after 10 half-lives, radiation negligible; dispose as ordinary waste remove/deface radiation hazard symbol for nondestructive disposal of low level waste separate dry and wet waste; separate different isotopes; do not mix with other chemicals or types of waste flush water soluble waste into sanitary sewer with generous water flow, as permitted; liquid scintillation materials not permitted label waste with date, type of isotope, the name of the person disposing it, etc disinfect biohazards chemically, not with autoclave; remove radioactive sources for separate disposal
Decontaminate exposed work areas and monitor for residual radiation	use decontaminating detergent daily periodically check with "wipe" test and/or Geiger counter
Keep accurate records for at least 3 years	reagent receipt, inventory, and disposal personnel dosimetry reports; share records of former employees with new employers so total dose can be tracked environmental radiation monitoring corrective actions taken
Appoint a radiation safety officer and/or committee	formation of institutional policy maintenance of records source of information before new procedure is implemented staff training on protocols; exam to verify is desirable

7

radioactive hazard symbol; access to these areas by unauthorized personnel should be avoided. Information on optimal storage conditions for reagents can be obtained from the manufacturer. Part of the reagent label should include the symbol for radioactive material hazards, which is bright purple on a yellow background; the NRC also accepts a symbol that is black on a yellow background. In addition to the symbol, the NRC Form 3 poster should be displayed prominently, which outlines the responsibilities of the NRC regarding employee rights, employer obligations, and the procedures to use if safety violations occur.

Secured storage. Radioactive materials are desirable to terrorists. Although medical laboratories seldom, if ever, have the amount and types of radioisotopes that would be desirable by criminals and terrorists, radioisotopes should still be secured so that only authorized personnel can access them. The lab must keep good inventory records of radioactive materials; the NRC takes very seriously the unexplained loss of radioisotopes. Materials must only be delivered to authorized recipients and kept in locked storage at all times.

Protective gear. Lab coats, safety glasses, and disposable gloves are essential protective gear and must be removed before leaving the laboratory. Gloves used to handle radioisotopes may require disposal into radioisotope trash rather than into the typical waste stream. Best practice is, after washing hands, to check them with a Geiger counter or another radiation detector.

Shielding. When working with significant amounts of radioactive material, protective shielding must be worn on the body or placed between the worker and the material. Because radiation is emitted in all directions, in some cases a 360° shield (eg, a box) must be placed around the isotope; shields made from lead are needed for γ emitters; plastic or Lucite for α and β emitters. Per CLSI, low energy β emitters (ie, ^{3}H, ^{14}C, ^{33}P, ^{35}S) do not require shielding, ^{32}P requires Lucite shielding, and γ emitters of significant quantity (ie, ^{57}Co and ^{125}I) require lead shielding. Lead *must not* be used for some high energy β emitters because the particles will interact with the lead and *cause* radioactive emissions (Bremsstrahlung radiation). Because certain β emitters (eg, ^{32}P) will attach to the silicon in glass, plastic test tubes and containers are preferred. In the unusual (and generally undesirable) event that a β emitter is mixed with a γ emitter, first place the material in a Lucite box to block the β radiation, and then place the Lucite box behind lead shielding.

Exposure monitoring. Appropriate body and hand dosimeters (devices that detect and measure accumulated radiation exposure) must be worn when radiation exposure is likely to exceed 10% of the maximum allowable yearly dose. Film badges are worn on a lab coat lapel or another external area of the clothing, while ring dosimeters are worn *under* gloves. These detection devices should only be worn during work with isotopes. When an employee moves to another company, his or her exposure should be tracked from the current institution to the next via record sharing. To facilitate this, employers should develop a form on which departing employees can indicate the institutions to which they have consented to have the necessary records released.

Good laboratory practice. Typical safety rules apply: no eating, drinking, smoking, inserting contact lenses, taking medication, applying cosmetics, or storing food and beverages in the laboratory. Inhalation and ingestion

 ISBN 978-089189-6463

are particularly hazardous ways to be exposed to radioisotopes, so performing activities involving hand to face contact without proper handwashing can be very dangerous.

Proper technique. Appropriate pipetting devices or automatic pipettes must always be employed; mouth pipetting is forbidden. Splashing and aerosol generation when pipetting or transferring materials must be avoided. Best practice is to use tongs, forceps, and other devices that help workers avoid touching radioisotopes and increase their distance from them, however modestly. If possible, devices used with radioisotopes should be separated from other devices.

Aerosol and vapor confinement. If permitted by local air pollution laws, ventilated hoods or glove boxes approved for radioactive materials should be used when materials have the potential to become volatile or airborne. Because radiation inhalation is so dangerous, the best first choice is to minimize or eliminate altogether aerosol generation.

Spill confinement. When working with large amounts of radioactive liquid, workers can confine spills by working in trays lined with absorbent material. In the event of a spill, air circulating equipment, eg, fans or air conditioning, should be turned off to prevent materials becoming airborne and contaminating other areas of the facility. Affected areas should be decontaminated with the appropriate detergent until a Geiger counter or other radiation detection device indicates residual radioactivity is gone. Radiation detection devices must be checked and calibrated at least annually to ensure their accuracy.

Proper disposal. Radioactive waste must be disposed of and stored in appropriately labeled containers. The licensing requirements and local regulations strictly specify how much radioactive waste can be discarded and in what manner. Institutional policy should reflect these requirements and should be obeyed accordingly. Medical laboratories usually handle material with very low levels of radioactivity, and typical protocols are as follows:

Dry waste and wet waste should be separated.

No other chemicals should be added, especially those that could cause gas to form, such as when iodine and bleach are combined.

Autoclaving should be avoided, as the vented steam could be radioactive and contaminate the interior of the autoclave. Chemical disinfection for biohazards with radioisotopes is preferred.

To allow quantification of residual radioactivity, waste should be labeled with date, time, isotope type, and the name of person discarding the waste. Best practice is to discard each isotope in a separate container.

Liquid scintillation fluids should be kept upright in their vials for disposal. If levels of radiation are low enough, isotopes in scintillation fluids may be treated as organic solvent chemical waste. These fluids are never appropriate for sink disposal.

Aqueous radioactive waste must be low level and water soluble for sanitary sewer disposal. The half-lives must be short, and the waste flushed with generous amounts of water to dilute the hazard. Wash water from glassware cleansing is also usually low level and acceptable for sink disposal, but local regulations must be checked. A sink dedicated to radioactive container washing is preferred.

Materials with short half-lives (eg, ^{125}I at 60 days) can be stored until no measurable

7

radioactivity exists. After 10 half-lives, <0.01% of radioactivity remains, and most jurisdictions allow the radioactive labeling to be removed and the waste to be discarded in the customary manner. This means that storage vessels for radioactive trash ("decay pigs") must be accurately dated and that mixing radioisotopes with different half-lives should be avoided. This disposal method is called "decay in storage" and is the method preferred by the NRC.

Certain solid wastes with low radioactivity can be sent directly for incineration or landfill burial, local regulations permitting. Radioactive labeling should be removed or defaced. For example, once a disposable plastic bottle containing radioactive reagent is emptied, the bottle can be triple rinsed; once its labeling is removed, it can usually enter the normal waste stream.

Human tissues implanted with radioactive pellets may require special disposal, depending on when they were implanted. The tissues themselves are biohazards, which must be decontaminated prior to final disposal; however, the pellets themselves must be handled as radioactive waste unless >10 half-lives have passed since they were implanted. Most pellets of this nature are γ emitters, so they should be placed in lead containers, and protocols should be developed for final disposal.

Decontamination. Work surfaces and items contacting radioisotopes should be decontaminated daily using the appropriate cleaning agents. Countertops made from nonporous material are vital in the lab, because when radiation spills into the pores of a porous material (eg, wood), it is impossible to remove. Spill kits should be posted in strategic locations.

Environmental monitoring. Laboratory equipment, glassware, and work areas must be regularly monitored and decontaminated to remove residual radiation. To verify adequate decontamination, a Geiger counter sweep and/or a "wipe" test should be performed, which consists of taking a sample from a work surface or piece of equipment by wiping it with a detergent soaked cotton swab. The swab is then inserted into the appropriate holder for a radiation scintillation counter. A "wipe" test sample should be taken from all laboratory work surfaces that are exposed to radioactive materials. Institutional guidelines may include a map of the laboratory indicating all areas from which samples should be taken **f7.4**; they should also include the acceptable levels of radiation for each area. Every test for radioactivity should be documented along with the action(s) taken if radioactivity levels were too high.

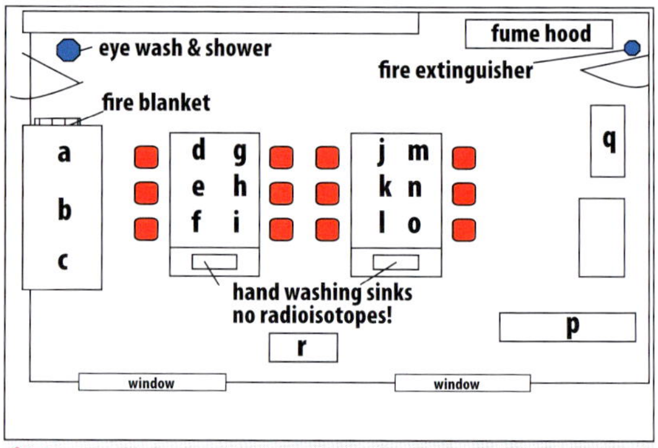

f7.4 Sample map of laboratory for "wipe" test; each contaminated sink (**a-c**), bench top (**d-p**) and refrigerator (**r**) should be wiped and tested with a Geiger counter on ^{125}I & ^{57}Co channels. Actions to be taken if radioactivity is too high in any area should be noted (counts may not exceed background by >20 cpm)

Records. Complete recordkeeping is essential when handling radioisotopes. Reagent receipt and inventory, personnel dosimetry reports, environmental radiation monitoring, and waste disposal records are absolutely critical to documenting compliance with licensing requirements.

 ISBN 978-089189-6463

All records related to radioisotope activities must be kept for a minimum of 3 years.

Safety officer/committee. Each institution should appoint a radiation safety officer, who is responsible for the maintenance of records as well as the formation and implementation of policy, and who should be consulted *before* new procedures are initiated. Standard operating procedures should incorporate a radiation control plan, and CLSI recommends that before working with radioisotopes, each employee pass a radiation safety exam related to institutional protocols.

As with any other hazard, protocols for use of radioisotopes must be customized to each laboratory. Appendix A of the CLSI publication GP12-A3 *Clinical Laboratory Safety; Approved Guideline Third Edition* (June 2012) is an excellent source for additional detail on radiation safety protocols.

Summary table: basic information on radioactive materials

Topic	Comments	
Types of decay	α particles	1. "naked" helium nuclei: 2 protons + 2 neutrons 2. strong positive charge; cause significant local internal damage when ingested, injected or inhaled 3. gloves, plastic, or Lucite provide external protection
	β particles	1. single negative or positive electrons 2. smaller than α particles, so penetrate more deeply 3. cause local internal damage when taken internally 4. gloves, plastic or Lucite provide external protection 5. β emitter ^{32}P interacts with lead, causing Bremsstrahlung radiation, and attaches to silicon in glass, so plastic preferred
	γ rays/X rays	1. forms of electromagnetic energy with no mass 2. lead used for shielding 3. cause internal and external hazard to humans
Radiation effects	cell death/mutation/malignancy, fetal damage/demise, cataracts, death	
Radiation exposure	1. measured in rems: 5 rems/year maximum permitted 2. special restrictions possible on pregnant workers 3. film badges or thermoluminescent detectors must be worn if exposure anticipated to be more than 10% of yearly allowable; usual sites on lab coat lapels or rings under gloves 4. thyroid scans to monitor radioactive iodine exposure 5. follow the "ALARA" principle, minimizing time with and distance from radioisotopes and using shielding 6. strict maintenance and testing for devices that contain radioactive sources	
Radiation licenses	1. NRC general license for most laboratories with low radioisotope handling 2. US has 36 "agreement states"; follow state regulations, which have been approved by NRC	
Radiation terms	1. rem (roentgen equivalent, man): unit of radiation exposure in US 2. Sievert: SI unit of exposure 3. Curie: unit of radiation amount in US 4. Becquerel: SI unit of radiation amount 5. half-life: amount of time radiation of radioisotope reduces by 50%	
Radiation symbol	dark purple or black with yellow background	

ALARA = as low as reasonably achievable; NRC=Nuclear Regulatory Commission

Self evaluation questions

1 Which type of radiation is the most penetrating?

- **a** β
- **b** α
- **c** γ
- **d** δ
- **e** all are equally penetrating

2 Matching. Only one answer is correct for each item.

rem _____

Curie _____

half-life _____

Lucite _____

lead _____

- **a** time it takes for isotope's radioactivity to decrease 50%
- **b** only type of shield that prevents γ penetration
- **c** only type of shield that prevents α penetration
- **d** can shield β penetration, but not γ
- **e** 50% of a reagent's shelf life
- **f** unit of amount of radioactivity
- **g** unit of exposure to radioactivity
- **h** unit of hazard for radioactivity
- **i** inventor of the Geiger counter

3 Which of the following is (are) a method to detect radiation exposure or contamination?

- **a** film badge
- **b** "wipe" test
- **c** TLD dosimeter
- **d** Geiger counter
- **e** all of the above

4 All of the following are responsible for granting radiation handling licenses or making legal requirements for the handling of radioactive isotopes and/or their waste EXCEPT:

- **a** NRC
- **b** EPA
- **c** radiation safety officer
- **d** local agencies (eg, county health department)
- **e** state agencies (eg, Department of the Environment)

 ISBN 978-089189-6463

5 Laboratory workers who use radioactive iodine may need monitoring of which gland?

- **a** thyroid
- **b** adrenal
- **c** pituitary
- **d** prostate
- **e** lacrimal

6 Fill in the blanks: If the isotope _____ is placed behind _____shielding, it can actually create additional radiation hazard.

- **a** ^{3}H/plastic
- **b** ^{14}C/Lucite
- **c** ^{32}P/lead
- **d** ^{35}S/polystyrene
- **e** ^{57}Co/polyethylene

7 When a laboratory uses a "decay in storage" program for its waste, the waste should be sealed, dated, and discarded after how many half-lives have passed?

- **a** 1
- **b** 5
- **c** 10
- **d** 20
- **e** 100

8 What does "ALARA" stand for, and what does it mean?

____

9 Fill in the blanks: The 3 most important variables in minimizing radiation exposure are _____, _____, and _____.

10 The radiation hazard symbol

- **a** is purple with a yellow background
- **b** should be displayed on a refrigerator storing radioisotopes
- **c** should be displayed on the entrance door of a laboratory using radioisotopes
- **d** should be removed or defaced from waste that is no longer radioactive
- **e** all of the above

11 You put 100 μCi of ^{33}P into the decay pig trash storage. When can you put this waste into normal trash?

- **a** 25 days
- **b** 250 days
- **c** 50 days
- **d** 500 days
- **e** never

12 Why is eye protection important when working with high levels of radiation?

- **a** radiation exposure can cause cataracts
- **b** contact lenses can absorb and hold radiation
- **c** splashes of radioactive materials can be absorbed through the eye
- **d** all of the above

13 A prostate gland containing ^{103}Pd seeds was removed. The seeds were implanted 2 months ago. Which of the following is correct regarding disposal of this specimen when work is complete?

- **a** autoclave the specimen and then discard it into the radioactive waste stream
- **b** discard the entire specimen, without additional treatment, into the radioactive waste stream
- **c** remove all the seeds for disposal into the radioactive waste stream; autoclave the prostate gland for disposal
- **d** remove all the seeds for disposal into the radioactive waste stream; chemically sterilize the prostate gland for disposal

14 β emitters are most dangerous if

- **a** handled without gloves
- **b** inhaled
- **c** stored in Lucite
- **d** stored in a refrigerator

15 Wearing radiation dosimeters is mandatory if the anticipated exposure is at what percent of the allowable amount per year?

- **a** 5
- **b** 10
- **c** 25
- **d** 30

16 Which of the following is (are) usually allowable?

- **a** washing empty containers that held a water soluble, low half-life isotope and allowing the wash water to enter the ordinary waste stream
- **b** dumping liquid scintillation fluid down the drain after 10 half-lives have passed
- **c** autoclaving blood that contains low level radioisotopes
- **d** bleaching serum that contains ^{125}I
- **e** all of the above

17 A radioisotope is spilled in the open lab. The air handlers and fans should be turned off.

- **a** true
- **b** false

 ISBN 978-089189-6463

18 A lab employee resigns from a job for which he handles radioisotopes, and takes a similar job at a new company. The new employer requests his radiation exposure records. What should the former employer do?

- **a** refuse to give the records, as employee health records are protected information
- **b** refuse to give the records, as employer records are the property of the employer
- **c** give the new employer the records if the employee has verified that it is, in fact, his new employer
- **d** give the new employer the records after the employee signs a release indicating who should receive the records that includes his new employer

19 The OSHA poster for employee rights while handling radioisotopes should be prominently displayed in the workplace.

- **a** true
- **b** false

7

Answers

1 c

2 rem: g; Curie: f; half-life: a;
Lucite: d; lead: b

3 e

4 c

5 a

6 c

7 c

8 "As low as reasonably achievable"; labs must use all available means to minimize exposure to radioisotopes

9 time, distance, shielding

10 e

11 b

12 d

13 d

14 b

15 b

16 a

17 a

18 d

19 b

 ISBN 978-089189-6463

Unit 8

Waste & waste management

The International Organization for Standardization standard ISO 15190 defines hazardous waste as "any waste that is potentially flammable, combustible, ignitable, corrosive, toxic, reactive, or injurious to people or the environment." This Unit describes hazardous waste categories and some general principles of good waste management. More detailed information on particular types of waste can be found in other units.

Hazardous waste categories

The first crucial task of waste management is to categorize a material that is about to be discarded by its hazards and proper means of disposal. The Environmental Protection Agency (EPA) puts hazardous waste in 4 categories: listed, characteristic, universal, and mixed. Listed wastes are chemicals or chemical sources specifically named on lists published by the EPA and state/local jurisdictions as being hazardous, regulated wastes. However, if a particular waste does not appear on these lists, that does *not* mean that it is unregulated or safe. Characteristic wastes are those regulated based on 1 or more of the following hazardous characteristics:

1. **ignitability:** flammable liquids with flash points <140°F (60°C), solids, and compressed gases that are susceptible to ignition and strong oxidizers
2. **corrosivity:** substances with pH <2.0 or >12.5, and substances capable of corroding steel >1/4 inch per year at 55°C
3. **reactivity:** highly reactive chemicals, chemicals unstable at ambient conditions, or chemicals capable of explosion or generating toxic gases on contact with water, acids, or bases
4. **toxicity:** chemicals that emit hazardous or irritating fumes and vapors or are harmful/fatal if ingested or absorbed; these include waste pharmaceuticals and substances harmful to the environment, microbes, or animals if released, including materials with a high biological oxygen demand, those which would kill significant numbers of micro-organisms, and waste with a temperature higher than 104°F (39°C)

The institution must determine whether its waste possesses any of these characteristics. It is important to remember that drugs are also chemicals, and some pharmaceuticals, particularly powerful antineoplastic drugs, are classified by the EPA as toxic waste. In order to simplify waste handling protocols, the EPA has designated the following 3 categories of commonly discarded waste as universal wastes:

- **batteries**
- **pesticides**
- **mercury containing items** available to the public such as equipment (eg, thermostats) and lamps (eg, fluorescent bulbs)

8

Methods for discarding universal waste and advice for the general public are available at www3.epa.gov/epawaste/hazard/wastetypes/universal/index.htm (accessed January 2, 2016). Finally, the fourth EPA category is mixed waste, which includes normally designated hazardous waste and radioactive waste.

The EPA is primarily concerned with chemical waste and does not have separate waste categories for biohazardous, sharp, and radioactive waste. This does not mean that the EPA does not retain authority to regulate other types of waste as it sees fit, only that other agencies often play a bigger role. Regulations for radioactive waste are generally handled by the Nuclear Regulatory Commission (NRC), while biohazardous and sharp wastes are handled by the Occupational Safety and Health Administration's (OSHA) Bloodborne Pathogens Standard, state regulations, and local mandates. Waste with multiple hazards poses special problems because it may be subject to conflicting regulations from different organizations. For example, the NRC recommends "decay in storage" for many radioisotopes, meaning that they should not be discarded for 10 half-lives, when they are no longer radioactive. However, large waste generators should not store waste any longer than 90 days, and this may not be long enough for some radioisotopes. Another reason to avoid multihazardous waste is that licensed waste handlers often refuse to accept it.

Regulations in the OSHA Bloodborne Pathogens Standard do not address all categories of infectious waste. Many states have categories such as "regulated medical waste" for materials that are likely to transmit an infection. The major categories of infectious waste are as follows:

- **live infectious agents:** cultures, stock cultures, and live or attenuated vaccines
- **anatomic and pathology waste:** body parts and fluids removed during surgery and biopsy (which may be multihazardous if preserved in chemicals such as formalin or if implanted with radioactive sources); stained and fixed tissues are no longer considered biohazardous and do not usually require special disposal
- **human blood and body fluids**
- **contaminated sharps**
- **animals, animal waste, animal blood/body fluids/tissues**
- **isolation waste:** tissues, blood, and body fluids from patients with highly communicable diseases

In general, most states require treatment of regulated medical waste so that it is noninfectious at final disposal. Some states require body parts to be unrecognizable at final disposal. The guiding principle for handling biohazardous and sharp waste is to treat it and/or segregate it to prevent transmission of infectious disease or physical injury. Sterilizing procedures, eg, autoclaving, incineration, and chemical treatment, are excellent ways to convert biohazards to nonbiohazards and permit routine disposal. That said, regulations vary widely regarding what is permissible to sterilize with incinerators and chemicals. A special problem of this waste category is limited storage capacity. Many pathogens continue to multiply in biohazardous waste, and much of the waste will decompose, producing dangerous organism loads, putrefaction products, and offensive odors.

Materials used in an infectious environment are not necessarily regulated medical waste. In some jurisdictions, fluids and excreta routinely handled by the sewer system can be flushed into the sewer system as waste. Urine, for example, can be flushed

 ISBN 978-089189-6463

down the sink; it is rarely considered a regulated waste unless it contains blood. In many states, objects such as gloves can be disposed into ordinary waste streams if contamination is not visible or likely to be sparse. Disposing of waste in this manner can be an important cost saving and environment sparing measure, but state laws about what is and is not considered regulated medical waste must be consulted. In some jurisdictions, if a material such as an empty specimen bag is not visibly contaminated with blood or debris, it can be disposed of by normal means, even if it displays the biohazard symbol; in others, it cannot unless the symbol is defaced. On another note, medical facilities must also ensure that protected personal health information on any discarded container is defaced or destroyed.

While it is permissible to discard some waste in the sink, which then travels to the sanitary sewer or landfills, laboratory staff must be very, very cautious about this practice because stiff fines can be levied for introducing pollutants into the environment. Solids or viscous materials that can cause plumbing and drainage system blockage are prohibited in most jurisdictions. If laboratory staff plans to use the sanitary sewer for disposal, the safety manager should consult with local wastewater management authorities for proper permits and procedures for monitoring discharge. It is illegal to use simple dilution as a waste treatment method. In general, no chemicals, radioisotopes, or organisms should be discarded into a septic tank, which would introduce the waste directly into the natural ground water. Additional information on waste management can be found elsewhere (chemical waste in Unit 3, sharp waste in Unit 4, biohazardous waste in Unit 5, and radioactive waste in Unit 7).

Handling hazardous waste

The categories of waste generators according to EPA regulations are based on the amount and type of hazardous waste generated, as shown in **t8.1**. While a medical laboratory could fall into the category of conditionally exempt small quantity generator if evaluated alone, it could fall under stricter guidelines if part of a larger institution assigned to a higher waste generating category.

Hazardous waste generated in the laboratory should be accumulated in a designated area within a short distance (ie, in sight) of workers. Workers should not have to go far with hazardous waste in their hands. In large facilities, these areas are called satellite waste sites. Each waste container must be closed or closable and

t8.1 EPA waste generator categories*

EPA waste generator category	Amount of waste generated	Waste storage limits
Large quantity generator	≥1,000 kg/month ≥1 kg/month acutely hazardous waste†	can accumulate waste up to 90 days
Small quantity generator	100-999 kg/month <1 kg/month acutely hazardous waste	can accumulate up to 6,000 kg in 180 days (270 days if transportation >200 miles)
Conditionally exempt small quantity generator	<100 kg/month <1 kg/month acutely hazardous waste	can accumulate up to 1,000 kg without time limits

**Determined by the highest amount of waste discarded in 1 month, rather than dividing the amount of waste discarded annually by 12*
†Examples of acutely hazardous waste include arsenic & mercury

f8.1 hazardous waste label

labeled as to the type of waste it holds. If the container is designated for storage, it should be labeled with the date waste was first placed inside it. All waste must be clearly marked, and labels on hazardous waste must indicate the exact nature of the hazard. **f8.1** shows an example of a label appropriate for hazardous waste. Waste must be kept separate by type unless it is certain they are compatible, and containers must never be overfilled.

Waste in satellite areas must be transferred regularly to the central waste storage area for final processing, which means that so many hazards in one room can make central waste sites quite dangerous. Personal protective protocols and engineering controls (eg, ventilation, sprinklers) for waste handling are identical to those for waste generation; waste also must be sealed and packaged for safe handling. All waste accumulation storage areas must be secured against unauthorized entry, and their contents documented in case of theft. Storage areas must be clearly marked, and provide good containment barriers and emergency equipment. Contact information for personnel who should be called on in case of emergency should be posted. Sites must be checked weekly to ensure containers are labeled and dated, and without damage or leaks; these inspections must be documented in a log. Waste should be promptly moved from storage to disposal, even if it is more costly to do so. While fewer, more widely spaced pickups save money, great care must be taken that waste is not kept longer and in amounts that exceed the limits set by the EPA. Because accumulated waste is evaluated by its oldest component, the date on any waste container should reflect when it was first put in use, not when it became full. One important exception to this is radioactive waste that is decaying in storage.

Air pollution from laboratories is usually low and not regulated by the EPA; however, no laboratory can assume that it is exempt, as in some jurisdictions, permits are required for emissions from chemical fume hoods. Evaporation of hazardous materials through a chemical fume hood is not an acceptable disposal method.

Many EPA regulations are based on the 1976 Resource Conservation and Recovery Act (RCRA), which covers waste from the moment it is generated until its ultimate disposal ("cradle to grave"). The following are best practices for waste handling:

- **Regulations.** Applicable federal, state, and local regulations must provide the foundation for any protocols. In the US, regulations can vary widely between states, so the first step must be to consult state policy in creating laboratory procedures. 2 very helpful websites in this regard are (both accessed January 2, 2016) www2.epa.gov/hwgenerators/table-noting-which-states-have-hazardous-waste-generator-categories-are-same-federal and www.envcap.org/statetools/hzrl/

 ISBN 978-089189-6463

- **EPA identification number/permits for onsite waste storage.** While not required of conditionally exempt small quantity waste generators, it is recommended for documentation and tracking purposes if licensed waste handlers are used. Laboratories must be able to validate their assigned waste generation categories by tracking all waste volumes.
- **Onsite waste treatment.** The decision to treat hazardous waste on site is not trivial. While onsite treatment can be very cost effective for large volumes of waste, it also incurs risk and liability for the laboratory. A laboratory must have an EPA permit to treat its own waste.
- **Waste minimization.** Laboratory protocols must be developed to minimize waste generation while maintaining work quality and staff safety.
- **Training.** Employees both inside and outside the laboratory must be knowledgeable about waste handling protocols associated with their job duties. Support staff members are also often part of the administrative and/or physical handling of waste and must be included in training to prevent mishandling of waste at any stage.
- **Licensed waste handlers.** It is important to verify the reputation and EPA credentials of anyone hired to remove waste from a site and transport it for final disposal. Waste generators are ultimately responsible for any errors made by licensed waste handlers, so selection of reliable companies is crucial.
- **Proper packaging and labeling for removal**
  - Waste being removed from a site must be labeled according to the Department of Transportation (DOT)/Globally Harmonized System (GHS) requirements (see Unit 1).
  - The EPA Hazardous Waste Manifest Form (Appendix 1.3), which documents the source and contents of the shipment, must be completed. The EPA requires this form to track all hazardous waste, and DOT requires that the form accompany all waste on public roads. The base form is identical for all 50 states, but each state may impose additional requirements. All local and state requirements should be consulted before waste is sent off site.
- **Tracking ultimate disposal**
  - The manifest must accompany the waste shipment at all times. When the waste is ultimately disposed of, the disposer completes the form, and a copy of the manifest is returned to the generator for records. This should occur within 60 days, or an exception report must be filed with the EPA.
  - Waste for which a manifest has not yet been received must be tracked. Follow-up must be consistent and prompt to ensure that hazardous waste is not in the wrong place.
  - Manifest records must be retained for at least 3 years to meet EPA requirements. A "paper trail" must document the waste's path beginning at its generation and ending at its ultimate disposal. Most experts recommend that manifests be kept longer than 3 years because labs can be held liable indefinitely for damage as a result of waste.

The numerous state and local regulations are beyond the scope of this Unit. Proper training is required for all staff involved in the waste handling process, including those outside the laboratory who may only perform administrative functions. Employees who maintain secure storage sites, inspect/package waste, prepare/sign manifests, release waste to outside handlers, archive documentation, or otherwise perform waste activities

8

for regulatory compliance must undergo documented training to perform those functions, which should also include DOT, EPA, NRC, OSHA, state, and local regulations. The person who signs manifests attests to compliance with all applicable regulations, so this is not to be taken lightly: "I hereby declare that the contents of this consignment are fully and accurately described above by the proper shipping name, and are classified, packaged, marked and labeled/placarded, and are in all respects in proper condition for transport according to applicable international and national government regulations" (EPA Form 8700-22, Rev. 3-05).

A serious concern in waste management is that some waste handlers with EPA licenses can be unscrupulous; still, the waste generator is still responsible for the waste even if the waste was out of its direct control. It is important, therefore, to investigate waste disposal companies and only use reputable handlers. Payment to a waste handler should never be given until the final copy of the manifest has been received. The generator is liable forever for hazardous waste, even if the documentation for final disposal is invalid.

Another problem in waste management is that manufacturers are not required to include chemicals on the safety data sheet (SDS) for a reagent if that chemical is <1% of the total. This means that laboratory staff may discard hazardous substances without knowing it. It may be necessary prior to disposal to request that a manufacturer certify that a particular reagent does not contain any of the hazardous chemicals listed in the regulations.

Accidental waste release

The preceding material discusses planning for waste generated under ordinary circumstances. Facilities must also plan for waste generated in emergencies such as chemical spills. Management of such waste is also covered by OSHA regulation 29 CFR 1910.120—Hazardous Waste Operations and Emergency Response, often referred to as "HAZWOPER." Medical laboratories that are conditionally exempt small quantity generators might not typically have waste emergencies on a large enough scale to fall under the requirements of HAZWOPER, unless they are part of a larger facility. Adequate emergency response procedures and plans for waste should be part of every laboratory's safety protocols to meet OSHA requirements. Major requirements are as follows:

- **Appoint an emergency response coordinator.** Ideally this person has HAZWOPER training and can develop an emergency plan, as well as direct staff and emergency responders to their appropriate duties during an event. This person should also be responsible for notifying the EPA National Emergency Response Center should an emergency occur that meets HAZWOPER criteria.
- **Train staff at a "first responder" awareness level.** Full HAZWOPER training may not be necessary for all staff, but designated staff members should be trained in the basic activity of a first responder in an emergency, the least of which are evacuation protocols and the location of emergency phone numbers; this also includes initial containment of a spill and providing communication (eg, posting signs) to prevent re-entry.

 ISBN 978-089189-6463

- **Prominently post emergency phone numbers**. These include, but are not limited to, local emergency contact information, such as that for first responders (911), the emergency response coordinator, and the EPA National Emergency Response Center (1-800-424-8802), which must be contacted when major accidental releases occur.

Basic waste management

Disposal of hazardous wastes is expensive, as licensed waste handlers often charge by weight and/or volume, and the more restrictive EPA categories for waste generators incur higher management costs. Minimizing hazardous wastes makes both economic and environmental sense. Some basic principles are as follows:

- **Planning.** Only the minimum quantity required of a particular substance should be purchased. While many chemicals have limited shelf lives, just because a chemical has expired does not mean its hazard is reduced. Planning chemical usage accurately should help minimize the amount of unused hazardous chemicals disposed of as waste.
- **Segregation.** Hazardous waste should always be separated from nonhazardous waste. When 1 mL of a hazard is mixed with 99 mL of water, the result is 100 mL of hazardous waste. Whenever possible, hazards should not be allowed to comingle with nonhazards. For example, a handwashing sink should have, in its vicinity, a biohazard disposal bag for contaminated gloves and a regular trash can for paper towels, rather than 1 trash can for both.
- **Reduce, reuse, recycle.** The "3 Rs" comprise 3 general ways to minimize hazardous waste from procedures. The RCRA requires a waste minimization program, and any 1 of these 3 techniques is acceptable.
    - **Waste reduction.** With careful planning, laboratories can avoid purchasing excess materials. Reducing the amount of materials used in a procedure to the minimum needed (eg, converting to microscale chemistry) and substitution of less harmful chemicals are both useful. Radioactive substances can be held until they decay to a safe level. Biohazardous waste can be sterilized. Some chemicals can be neutralized (eg, bases neutralize acids) or rendered less harmful (eg, mercury treatment).
    - **Waste reuse and recycling.** Some procedures tolerate reuse of chemicals, which should be maximized. Procedures requiring very pure chemical can be paired with procedures requiring the same chemical at a lower level of purity. Some chemicals can be recycled either on site or at recycling facilities. For example, some facilities that generate a lot of waste have found that, when compared to the cost of storage and disposal of waste, onsite solvent distillation facilities are more cost effective. This process has been highly successful in the disposal of formalin and xylene, particularly in histology laboratories. Good training is essential, however, in the use of recycling instruments, to ensure the safety of the operation and the quality of the product. For example, if formalin contains picric acid (in Bouin fixative) or colloidin (nitrocellulose), it cannot be distilled because these materials are explosive when heated.

8

- **Disposal.** Final disposal methods must be chosen carefully after all options have been investigated. A balance of compliance, environmental impact, worker safety, liability, cost, and convenience must be achieved when deciding between options, eg, onsite waste treatment vs outside contractors. The correct options may vary widely between laboratories since the needs of the entire operation must be considered.

Only the basic information on waste management has been reviewed in this Unit. For more detailed information, the reader should refer to EPA/state/local regulations, SDSs, the information on biohazards from the Centers for Disease Control and Prevention, and NRC regulations for radioactivity. The CLSI publication GP05-A3. "Clinical Laboratory Waste Management" is an excellent resource to use as the foundation of hazardous waste protocols as well as the following 2 websites (both accessed January 2, 2016): www2.epa.gov/hwgenerators and hercenter.org.

 ISBN 978-089189-6463

Summary table: waste & waste management

Topic	Comment
Identification of hazardous waste	1. EPA lists of specific chemicals 2. wastes with characteristics specified by the EPA: toxicity, corrosivity, reactivity, ignitibility 3. EPA universal waste: eg, batteries, pesticides, mercury-containing devices 4. regulated at a state level: medical waste, sharps, biohazards; exceptions are provisions of OSHA Bloodborne Pathogens Standard 5. radioisotopes as identified by NRC 6. mixed waste: to be avoided
Hazardous waste regulations	1. EPA permits, licenses, and regulatory level based on amount/type of waste a. time limits on holding hazardous waste b. permits required to generate, store, transport, and treat waste unless exempt c. primary authority from RCRA: waste generators responsible for waste forever ("cradle to grave") and must minimize waste 2. variance of regulations at federal, state, and local level for sharps and biohazards; customize policy to location 3. NRC: "decay in storage" disposal preferred 4. DOT: transport of waste 5. state and local regulations could be added to above
Waste accumulation sites	1. satellite accumulation areas very close to workers 2. central accumulation sites for final disposal a. locked storage with hazard signs posted b. proper ventilation, protective equipment, emergency devices, etc c. inspect weekly for leaks, damage, theft, etc d. waste dated so not held too long e. incompatible wastes segregated 3. proper labels on all waste (identity, date, generator, etc)
Disposal methods	1. sanitary sewer a. some normal human waste; states vary for blood/body fluids b. low risk chemicals and radioisotopes as permitted c. not permitted for flammables, strong corrosives, toxics (including drugs), unstable chemicals, very hot materials, viscous materials, chemicals that generate noxious fumes/odors, environmentally damaging materials 2. septic tank: virtually never; waste enters ground water 3. biohazards and sharps: incineration, autoclaving, chemical treatment a. made noninfectious at final disposal b. some exceptions for low risk items not visually c. biohazard symbol may require defacing 4. radioisotopes a. discard as ordinary waste after 10 half-lives have passed b. sewer disposal as permitted if low level and water soluble c. disposal varies with isotope, half-life and amount d. radiation symbol may require defacing 5. licensed waste handler: must have EPA license a. package waste per DOT/GHS specifications b. track waste with mandatory EPA manifest c. do not pay handler until completed manifest is received (60 days or less) d. keep completed manifests at least 3 years e. only use reputable companies

Summary table: waste & waste management (continued)

Topic	Comment
Training	1. workers handling waste at any phase know applicable regulations to their job 2. applicable regulations may include EPA, NRC, OSHA, DOT, state, and local 3. "first responder" training for HAZWOPER accident protocols 4. full HAZWOPER training for emergency response coordinator a. coordinator develops emergency plan and trains first responders b. prominent posting of contacts for emergency c. coordinator organizes emergency response during event
Basic waste management	1. plan ordering, storage, usage to minimize waste generation 2. waste segregation a. hazardous from nonhazardous wastes b. separate by hazardous waste category c. clear labeling, physical separation, sealed and secured 3. waste minimization required by RCRA— "reduce, reuse, recycle" 4. disposal choices a. regulatory compliance first priority b. balance between cost, convenience, risk, liability, environmental impact, worker safety and needs of the larger organization

 ISBN 978-089189-6463

Self evaluation questions

1 Concentrated hydrochloric acid (pH 1.2) does not appear on the EPA list as a hazardous regulated waste. Which of the following is true?

- **a** EPA regulates it as a corrosive hazard
- **b** it is acceptable to discard it into a landfill
- **c** it is acceptable to flush it in the sanitary sewer
- **d** it is acceptable to evaporate it in a chemical fume hood
- **e** state and local regulations should be consulted prior to disposal

2 A facility generates arsenic waste. Its responsibility for maintaining documentation on this waste ends when

- **a** they receive the completed disposal manifest from the disposal site
- **b** the completed disposal manifest has been held for at least 3 years
- **c** the waste is sealed into impermeable bags and labeled with DOT codes
- **d** the waste and a completed manifest are given to a licensed waste handler
- **e** 30 years after the staff who were exposed to the waste have completed employment

3 A facility generates arsenic waste. Its responsibility for any harm that this waste causes

- **a** never ends
- **b** ends when the disposal facility destroys the waste
- **c** ends when the waste is removed from the facility by a licensed waste handler
- **d** ends when the disposal facility sends the completed manifest documenting disposal
- **e** ends when the disposal facility completes the waste treatment process and the waste is deposited at its final disposal site

4 The EPA regulates the following categories of hazardous waste: chemical, radioactive, biohazard/medical.

- **a** true
- **b** false

5 Which of the following can usually be discarded in the sanitary sewer?

- **a** batteries
- **b** bag of human blood for transfusion
- **c** cancer chemotherapeutic agent
- **d** all of the above
- **e** none of the above

6 The EPA allows each of the 50 states to develop and require its own hazardous waste manifest form.

- **a** true
- **b** false

7 State the 3 "Rs" of minimizing waste.

__________________, __________________, __________________

8 When should a licensed waste handler receive payment?

____

9 The authority for regulating laboratory waste by the EPA comes primarily from the

- a Clean Air Act
- b Medical Waste Tracking Act
- c OSHA "Right to Know" Standard
- d Resource Recovery and Conservation Act
- e OSHA Hazard Chemicals in Laboratories Standard

10 Fill in the blanks: The NRC prefers that radioactive waste is eliminated by a ________________________ program that discards radioactive waste after it has been in storage for more than _____ half-lives.

11 What happens if nonhazardous waste is mixed with hazardous waste?

- a the whole mixture is classified as hazardous and disposed accordingly
- b the 2 wastes should be separated and disposed of in the normal manner
- c the SDSs for both wastes should be consulted to see if disposal techniques for each are compatible
- d the mixture should be taken to the safety officer for an assessment of how or if they can be separated
- e the hazardous waste should be washed off of the nonhazardous waste and then each disposed in the normal manner

12 A pesticide used on a farm

- a is not classified as hazardous waste
- b is classified as EPA listed waste
- c is classified as EPA characteristic waste
- d is classified as EPA universal waste

13 Which of the following is acceptable?

- a discharging boiling water into a river
- b an EPA small quantity generator located 400 miles from the nearest disposal site holding waste for 200 days before pickup
- c pouring the blood from an Ebola patient into the sanitary sewer
- d discarding urine container cups with the patient's name still visible into the municipal trash
- e none of the above

8

 ISBN 978-089189-6463

14 Central waste accumulation sites

- **a** must be inspected monthly and the inspection dates documented
- **b** do not need as much ventilation as the laboratory if containers are sealed
- **c** must have signs indicating whom to contact in an emergency
- **d** must be locked when everyone goes home for the day
- **e** all of the above

15 "First responder" HAZWOPER training should include all of the following **EXCEPT**:

- **a** training and rehearsal of evacuation protocols
- **b** knowledge of contact information and communication protocol with emergency response coordinator
- **c** thorough explanations of all HAZWOPER requirements
- **d** training on primary spill containment protocols

16 An EPA permit is required

- **a** to store hazardous waste
- **b** to dispose of hazardous waste
- **c** a & b
- **d** depending on how much waste a facility generates

8

 ISBN 978-089189-6463

Answers

1 **a**

2 **b**

3 **a**

4 **b**

5 **e**

6 **b**

7 **reduce, reuse, recycle**

8 **when completed manifest is received**

9 **d**

10 **decay in storage; 10**

11 **a**

12 **d**

13 **b**

14 **c**

15 **c**

16 **d**

 ISBN 978-089189-6463

Unit 9

Identify hazards

Examine the following photos and identify as many hazards as you can. Answers start on p 187.

f9.1 ______________________________

____

____

____

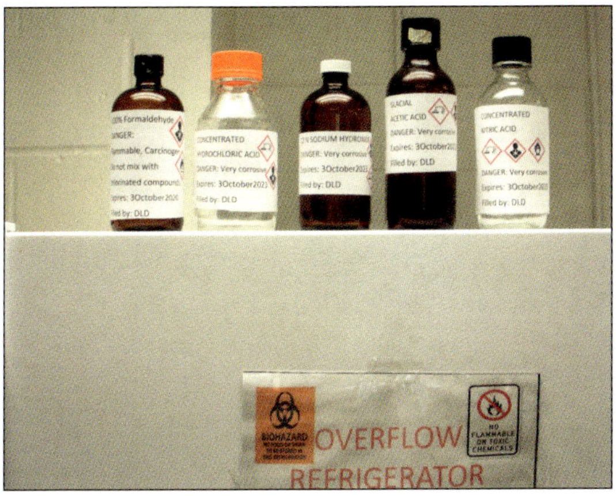

f9.2 ______________________________

____

____

____

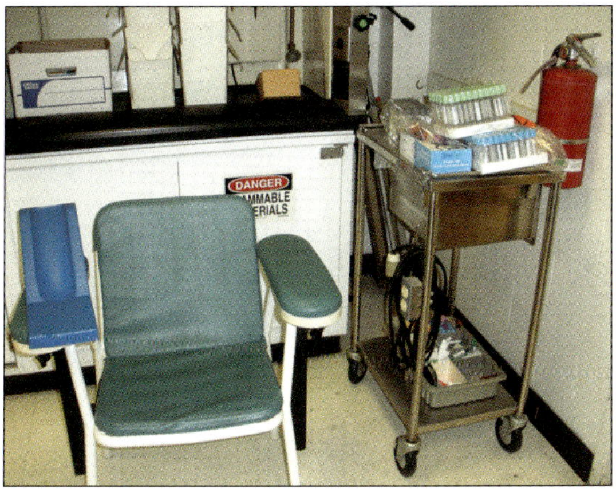

f9.3 ______________________________

____

____

____

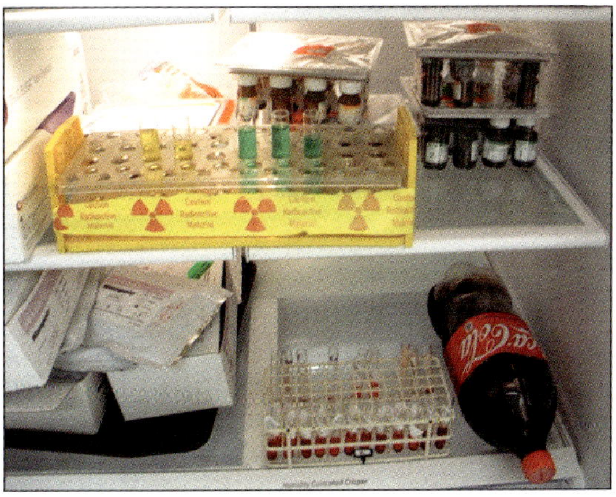

f9.4 ______________________________

____

____

____

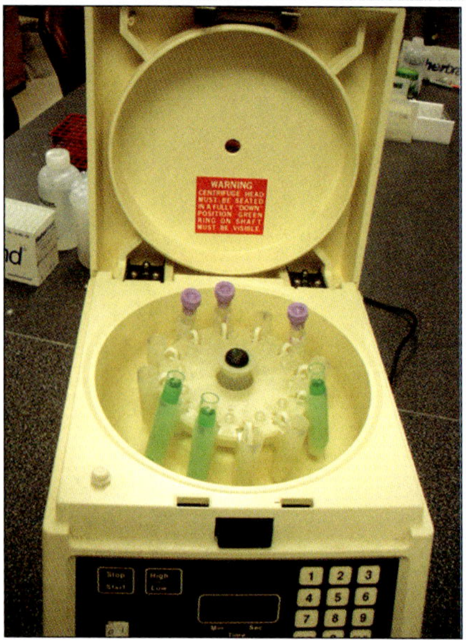

f9.5 ______________________________

____

____

____

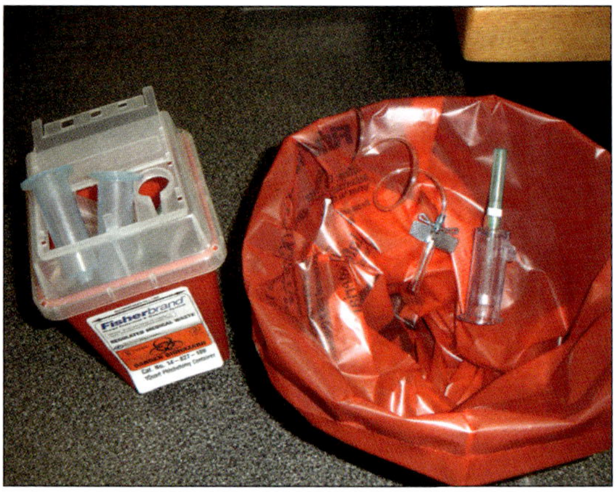

f9.6 ______________________________

____

____

____

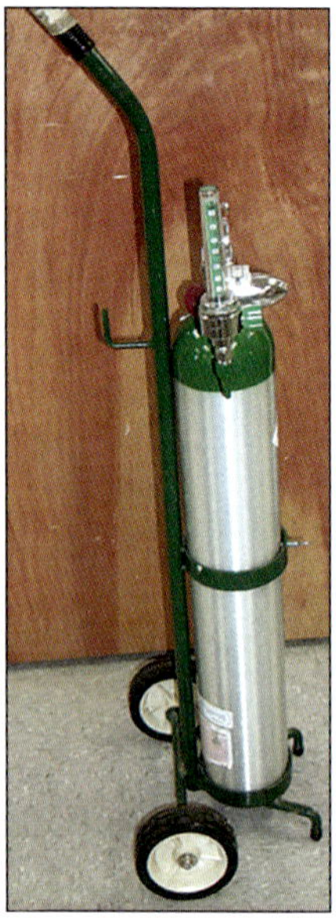

f9.7 ______________________________

____

____

____

f9.8 ______________________________

____

____

____

ISBN 978-089189-6463

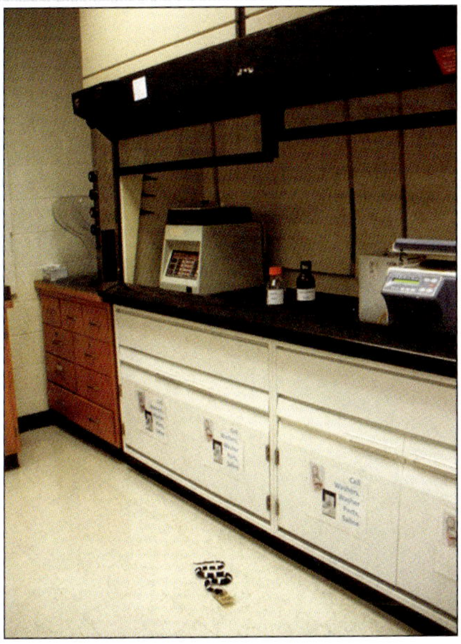

f9.9 ______________________________

f9.10 ______________________________

f9.11 ______________________________

f9.12 ______________________________

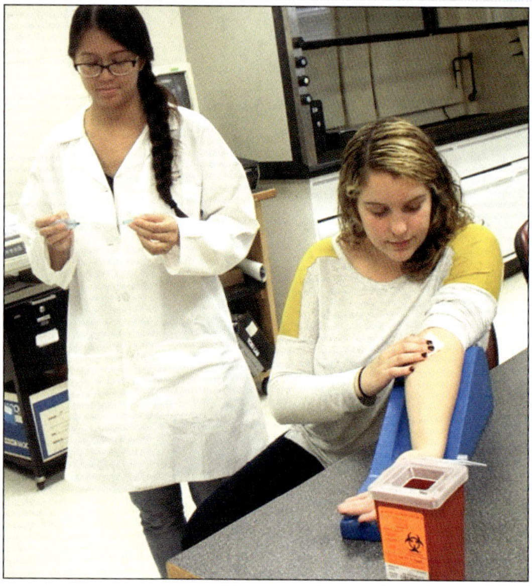

f9.13 ______________________________

f9.14 ______________________________

f9.15 ______________________________

f9.16 ______________________________

ISBN 978-089189-6463

f9.1
LABCONCO PURIFIER CLASS II BIOSAFETY CABINET
signs & notes should not be placed on hood sashes
Do your quality control FIRST! Anyone found not doing QC will be written up!
pipette/suction devices should not stick out of containers
waste should not be overfilled
nothing should block the biosafety cabinet air intake

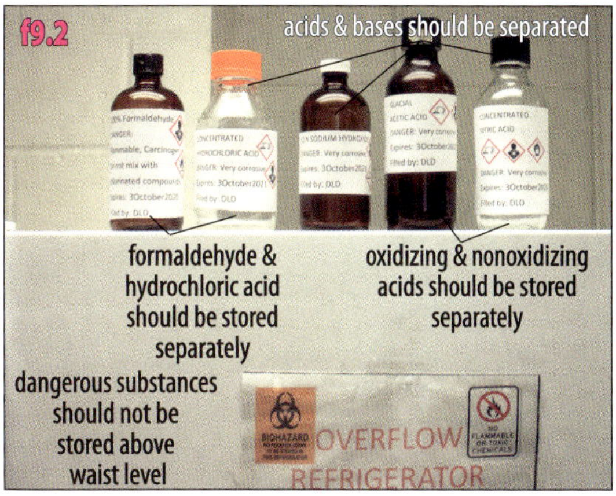
f9.2
acids & bases should be separated
formaldehyde & hydrochloric acid should be stored separately
oxidizing & nonoxidizing acids should be stored separately
dangerous substances should not be stored above waist level
OVERFLOW
REFRIGERATOR

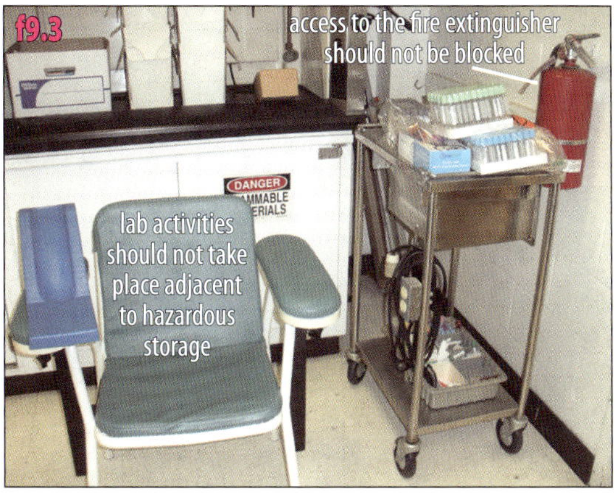
f9.3
access to the fire extinguisher should not be blocked
DANGER
lab activities should not take place adjacent to hazardous storage

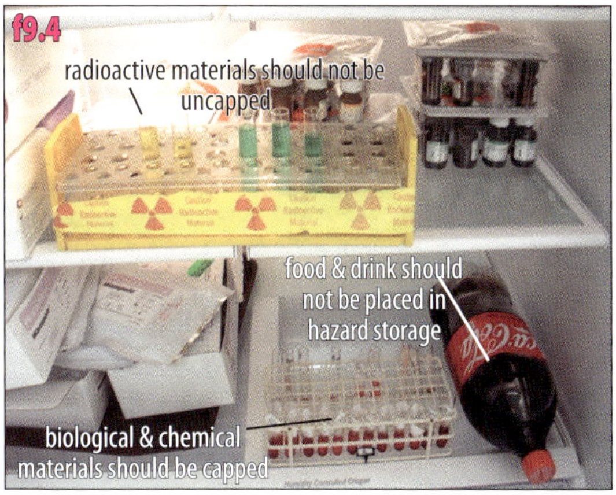
f9.4
radioactive materials should not be uncapped
food & drink should not be placed in hazard storage
biological & chemical materials should be capped

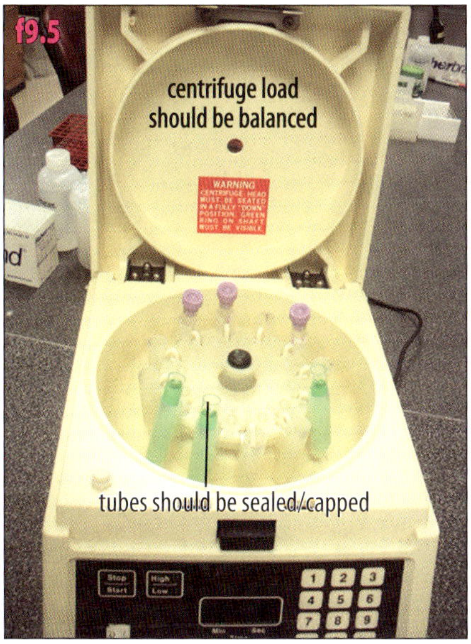
f9.5
centrifuge load should be balanced
WARNING
tubes should be sealed/capped

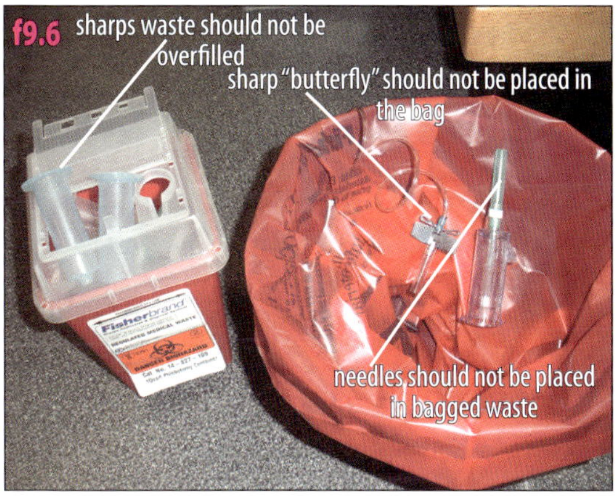
f9.6
sharps waste should not be overfilled
sharp "butterfly" should not be placed in the bag
needles should not be placed in bagged waste

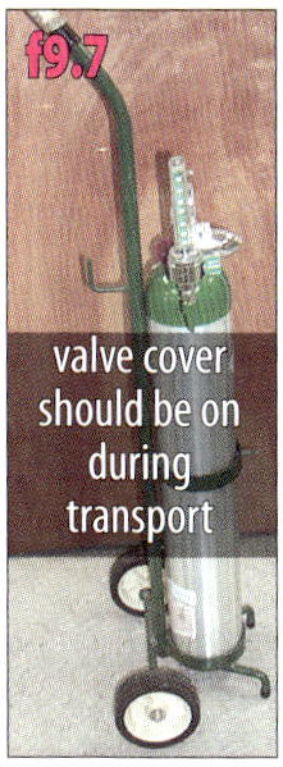
f9.7
valve cover should be on during transport

f9.8
bucket & trash should not block the electrical breaker box

 ISBN 978-089189-6463

f9.9
fume hood opening should not be more than 12 inches
vermin & insects must be controlled

f9.10
should stand to the side & turn face away when opening the autoclave
gloves & a lab coat should be worn, and hair tied back

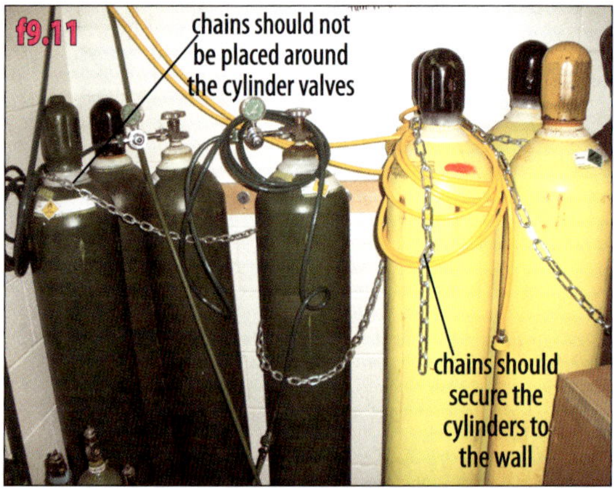
f9.11
chains should not be placed around the cylinder valves
chains should secure the cylinders to the wall

f9.12
extension cords & multiple plugs should be avoided as an electrical & trip hazard

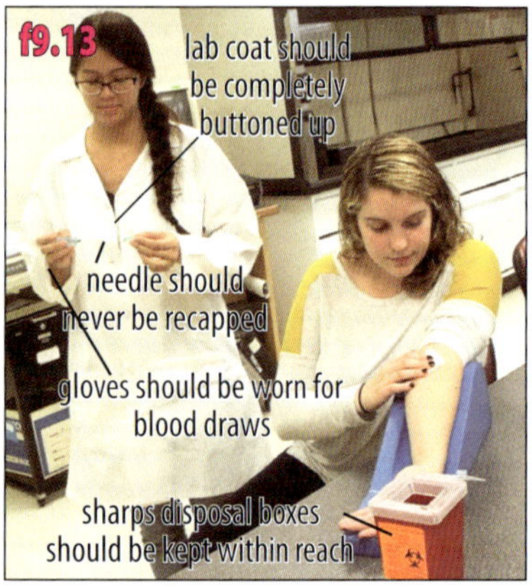
f9.13
lab coat should be completely buttoned up
needle should never be recapped
gloves should be worn for blood draws
sharps disposal boxes should be kept within reach

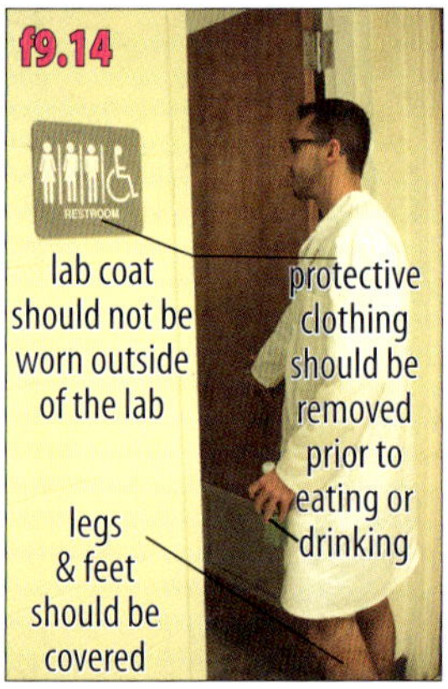
f9.14
lab coat should not be worn outside of the lab
protective clothing should be removed prior to eating or drinking
legs & feet should be covered

f9.15
never pipette by mouth
gloves should be worn to handle chemicals
containers should be held by the bottom, not the "neck"
container should be labeled

f9.16
should not blow the nose with gloves on
gloves should not make contact with the face or be used with cell phones
electronic devices or ear buds should not be used in the laboratory
gloves should be worn when handling biohazards

 ISBN 978-089189-6463

Unit 10

Work practices & safety equipment

Various types of safety equipment used in laboratories, referred to as "engineering controls" by the Occupational Safety and Health Administration (OSHA), have been discussed in previous units. Safety equipment must be on a regular inspection and maintenance schedule so that it is able to function in the event of an emergency, and all workers should know where this equipment is and how to use it. This Unit also outlines general concepts of safe laboratory behavior, referred to by OSHA as "work practice controls," and proper use of personal protective equipment (PPE).

Basic work practices

Although customized safety rules are essential for specific tasks, some rules are basic to virtually any laboratory procedure. These include the following:

- Smoking, eating, drinking, applying cosmetics, taking medication, inserting contact lenses, blowing the nose, or any other activity involving hand to face contact is forbidden. This includes mouth pipetting.
- The appropriate personal protective clothing and equipment must be worn correctly and at all times during laboratory procedures. (Additional information is below.) At a minimum, all workers must wear long sleeved, buttoned laboratory coats and sturdy closed toe shoes. If personal protective clothing and equipment are grossly contaminated during a procedure, they should be replaced.
- Contaminated clothing and equipment must never be worn outside the laboratory. Laboratory coats and protective equipment should be hung on designated hooks within the laboratory, and storage facilities outside the laboratory should be used for personal items such as hats, purses, coats, and medicines.
- Personal grooming and ornamentation must be compatible with lab safety. Hair must be tied back and jewelry must not interfere with protective equipment or procedures. Because long fingernails, particularly artificial nails, have been shown to harbor micro-organisms, they are also forbidden in many workplaces. Increased contamination of the hands due to jewelry and chipped nail polish has been documented and long nails also poke holes in gloves. For all these reasons, fingernails should be kept clean and short.
- Serious and professional behavior is required at all times. No one should work in the laboratory under the influence of drugs or alcohol, and evidence of such must be cause for dismissal. Workers should be undistracted and focused on the tasks at hand.

- Work areas and laboratory equipment should be kept clean and uncluttered. Laboratory workers, not housekeepers, are responsible for cleaning and decontaminating their own materials and area at the end of the work session. Hazardous materials should not be left for laboratory assistants or janitorial staff to clean up unless they have been properly trained. Trash should not accumulate and must be picked up regularly.
- Keeping plants and festive decorations, including lights, should be avoided in a laboratory. Consumer electrical devices may be unsafe around laboratory materials, and many decorations would be difficult to decontaminate. Similarly, plants cannot be decontaminated and could potentially harbor fungi, mold, and insects. The only animals in a laboratory should be those involved in the work.
- All workers must wash their hands when they are contaminated, when gloves are removed, and at the end of a work session. While applying cosmetics is forbidden, use of approved hand cream to prevent chapped hands from frequent hand washing is encouraged.
- Working alone in the laboratory or engaging unattended laboratory operations is to be avoided; some procedures are sufficiently hazardous as to have such practices forbidden. If someone does work alone in the laboratory, a second person should be made aware. A procedure should be put in place for how and when workers are alone in the laboratory and who should be informed. A sample form for unattended operations is shown in Appendix 10.1, page 210.
- Workers should strictly adhere to established safety protocols and expect disciplinary action when they do not. Many laboratories are busy and generate results that are time sensitive, but rushing procedures or taking unsafe shortcuts should not be tolerated. Both worker safety and work quality are compromised.
- Laboratory procedures should be planned carefully ahead of time to ensure that all equipment is in good working order and all safety requirements are met before the procedure begins.
- Electronic devices (eg, cell phones, ear buds, tablets) that are used outside of the laboratory should not be used inside the laboratory, especially not while wearing gloves. These devices can be contaminated and transmit hazards outside the laboratory. Many of them require hand to face contact, which is also a problem. Further, the use of ear buds can impair the ability to hear alarms or emergency communication. Also, protected information could be captured by these devices and removed from the laboratory. However, it may be permissible to use such devices if they are permanently left in the laboratory (eg, computers, radios).
- Laboratory staff is responsible for keeping exits unobstructed and the areas around emergency equipment clear. Shipments of new supplies must be put away promptly, and the permanent arrangement of furniture and equipment must not block access to fire extinguishers, eye washes, showers, etc.
- Devices, furnishings, and work surfaces must be kept clean and in good repair. Hazards can accumulate in cracked countertops, torn upholstery, broken instrument covers, etc, making it difficult or impossible to decontaminate them.

 ISBN 978-089189-6463

Signage

All pertinent hazard information and safety instructions must be clearly posted on entrance doors, storage units (eg, cabinets, refrigerators), equipment, and laboratory walls. Signs should be positioned to be accessible by both normal and handicapped individuals, including those who are sight impaired. Some important items to post would include the following:

- universal hazard symbols as appropriate, such as symbols for biohazard and radiation hazards **f10.1**, **f10.2**
- laboratory area designations such as "Clean area—no gloves" or "Dirty sink—no hand washing"
- refrigerator information such as "For food only," "No food permitted," "Explosion proof"
- standard danger signs (black, red, and white), such as the GHS symbol for explosive **f10.3** or caution signs (yellow & black) such as the electricity warning symbol **f10.4**
- safety instructions (green & white), such as signs for eyewashes and shower; symbols shown in **f10.5** and **f10.6**
- fire evacuation routes and exit signs
- temporary signs or barrier tapes, such as spill area warnings and evacuation routes to be used during construction

Telephone

A laboratory should have a telephone and/or other dependable means of communication. Emergency numbers should be clearly posted, especially if the laboratory is an area that does not have the "911" emergency system. Emergency numbers to consider posting include the fire department, ambulance, police, poison control, security, supervisor(s), and safety officer(s).

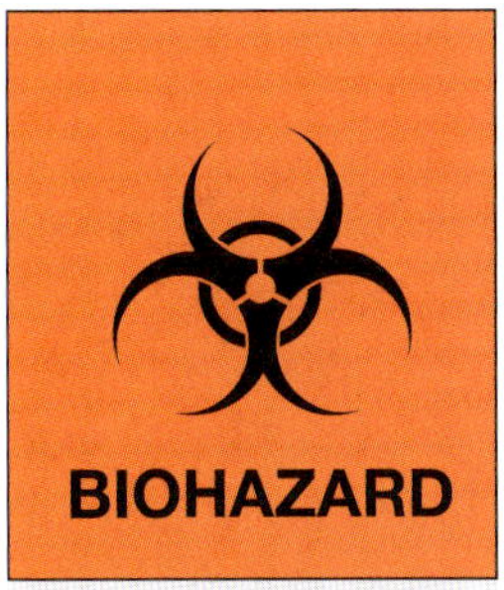

f10.1 biohazard symbol

f10.2 radiation hazard

f10.3 explosive symbol

f10.4 electricity warning

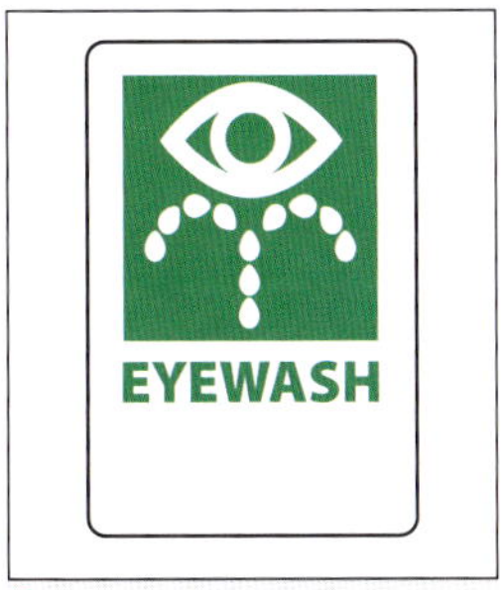

f10.5 eyewash sign

f10.6 deluge shower sign

Fire safety equipment

No laboratory should be without the means to extinguish fires. In general, 1 or more fire extinguishers, a fire blanket, a fire hose, a sand bucket, and a fire alarm system are required. Local fire codes outline amount and types of fire equipment that must be on hand in any laboratory area. These codes take into account the type of work being conducted in a specific area. (More information on types and usage of fire safety equipment is in Unit 2.)

Safety shower

Safety showers and eyewashes are necessary to cleanse chemical, biohazardous, and radioactive body spills. They can also extinguish clothing fires. Showers and eyewashes should be well marked, centrally located, and easy to access directly, because a person who has sustained a spill or is on fire may be disoriented and upset. A "lockout/tagout" mechanism should be in place to be sure that the water supply lines to these devices are never turned off except for brief periods of maintenance. The American National Standards Institute (ANSI) publication ANSI Z358.1 (summary available at www.eyewashdirect.com/v/vspfiles/pdf-new/eyewash-ansi-2015.pdf, accessed January 4, 2016) contains specifications for safety showers and eyewashes. Although this is not an OSHA standard, OSHA typically follows its recommendations.

Safety showers should be located <55 feet or 10 seconds traveling time from the farthest point in the laboratory, but still well away from instruments and electrical sources. The path to the safety shower must be unobstructed and on the same building level. If the hazard is corrosive, the access route must not include doors, which could need opening or accidentally be locked. For noncorrosive hazards, one door opening in the direction of the safety shower is permissible; the door should be kept open. Safety showers should be inspected and tagged at least once a month to verify that an uninterrupted flow of 20 gallons of temperate water per minute at a pressure of 30 pounds per square inch (psi) is available. ANSI recommends testing weekly to verify flow, but since this is not an OSHA standard, many laboratories find it difficult to justify this given the mess that shower testing can cause. Showers should not be connected solely to cold water because this could cause shock in the victim. Laboratories in geographic regions where water can get very cold may or very hot may need to use devices such as thermostatic mixing valves to ensure proper water temperature.

Safety showers are activated by pulling on the large handle ring to release the water, which must be low enough for a person in a wheelchair to reach. In addition, pull rings must not require undue force so that disabled or injured users can activate the shower. Many plumbing devices can ensure appropriate accessibility, and laboratories should evaluate whether they should replace older equipment to comply with accessibility laws such as the Americans with Disabilities Act.

Spill victims should remain under the shower for 15 minutes until all the material is removed. Some showers remain activated for 15 minutes even if the ring is released, while others turn off once the ring is released. Continuous water flow without operator intervention is considered superior and recommended by ANSI. Victims of body fires and spills serious enough to warrant a deluge shower must always be given medical attention. Even if the victim appears to be suffering no ill effects, delayed reactions may occur and it is safer to obtain medical treatment.

Safety shower drains can be problematic because they are used infrequently so insects and debris can accumulate in them, thus necessitating periodic flushing. In addition, chemicals that should not enter the sanitary sewer system cannot be allowed to enter safety shower drains. If drain covers are used to minimize these problems, they must be removed when the deluge shower is activated. In some jurisdictions, emergency showers simply do not have floor drains.

 ISBN 978-089189-6463

A deluge shower and eyewash in a combination unit is often preferred as it allows for less involved plumbing and the ability to flush the eyes and body simultaneously. ANSI recommends against such devices that do not allow simultaneous operation of both the shower and the eyewash.

Eyewash

Squirt bottles and portable eyewashes are not permitted by OSHA except in field conditions. If they are present, the bottles must be replaced when the solution inside is past the expiration date. Eyewashes should be plumbed to a continuous source of tepid water (60°-100°F, or 16°-38°C, per ANSI) and should be accessible to the handicapped. Water temperatures in excess of 100°F actually hasten chemical reactions in the eye and must be avoided. Annual verification of water temperature is usually sufficient. As with safety showers, paths to eyewashes must be unobstructed and <55 feet or 10 seconds from all users, well away from instruments and electrical sources. Eyewashes should require minimal effort (1 hand) to activate and should remain on for 15 minutes without intervention. 2 jet eyewashes (1 for each eye) are preferred over single nozzle washes. Eyewash stations should be inspected often to verify that clean, temperate, aerated water freely flows from it at a rate of at least 0.4 gallons (1.5 L) per minute at a pressure of 30 psi. According to ANSI, eyewash plumbing should be flushed for ~3 minutes each week to decrease microbial growth in the lines and minimize possible infections from eyewash use. Many eyewashes are associated with sinks, so they are regularly exposed to contamination. Eyewash nozzles should be kept covered when not in use, and both the nozzles and covers should be disinfected weekly. OSHA has published detailed information about infections associated with eyewashes (www.osha.gov/Publications/OSHA3818.pdf, accessed January 4, 2016), and many are serious.

When the eyes are splashed with chemicals, especially corrosives, they should be rinsed for 15 minutes in an eyewash or under a faucet if an eyewash is unavailable. To use an eyewash, the victim should place his or her eyes directly in the flow of water and keep them open while roll the eyeballs around so water covers all surfaces and under the eyelids. If contact lenses are in the injured eye, they should be removed as soon as possible. An eyewash is much less effective in the presence of contact lenses, and liquids that get trapped under them can cause severe damage. If the eye is sufficiently damaged that the contact lens is "stuck," it should not be forced from the eye. Again, the victim should receive medical attention.

Protective wearing apparel

The OSHA Personal Protective Equipment (PPE) standard (29 CFR 1910.132) requires that employers assess workplace hazards and provide PPE at no charge to staff who will encounter hazards. When PPE is necessary, there is a reasonable chance that a worker could come in contact with a hazard; therefore, minimizing or eliminating possible contact with a hazard by engineering and/or administrative controls must first be attempted. For example, if a chemical with hazardous fumes must be used, the first protective steps include ensuring good ventilation in the laboratory with functional fume hoods—not requiring all workers wear respirators. Protective wearing apparel, while certainly necessary in many situations, should only be used

 ISBN 978-089189-6463

when all other means to reduce a hazard have been exhausted. Once it has been determined that PPE is necessary, staff must be trained when and how to use and care for it properly. PPE must fit well and be as comfortable as possible.

Protective wearing apparel ranges from the common laboratory coat to the positive pressure "space suit" worn in biohazardous containment facilities. The correct apparel to be worn in any situation should be determined by the following:

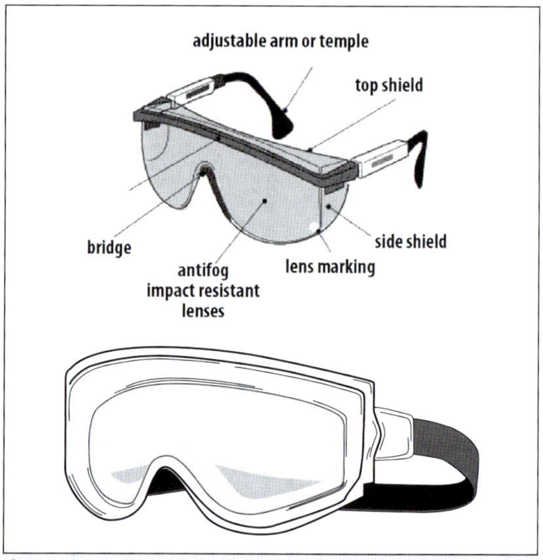

f10.7 safety glasses (above) & safety goggles (below)

Type of procedure

Will it be difficult to minimize splashing? Will it be difficult to minimize aerosols? Will there be extreme heat or extreme cold? Are there sharp objects? Is electrical equipment involved? Everyone who is involved in a process must be fully informed of the inherent risks and be given the opportunity to voice concerns.

Degree of risk

Are the chemicals corrosive, flammable, toxic, explosive, radioactive, or reactive? What is the appropriate biosafety level for the organisms/animals/specimens used in the laboratory? Are compressed gases, cryogens, or radioisotopes involved?

Condition of the worker

Is the worker ill or immunocompromised? Does the worker have cuts, abrasions, or skin conditions that might make him or her more vulnerable to a hazard? Does the worker wear glasses or contact lenses? Is the worker pregnant? Is the worker's respiratory system compromised? Does the worker need accessibility or ergonomic accommodations?

Given these considerations, some broad categories of protective wearing apparel are commonly used. An important aspect of all PPE is that it should fit the user properly so that safety is not compromised. This may mean, for example, that employers have to purchase many sizes of laboratory coats, rather than several "one size fits all" models. As a general rule, gloves should be put on last so the worker has maximum dexterity when putting on the other PPE, eg, masks, gowns. When work is complete, gloves must be taken off first, allowing the removal of other PPE by grasping the noncontaminated areas against the body. When possible, protective apparel is turned inside out when removed to keep the contaminated areas contained. PPE should not be worn outside the lab in clean, public areas. Conversely, PPE may be needed by anyone in a laboratory, even those not directly involved in work (eg, janitors, maintenance workers).

Face protection

This type of protection includes eye goggles, masks and face shields. 2 types of eye protection are shown in **f10.7**. Safety glasses should have side shields and be made of shatterproof glass or plastic. However, safety glasses protect only against minimal

 ISBN 978-089189-6463

f10.8 laser hazard warning

aerosol formation as they do not make a seal around the eyes the way goggles do, which is why prescription eyeglasses are sometimes considered adequate protection in low risk procedures. Any procedure with a high risk of splashing, especially of corrosives, requires the use of eye goggles and a face shield. Face shields are *not* considered adequate eye protection, and goggles or glasses may have to be worn under the face shield. Also available are special goggles that protect the eyes from ultraviolet (UV) and laser light. Even with these goggles on, however, one should *never* look directly at UV and laser light sources. **f10.8** shows hazard labeling for lasers.

Respiratory protection

The National Institute for Occupational Safety and Health (NIOSH) lists 5 categories of respiratory protection (www.cdc.gov/niosh/npptl/topics/respirators/factsheets/respfact.html, accessed January 4, 2016):

1. **Escape respirators** are designed for emergency evacuations for very short term (<1 hour) of protection; they should never be used for long term work with hazards.
2. **Particulate respirators** fit over the nose and mouth and useful for respiratory protection from particulate hazards, such as spores from a mold, but are not effective against chemical fumes. Masks worn for protection against bacteria such as *Mycobacterium tuberculosis* are particulate respirators and are required to filter particles of 1 μm in size with >95% efficiency (N95 particulate masks).
3. **Chemical cartridge respirators** ("gas masks") filter out volatile chemicals through interchangeable and replaceable cartridges. Depending on the cartridge, some protection against particles may also be possible. Unfortunately, no filter/cartridge works against all potential chemical or biologic hazards. Effective use of these respirators requires some knowledge and certainty of the hazard so that the correct cartridge can be put in place. Cartridges have expiration dates, so if these are kept for emergency use only, they should be inspected routinely so that they can be replaced when necessary.
4. **Powered air purifying respirators** are similar to chemical cartridge respirators except that a fan blows air through the filter, making it easier for the user to breathe.
5. **Self contained breathing apparatus** provides air through a tank that the user must carry. The air tanks are heavy and these devices require special training to use. These are used by fire fighters and would rarely be encountered in a laboratory.

No respirator is effective if it does not fit properly and contaminated air can get to the user. Facilities must conduct fit testing to verify that staff can don respiratory protection properly and make a facial seal that does not leak. Fit testing is usually

10

 ISBN 978-089189-6463

repeated annually and/or if a wearer has undergone significant facial change such as that which results from weight loss. Detailed fit testing procedures are beyond the scope of this text and available on the OSHA website. Beards or facial anomalies can prevent respirators from fitting properly, so some staff may not be eligible for tasks requiring full respiratory protection. Staff with compromised pulmonary function also may not be eligible to wear respirators because of the extra effort required to breathe through filters. Before fit testing, users should complete a medical questionnaire to ensure that they are healthy enough to wear a respirator.

Respirators are not the same things as face masks. Face masks do not seal tightly against the wearer's face, so they do not provide protection to the wearer from small particles. Face masks are put on infectious patients to reduce the number of infectious particles released by them. Healthcare workers wear face masks to protect patients and to provide protection to themselves against large droplets from patients. Good discussions regarding the nature of respiratory protection and selection of devices for various purposes can be found at www.cdc.gov/niosh/npptl/topics/respirators/disp_part/ and www.osha.gov/Publications/OSHA3767.pdf, both accessed January 4, 2016).

Hand protection

Hands are protected with various types of gloves. Insulated gloves should be used to handle hot objects such as items that have been autoclaved or cold items such as dry ice. For handling animals, performing autopsies, using bone saws, and changing cryostat or microtome blades, metal mesh gloves should be used, which are puncture resistant but do not always prevent injury. Infectious materials require the use of latex, nitrile, or vinyl gloves. Correct glove composition (eg, nitrile, neoprene, or rubber) for chemicals may require evaluation on a case by case basis because many chemicals dissolve latex and vinyl. In 1997, dimethyl-mercury penetrating a latex glove led to a fatality; it only takes ~10 minutes for 100% isopropanol to penetrate latex or vinyl. A chemical that can penetrate a glove will come in close contact with the hand and is more likely to cause damage than not wearing any glove. Many safety data sheets (SDSs) will specify the optimal composition of gloves for a chemical, including toxic antineoplastic drugs. Thicker gloves can be used longer without chemical breakthrough, but they may impair grip and dexterity, which is potentially hazardous. Therefore, evaluation of the task to be performed as well as the chemical involved influences glove selection. Many gloves are intended to be single use, and should never be washed and reused because cleaning chemicals can compromise the integrity of the glove. Heavy duty gloves resistant to chemicals may be intended for multiple uses, so staff should be trained to inspect reusable gloves before and after use, and follow decontamination protocols carefully. Even highly resistant gloves may be penetrated after multiple uses, so it may be necessary to track when such gloves were put in service. Chemical resistance charts for reusable gloves are typically provided by the manufacturers.

Some people have serious allergic reactions to products that contain latex, including skin rashes, hives, flushing, itching, nasal/eye/sinus symptoms, asthma, shock, and death. Therefore, a person with severe allergies may wear a medic alert bracelet. There are 4 types of hypersensitivity reactions, and latex tends to cause Type 1 or Type 4. Type 1 reactions are usually from the latex itself and can

 ISBN 978-089189-6463

involve the more dangerous manifestations such as asthma and shock, while Type 4 reactions are usually from the chemicals used in latex processing; contact dermatitis is more likely. Irritation and inflammation associated with glove use should always be investigated and corrected. An occupational health physician may choose to test an affected person to determine the exact root cause so that allergenic materials can be avoided.

Sometimes a reaction to latex is enhanced by the powder in the gloves because the powder can make latex particles airborne for up to 5 hours and cause respiratory reactions in sensitized individuals. These airborne allergens can affect anyone in the room, even individuals not wearing the latex gloves. Corrective options include the use of glove liners, powder free gloves, and latex free gloves. When unnecessary, powdered and latex gloves should simply be avoided. Because some hypoallergenic gloves are more expensive, it is usually financially smart to determine the exact cause of an allergy. Chronic skin inflammation, however, is never acceptable because broken skin makes workers more vulnerable to hazards.

The choice of glove material is not simple. An excellent review of glove materials by Ozanne is listed in the references and is summarized here. Vinyl gloves are more likely to leak than nitrile or latex gloves, and they generally show higher rates of penetration by biologic agents. Nitrile gloves are more resistant to perforation than latex gloves, but when they are perforated, the perforations enlarge faster, which also holds the benefit of more quickly drawing attention to the hole. All 3 materials have different profiles regarding chemical penetration, so the SDS for a chemical should typically be checked for recommended glove type. In some cases, using double gloves may be recommended. Of the 3 materials, vinyl is generally least preferable in terms of barrier protection, but usually the least expensive. Regardless of material, jewelry and long fingernails are more likely to compromise gloves and should be avoided.

Because latex is so widely used, it is important to note ways in which latex is compromised. Latex becomes more porous upon exposure to ozone (generated by electrophoresis, for example), certain hand lotions, X rays, UV light, temperatures higher than 33°C, and humidity >40%. Therefore, latex gloves should be stored in cool, dry places away from electrical equipment and without exposure to light sources, including UV. Boxes should be dated when they are opened and any remaining gloves discarded after 3 months; the only hand lotions permitted in the laboratory should be compatible with latex. Users should be trained to briefly inspect all gloves for imperfections when they are first worn and to change gloves after 30 minutes of work. Studies have shown that the combination of sweat and heat in a latex glove can make it permeable to the human immunodeficiency virus (HIV) and hepatitis B virus (HBV) after ~50 minutes. Gloves must also be changed between patient contacts or if they become significantly contaminated.

Even when the correct type of glove has been selected, stored, and donned properly, users cannot assume that even new gloves are without defects. Until its new standards were published in 2006, the Food and Drug Administration (FDA) accepted defect rates of 4.0% and 2.5% for patient examination medical gloves and surgeons' gloves, respectively. Although they lowered the new acceptable defect rates to 2.5% and 1.5%, respectively, the expected defect rate is still not 0. The FDA estimates that at the old defect rates, 2.4 cases of HIV and 2.4 cases of HBV were transmitted annually because of faulty gloves alone. Therefore, although it will not guarantee the absence of a defect, users must be responsible for visually inspecting each glove they don before beginning work. For reusable gloves, performing an "air test" is useful to detect leaks, during which the glove is rolled up at the cuff end so that air is trapped, and the rolled toward the fingers to reveal any significant leaks.

Body protection

Laboratory coats come in various materials, including fluid resistant and fluidproof substances that minimize penetration of liquids and aerosols. Laboratory coats should be worn completely buttoned during work; if open, a laboratory coat is not much of a barrier and can easily snag objects. The barrier provided by many fluid resistant/fluidproof laboratory coats makes them hot to wear and can create static electricity, so tasks should be analyzed to be sure that such garments are absolutely necessary. Some manufacturers weave in a black static discharge carbon thread to minimize static electricity, and others provide barrier protection only in front, making the back more porous so as to make the user more comfortable. Knitted cuffs that allow workers to pull gloves over them for additional protection are preferred. For more coverage, removable sleeve protectors are useful for working inside contaminated hoods.

Knee length lab coats are a better option than hip length. The best policy is also to require that legs be covered with pants or an ankle length skirt. Bare legs or leg coverings that are very thin (such as pantyhose) are not nearly as safe. Per the Clinical and Laboratory Standards Institute (CLSI), garments should be 1-1.5 inches above the floor and should not touch the floor. Lab coats must also cover the full arm and never be pulled up or rolled up.

For staff engaged in work with blood and body fluids, OSHA requires fluid resistant laboratory coats that would not, under "normal conditions of use," permit the passage of infectious fluids. OSHA specifies the following qualities as tested by the American Society for Testing and Materials (ASTM): (1) spray rating of 90 or higher (water repellence), (2) air porosity rating of 10 or higher (for comfort), and (3) Suter resistance of 340 or more to fluid pressure. Because of the various laboratory coats available on the market, these criteria may be helpful in selecting what to purchase and understanding cost differences.

"Fluid resistant" means that if the lab coat sustains a spill, penentration can occur, but the wearer likely has time to remove it before that happens. Therefore, these lab coats must be removed if there is a significant spill of any kind. If lab staff members are part of procedures at a patient's bedside in hospital settings, they must wear clean lab coats that have not be used, even if no previous spill on a coat is suspected.

Fluid impermeable lab coats and plastic aprons are useful to completely prevent penetration from liquid spills. Plastic can collect static electricity, so caution is needed around flammables. Barriers such as work shields may be a better choice, but gloves and laboratory coats must still be worn because the arms must reach around the shield to perform the work. Sleeve protectors may be useful when working with shields.

Since gloves are typically removed first, lab coats must be removed to prevent hand contamination. Sleeves should be pulled

 ISBN 978-089189-6463

off by grabbing under the cuff, and this is often easier when reaching from behind. Dirty lab coats should be folded so that the outside is facing inward. A video showing proper lab coat donning and removal is located at www.youtube.com/watch?v=fqViWi1L3gE (accessed November 7, 2015). Significantly contaminated clothing must be removed immediately and bagged in leak resistant material. Ideally, decontamination happens on site, but cleaning can be performed by an outside contractor. Workers should never take contaminated clothing home.

If no spill is evident on a lab coat, it can be reworn in many circumstances. Best practice is to have a set of hanging hooks for lab coats to be stored when not in use and separate storage for street clothes. If a lab coat is worn daily, it should be laundered every week or 2, even if no contamination is evident. Per CLSI, lab coats should be laundered in hot water (>160° F, or 71° C) or in cold water with bleach. Fabric softener should not be used as it compromises fluid resistance. The Centers for Disease Control (CDC) have additional information about laundering reusable materials (www.cdc.gov/HAI/prevent/laundry.html, accessed January 4, 2016).

Proper footwear is essential to prevent damage from spills. Sturdy, fluid impermeable shoes with nonslip soles that cover the entire foot are required. Additional disposable "bootie" foot covers can be worn over shoes, if necessary. Canvas shoes, open toed shoes, open backed "clogs," and sandals are strictly forbidden. Shoe covers are never acceptable substitutes for inappropriate footwear; the purpose of shoe covers is to provide a disposable barrier to contamination since shoes are difficult to decontaminate and expensive to discard. When handling heavy objects such as compressed gas cylinders, steel toed shoes are ideal.

Ear muffs or ear plugs are required if the noise level exceeds the 85 decibel (dB) time weighted average for an 8 hour day specified by OSHA. Noise levels in excess of 100 dB even for a short time require ear protection. According the American Speech and Hearing Association, the noise level of an ordinary conversation is ~60 dB, a blow dryer is ~85 dB, and a hand drill is ~100 dB. Although a noise dosimeter would be required to make actual measurements, a good rule of thumb is that any time an ordinary speaking voice cannot be heard, the area is too loud. Some laboratory equipment is very noisy (eg, tissue homogenizers and sonicators) and should not be used without ear protection. Soft foam ear plugs are generally adequate. Users should wash their hands to avoid introducing contaminants into the ear canal and then roll the foam into a narrow cylinder that is easily inserted and shaped to the ear canal. If the top of the ear is pulled up and back so the ear canal straightens, the rolled plug should be easy to insert. Proper fit can be verified by comparing how muffled the sound is when the hands are placed over the ears and when the ears are uncovered. Good fit into the ear canal is essential, as an ear plug that is simply sitting outside of the ear canal creates a channel into the ear that actually amplifies sound instead of reducing it.

Personal

If name badges are worn on lanyards, the lanyards should be either the "break away" kind that split apart if the badge is caught or the retractable kind so that the badge is pulled rather than the person. Ideally badges are worn under lab coats, but if identification is necessary, the badge should be positioned high on the collar in an area unlikely to be contaminated.

Jewelry of all kinds should be avoided. Dangling jewelry can get caught in equipment, and hazardous substances can accumulate underneath rings. Many rings can also puncture gloves and eliminate the barrier that they provide. Hair should be worn tied back away from the face, and beards should be neat and closely trimmed. For some types of sterile work, beard covers may be necessary. OSHA does not specifically regulate contact lens use in the PPE standard because of evidence that eye protection worn over the eyes is sufficient. However, in many situations, contact lenses should be avoided because liquids and chemical fumes can be trapped under the lenses and greatly intensify the amount of hazardous material contacting the eye. In addition, contact lenses can be permeable to chemical vapors and radioisotopes, so in some cases they are categorically forbidden.

Hoods

All chemicals that release harmful or combustible vapors must be handled under a chemical fume hood. Airborne biological hazards must be handled in biological safety cabinets. Class 1 and 2 biohoods are for organisms up to CDC biosafety level (BSL) 3, and the sealed class 3 hoods are for BSL 4. Laboratories handling any of these hazards must use hoods. More information on hoods is in Units 3 and 5.

Sharps

All sharp objects must remain covered when not in use and disposed of in puncture resistant containers when work is complete. Special receptacles, not ordinary trash bins, are required for broken glass. Brooms, dustpans, tongs, and other devices must be available so that broken glass or sharp objects do not have to be picked up by hand. Puncture resistant gloves should be used when managing sharp exposure is difficult to predict, such as when handling animals. More information is in Unit 4.

First aid & spill containment supplies

Sand, kitty litter or vermiculite, spill pillows, or general absorbents can be used for almost any spill; a broom and dustpan are useful to clear them away. Disinfectants for biohazards should be readily available. Barrier tape is helpful to prevent inadvertent traffic through the spill area. PPE such as chemical resistant gloves, goggles, and shoe covers should be available for spill cleanup. Other necessities, eg, acid/base neutralizing agents, mercury spill kits, and radiation decontaminating solution, are based on the types of hazards present. Unit 12 contains information on first aid and first aid supplies.

Hand hygiene

The World Health Organization (WHO) has published an exhaustive review of hand hygiene research and effective protocols (WHO Guidelines on Hand Hygiene in Health Care, apps.who.int/iris/bitstream/10665/44102/1/9789241597906_eng.pdf, accessed January 4, 2016). The reader is referred to this document for additional information and detailed protocols to establish hand hygiene programs.

Laboratory staff must wash their hands whenever they complete work and/or leave the laboratory. This is particularly important before making hand to face contact and eating or drinking. Proper hand washing is absolutely critical for preventing the spread of infectious disease, and eliminating chemical and radiation contamination from workers. Gloves are effective barriers, but they can easily get small tears and holes that cause hand contamination. Studies have also shown that a small percentage of

ISBN 978-089189-6463

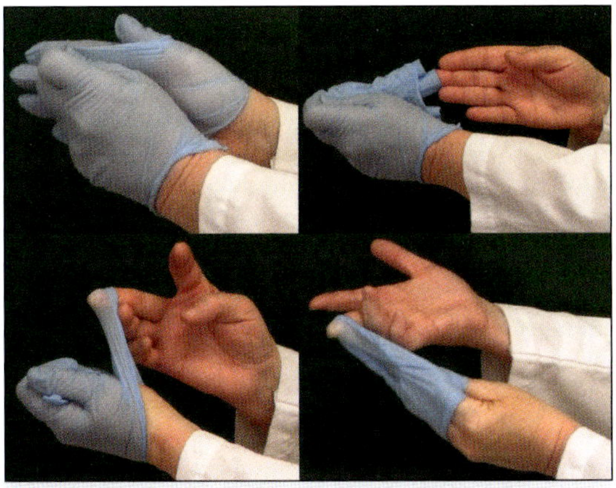

f10.9 proper glove removal

f10.10 top: label on sink used for hazards; bottom: label on sink to be kept clean

new gloves also have small holes. Therefore, a thorough cleansing of the hands is vital upon completion of any work, even if gloves are worn.

f10.9 illustrates proper glove removal to prevent hand contamination. Gloves must be removed one at a time. The first glove can be pulled off by grabbing the fingers and pulling. The second glove cannot be removed this way, however, because the other hand is now bare and should not contact the contaminated surface. The correct procedure is to grab the first glove with the remaining gloved hand, slide the bare hand under the cuff where there is no contamination and roll the second glove off so that it becomes "inside out" and creates a noncontaminated surface around the 2 gloves. The gloves can then be discarded with bare hands because the contaminated surfaces are inside. During training, it is useful to cover the outside of gloves with shaving cream and ask trainees to remove them properly. If they are successful, no shaving cream will get on their hands.

The hands must then be washed with soap in warm, not hot, water. Hand washing sinks should be dedicated to noncontaminated processes only and should be separate from sinks with contaminated functions such as waste disposal, instrument cleaning, and Gram staining. Control of the sink with foot pedals is ideal. If separate sinks are not possible, sinks must be decontaminated before being used for hand washing. **f10.10** shows sample labels for laboratory sinks. Hand washing sinks accessible to the handicapped must also be provided.

Hands should be wet before soap is applied. Plenty of soap should be used, and the hands should be washed in a downward motion for at least 1 full minute. Studies show that the average duration of hand washing for healthcare workers is <15 seconds, so a sustained conscious effort is likely required so hands are washed for long enough. Particular attention should be given to the nails, between the fingers, and under rings, which are frequent sites of contamination. A nail brush is useful to ensure that the areas under and around the nails are clean. The water should be left *on* while the hands are being dried, and the paper towel used to dry the hands should be used to turn off the water to prevent recontamination of the hands from the

10

dirty faucets. Reusable cloth drying towels are not recommended.

An antiseptic soap is helpful, but most of the decontamination occurs from the physical actions of scrubbing and rinsing, and bacterial spores are highly resistant to most antiseptics. Laboratory workers may wash their hands many times during a work day, but repeated use of harsh soaps may cause irritation and chapping, so these products must be used cautiously. When skin is irritated by loss of skin cells and/or lipids, its barrier function is reduced and only recovered by 50%-60% after 6 hours. Full recovery can take up to 5-6 days, and often normal skin flora is deranged, which can promote colonization by undesirable organisms. Skin irritation is more likely if the water used is too hot, the humidity is low, the towels are too rough, and hand cream is not used after washing. Hand cream should not be petroleum based because glove integrity could be compromised.

Many facilities have installed dispensers of alcohol based hand gels for biohazard decontamination because they are equal or better germicides than soaps. The addition of emollients to alcohol reduces skin irritation compared to normal hand washing with antimicrobial soaps, and fewer allergic reactions to alcohol have been reported. Use of alcohol gels does not preclude hand washing, which is still necessary when the hands are visibly soiled because alcohols poorly penetrate proteinaceous material. In addition, alcohols have no residual antimicrobial activity on skin once they have evaporated, while many antiseptic products, such as chlorhexidine, remain on the skin and have residual activity.

Various alcohols (isopropanol, ethanol, n-propanol) alone and in combination have good germicidal activity against Gram+ and Gram– vegetative bacteria (including multidrug resistant forms), *Mycobacterium tuberculosis,* many fungi, and enveloped viruses (such as human immunodeficiency virus, respiratory syncytial virus, herpes simplex virus, influenza, hepatitis B virus, and hepatitis C virus). Infectivity of many nonenveloped viruses (such as rotavirus, adenovirus, rhinovirus, poliovirus, and hepatitis A virus) is reduced by alcohols. Alcohols kill microbes by protein denaturation, and minimum alcohol concentrations are ~60% for this effect. Maximum alcohol concentrations are ~80% for killing because water is also necessary to denature the proteins. It is important to note that prions and some organisms are *not* reliably killed by alcohol hand gels. According to the CDC, some organisms not susceptible to alcohol gels include:

- Spore forming bacteria: *Bacillus anthracis, Bacillus cereus, Clostridium botulinum, Clostridium tetani, Clostridium perfringens, Clostridium difficile*
- Certain parasites: cysts of amoebic dysentery, *Giardia lamblia, Cryptosporidium*
- Certain viruses: norovirus, calicivirus, picornavirus, parvovirus, some nonenveloped viruses

If the presence of these organisms is suspected, hand washing is absolutely essential. Iodophor antiseptics are the most likely to be effective against spores, but they are generally too irritating to the skin to be used routinely, so they must be used sparingly.

Another problem with alcohol hand gels is that the alcohols used typically have flash points at room temperature. Incidents have been documented in which residual gel on the hand has caught fire when exposed to a spark as small as that from a light switch or electrical outlet, an effect made worse by low humidity. The simplest means to prevent fire is to apply the minimal amount

ISBN 978-089189-6463

of gel necessary to coat the hands and let it air dry before doing anything else.

The correct amount of alcohol gel varies with the concentration of alcohol in a particular product, and many products have dispensers that provide the optimal amount of gel. Dispensers should be maintained regularly because over time they can develop blockages that prevent an adequate amount of gel from being dispensed. A rough guideline is that if the gel dose dries on the hands in <15 seconds, an adequate amount may not have been present. Users should dispense the gel and spread it thoroughly over all the surfaces of both hands and not proceed with work until the gel is dry.

Studies have shown that some organisms, including methicillin resistant *Staphylococcus aureus,* are developing resistance to antiseptics used in hand washing products. There is currently no cause for alarm because the resistance is minimal compared with the concentrations of antiseptics currently in use. However, as soap dispensers are used, the bacterial load in the soap will increase, so it is a good practice to replace disposable dispensers rather than "topping them off" with additional soap.

The CDC recommends that for routine hand hygiene a worker should choose an alcohol gel or hand washing, but not both at the same time, to minimize skin irritation. For this reason, placement of alcohol gels near hand washing sinks is not recommended. If the hands are not visibly soiled, the use of alcohol is usually superior unless numerous (5-10) applications have resulted in a buildup of emollient. The CDC also provides data to show that the increased cost of the gel products is easily justified with expected reductions in the rates of infections, particularly if gel dispensers are placed strategically at points of contact where hand washing sinks are far away or inconvenient. Placement of dispensers must be considered carefully to ensure that they are easily accessed, but since alcohols are poisonous, they must also be out of the reach of children. Pocket sized bottles of gel have also been shown to increase compliance with hand hygiene protocols. Further, because gels can be properly used in as little as 15 seconds, while correct hand washing takes 60 seconds, decontamination is likely more effective with gels.

Many studies have shown increased bacterial burden with long nails (natural & artificial), chipped nail polish, and jewelry. Even with the use of gloves, bacteria colonizing patients have been detected in up to 30% of healthcare workers with long nails, and infection has been shown to be transmitted from long fingernails. Current CDC recommendations are that workers with artificial nails should not be allowed around high risk patients and that natural nails be 1/4 inch long or less. Data are insufficient to forbid jewelry and nail polish, but the aforementioned risk factors might be considered in outbreak situations when every possible avenue of transmission must be controlled. In addition, nails and jewelry could harbor chemical and radioactive contaminants, so workers should consider these factors carefully when engaging in any method of hand hygiene.

Many hospitals have adopted a "sanitize in, sanitize out" policy in which staff must perform hand hygiene both before and after patient contact. Hands must be cleaned even if gloves were worn. While these are excellent policies to minimize the spread of micro-organisms, intelligent selection of soaps and disinfecting gels is required to minimize skin irritation from so many daily cleansing episodes.

Summary table: safety equipment & safe work practices

Topic	Comments
Basic work practices	1. no hand to face activities permitted 2. wash or sanitize hands after removing gloves and when leaving lab 3. no jokes or horseplay, and no alcohol or drugs; discipline workers who ignore safety protocols 4. avoid having personnel in the laboratory alone or engaging unattended lab operations 5. keep work areas clean and uncluttered; decontaminate after procedures 6. wear appropriate personal protective equipment (see below); remove PPE when leaving lab 7. plan procedures carefully so proper safety equipment is available 8. no electronic devices, plants, pets, decorations 9. full attention to work; no rushing or shortcuts 10. no obstruction of safety equipment or evacuation routes 11. keep equipment, work surfaces, chairs, etc in good repair
Signage	appropriate signs on entrances, storage units, equipment, and walls
Inspection & maintenance	1. all equipment should be on maintenance and inspection schedule 2. all equipment should bear tag stating latest inspection and/or maintenance
Telephone	1. all labs should have easy access to a reliable telephone 2. post emergency numbers beside telephone
Fire safety	1. know local fire codes and fire procedures for your laboratory 2. fire equipment: extinguishers, blanket, hose, sand bucket, alarm system
Safety showers and eyewashes	1. well marked, easily located, unobstructed within 55 feet or 10 seconds of worker 2. 20 gallons per minute of temperate water at 30 psi for showers; 0.4 gallons per minute of temperate, aerated water at 30 psi for eyewashes 3. flush victim/eyes 15 minutes; if possible, remove contact lenses 4. seek immediate medical attention even if victim feels fine 5. position for wheelchair access 6. flush eyewashes 3 minutes weekly to minimize bacterial growth; bleach nozzles and covers 7. lockout/tagout protocol to prevent water supply from being shut off
Hoods	1. chemical fume hoods for chemicals with harmful or combustible vapors 2. biohazard cabinets: Class 1 & 3 for BSL 1-3; Class 3 for BSL 4
Sharps	1. puncture resistant containers and special needle boxes 2. broken glass boxes 3. brooms and dustpans to clean broken glass 4. cover sharp at all times it is not in use
Spill equipment & first aid supplies	1. general purpose absorbents, eg, sand, kitty litter, spill pillows 2. treatments for hazards present: neutralizing agents for acids and bases; radiation decontaminating solutions; mercury spill kits; disinfectants for biohazards 3. see Unit 12 for first aid supplies

 ISBN 978-089189-6463

Summary table: safety equipment & safe work practices (continued)

Topic	Comments
Protective apparel	1. criteria for determining type needed: type of procedure, degree of risk, condition of worker 2. learn to wear and remove PPE properly 3. face/eye protection a. goggles make seal around eyes; better than side shield glasses b. face shields and prescription glasses NOT adequate eye protection c. special glasses for UV and laser lights 4. respiratory protection a. particulate respirators for microbes: N95 is minimum b. chemical cartridge respirators: must choose proper cartridge for hazard c. masks only protect against large droplets; respirators must be fit tested so they don't leak 5. gloves a. latex, nitrile, vinyl, or chemical resistant b. insulated for heat and cold c. puncture resistant (animal handling) d. hypoallergenic/powderless gloves or glove liners for allergies e. compromised gloves: don't reuse or decontaminate single use glove or use gloves that are too old f. latex damage: heat, light, humidity, certain hand lotions, ozone 6. body protection a. lab coats & fluid resistant lab coats buttoned; plastic aprons b. proper footwear: no open toes or heels, sandals, or canvas shoes c. legs covered d. ear muffs or ear plugs if conversation can't be heard e. disposable sleeve and foot covers 7. personal a. avoid jewelry, long nails, artificial nails, nail polish b. hair pulled back; beards may need beard covers c. avoid contact lenses d. breakaway or retractable lanyards e. identification badges under PPE
Hand hygiene	1. remove gloves without hand contamination (see text) 2. antiseptic soap for at least 1 minute; check nails and between fingers 3. turn water off with paper towel, not the clean hand 4. minimize irritation with mild temp and gentle towels; use nonpetroleum hand cream 5. alcohol a. disinfects most biohazards but not prions and spores; some viruses and parasites not susceptible b. let dry on hands to remove flammable hazard c. not a hand washing substitute; wash hands when visibly soiled 6. designate dirty sinks and clean hand washing sinks

Self evaluation questions

1 You are about to handle a chemical with toxic fumes. You should (choose the BEST answer)

- **a** wear a respirator
- **b** wear a face shield
- **c** work under a fume hood
- **d** wear a self contained breathing apparatus
- **e** wear a mask that covers your mouth and nose

2 You are about to perform a procedure which requires vigorous mixing of a hazardous micro-organism, and it is difficult to prevent splashing and aerosols. The minimum you should do is

- **a** wear laboratory coat and gloves
- **b** wear laboratory coat, gloves, and face shield
- **c** wear laboratory coat and gloves, and work in a biological safety cabinet
- **d** wear gloves and laboratory coat with disposable sleeve protectors, and work in a biological safety cabinet

3 Fill in the blank. If a chemical is splashed into someone's eye it is important to be sure that __________ are removed from the eyes as soon as possible.

4 Fill in the blank. The safety equipment used to remove large spills from a person's body is a ______________.

5 Fill in the blank. Chemical spills on a person's body or in the eyes should be flushed with water for at least __________.

6 If a corrosive chemical has been completely washed off a victim, and he or she is feeling well, it is not necessary to seek medical attention.

- **a** true
- **b** false

7 Which of the following should be avoided when working in the laboratory?

- **a** unconfined long hair
- **b** wearing wedding rings
- **c** wearing contact lenses
- **d** canvas shoes
- **e** all of the above

8 You come upon a spill in the laboratory and do not know what it is. Which of the choices below is the BEST to treat the spill at this time?

- **a** sprinkle kitty litter around the perimeter of the spill
- **b** spread paper towels over the top of the spill
- **c** soak the spill in 10% bleach
- **d** soak the spill with chemical neutralizers

 ISBN 978-089189-6463

9 Which step(s) in the following sequence is(are) incorrect or missing?
i. remove first glove by tugging on contaminated surface
ii. remove second glove by inserting bare hand under cuff
iii. roll second glove off uncontaminated side out around first glove
iv. dispose of gloves
v. turn water on and wash with antiseptic soap for at least 15 seconds
vi. inspect nails and between fingers, use nail brush if necessary
vii. turn the water off
viii. dry hands with a cloth towel

a at the conclusion of hand washing, water should be left on and turned off using a paper towel
b A cloth towel should not be used to dry hands unless it has not been used by someone else first
c the hands should be wet before soap is applied
d hands should be washed a minimum of 60 seconds
e all of the above

10 A worker develops a severe case of hives on her hands, and she is having some symptoms of asthma. What is the most likely concern to investigate first?

a latex allergy
b allergy to a chemical used in the laboratory
c allergy to a micro-organism being cultured in the laboratory
d infection from a micro-organism being cultured in the laboratory
e infection from a micro-organism contaminating the laboratory environment

11 All of these substances may compromise latex gloves in a relatively short time period EXCEPT

a light
b hand lotion
c isopropanol
d heat and sweat
e radioactive isotopes

12 Fill in the blanks. Safety showers and eyewashes must be located __________ feet or __________ seconds away from the worker. Eyewashes should be flushed __________minutes every __________ to reduce microbial growth.

13 Which situation below describes the correct use of alcohol hand gel?

a put on hands and work begins while gel is still wet
b used to wash off visible blood contamination from hands
c used when gloves are removed and hands look clean
d used after doing venipuncture on a patient with *Clostridium difficile*
e all of the above

 ISBN 978-089189-6463

14 Matching each item to the correct answer below.

Chemical fume hood _____

Pipetting liquids to minimize splashing _____

Scheduling pregnant woman in nonradioisotope laboratory rather than radioisotope laboratory _____

- **a** OSHA administrative control
- **b** OSHA work practice control
- **c** OSHA engineering control

15 Which situation below is acceptable?

- **a** cell phone is used in speaker mode in biohazard laboratory where hard line doesn't exist; phone is kept in the laboratory permanently
- **b** radio is brought into laboratory daily and decontaminated prior to being returned to the employee's locker
- **c** plants that maintain air quality are kept by the window in a chemical laboratory
- **d** holiday decorations are put on the laboratory instruments and discarded into contaminated trash when the holidays are over
- **e** all of the above
- **f** none of the above

16 Signs that communicate safety instructions or device locations should be

- **a** red, black and white
- **b** yellow and black
- **c** green and white
- **d** orange
- **e** yellow and purple

17 A safety shower is inspected for a biohazard laboratory. Which finding below does not meet ANSI standards?

- **a** water pressure 30 psi
- **b** water temperature 75° F
- **c** shower 40 feet from workers through a door that opens toward the shower
- **d** shower located next to electrical instruments used to process specimens
- **e** shower flow is 20 gallons per minute

18 OSHA requires that all safety showers have drains.

- **a** true
- **b** false

 ISBN 978-089189-6463

19 You have a chemical cartridge respirator. What kind of cartridges should you buy?

- **a** all purpose cartridges that are effective against almost all chemicals
- **b** cartridges that guard against flammables and corrosives
- **c** cartridges that filter both particles and chemicals
- **d** cannot answer without doing assessment of hazards in the laboratory

20 Even if you are wearing the proper eye protection you should not look directly at a laser.

- **a** true
- **b** false

21 Which of the following is required for respiratory protection against a BSL 3 respiratory pathogen?

- **a** if respirators are to be worn, wearers must be medically fit enough
- **b** N-95 particulate respirators are the minimum to be used
- **c** people with sideburns or beards who cannot get a good respirator fit are ineligible to work in an environment that requires respirators
- **d** all of the above

 ISBN 978-089189-6463

Answers

1 c

2 d

3 contact lenses

4 safety shower

5 15

6 b

7 e

8 a

9 e

10 a

11 e

12 55 feet; 10 seconds; 3 minutes; week

13 c

14 chemical fume hood: c
pipetting liquids to minimize splashing: b
scheduling pregnant woman in nonradioisotope: a

15 a

16 c

17 d

18 b

19 d

20 a

21 d

Appendix 10.1: sample notice of unattended laboratory operation

(modified from Wayne Shelton, Salisbury University, with permission)

CAUTION

This procedure, ____________________, is being performed in this laboratory beginning ____________________ and will end on (or about) ____________________. The procedure is continuous and may not always be attended.

SDSs available for reaction ____________________
LOCATION OF SDS

Associated hazards and precautions: ____________________,

In the event of a failure of the exhaust system, potential emissions from the procedure primarily consist of: ____________________,

SPECIAL INSTRUCTIONS:

If exhaust system fails: ____________________

If water supply fails: ____________________

If electricity fails for extended period: ____________________

If there are concerns, contact the Principal Investigator: ____________________

Office room number: ____________________ Office Extension: ____________________

Emergency phone numbers: ____________________ ____________________

Phone number for maintenance and repair: ____________________

Evening/weekend contact number: ____________________

Safety Officer contact number: ____________________

Unit 11

Locating safety equipment, signs & documents

No piece of safety equipment or document can be used unless the laboratory worker can identify signage and learn its location.

The purpose of this Unit is to ensure that the reader

1. knows what each piece of safety equipment and signage looks like
2. knows the location of all of the safety equipment and documents in a laboratory in which she or he works
3. can correctly operate, when possible, the safety equipment available in a relevant laboratory
4. can identify when new safety equipment needs to be incorporated

Materials needed

- ☐ access to a laboratory (preferably a laboratory in which the reader will be working/learning) and its safety equipment
- ☐ access to standard operating procedures and documents related to laboratory safety
- ☐ instructions or procedures, as appropriate, for each piece of safety equipment
- ☐ map of the laboratory indicating exits and locations of all hazards and safety equipment with which the reader should become familiar
- ☐ safety data sheets (SDSs) and chemical inventory

11: Locating safety equipment, signs & documents

Instructions

1. Attempt to locate the various pieces of equipment, signage, and/or documents in the list that follows. They may be found in more than 1 laboratory area, and everything listed may not be found in all laboratories. Note the equipment and/or documents that are available but not listed below, as well as what may be absent but necessary.

☐ accident report forms

☐ accident and emergency procedures

☐ biological safety cabinets, including inspection tags

☐ bloodborne pathogens exposure control plan

☐ broken glass containers

☐ brooms/dustpans/devices to clean broken glass and sharps

☐ buckets and carts for carrying chemicals

☐ chemical aprons and chemical resistant gloves

☐ chemical fume hood, including inspection tags

☐ chemical hygiene plan

☐ chemical and compressed gas inventory

☐ chemical labels/signs and transition to GHS labels

☐ chemical and compressed gas storage facilities

☐ chemical cabinets for defined hazards (eg, acid, flammable, vented)

☐ compressed gas storage collars, chains, hand trucks, valve covers

☐ compressed gas regulators and labels

☐ cryogen storage and containers for use

☐ decontaminating and disinfecting solutions

☐ ear protection

☐ electrical breaker box

☐ electrical emergency protocols: lockout/tagout procedures, generators, uninterruptable power supplies

☐ ergonomically designed equipment and procedures for reducing skeletomuscular injury

☐ eyewash, including inspection tags

☐ fire alarm code

☐ fire alarms

☐ fire blankets, including inspection tags

☐ fire evacuation route, assembly points, procedures, and posted instructions

 ISBN 978-089189-6463

11: Locating safety equipment, signs & documents

- ☐ fire extinguishers and hoses, including inspection tags
- ☐ first aid kit
- ☐ flammable safety can
- ☐ fume hoods, including inspection tags
- ☐ gas shutoff valves
- ☐ Geiger counter and radiation detection devices
- ☐ glasses for special hazards: UV light, lasers
- ☐ gloves: latex, nitrile, thermal, chemical resistant, puncture resistant
- ☐ goggles/face shields/work shields
- ☐ hand sanitizer locations
- ☐ hazardous waste protocols, disposal containers, segregation and storage areas
- ☐ locked doors or restricted access points
- ☐ pipettor bulbs and suction devices
- ☐ radioisotope shields and storage
- ☐ respiratory protection/masks
- ☐ safety data sheets: access by paper, electronic and/or commercial service
- ☐ safety engineered sharps
- ☐ safety shower, including inspection tags
- ☐ safety apparel: lab coats, sleeve protectors, head covers, shoe covers
- ☐ sand bucket
- ☐ signs: eg, biohazard, radiation, chemical, laser, electrical, exits, emergency equipment
- ☐ sharps disposal and puncture resistant containers
- ☐ shipping instructions for hazardous materials
- ☐ sinks: contaminated and noncontaminated for handwashing
- ☐ spill containment materials (eg, kitty litter)
- ☐ special spill kits (eg, those for mercury and formaldehyde)
- ☐ standard operating procedures for safety and waste disposal
- ☐ steam sterilizer/autoclave, quality control organisms, and heat indicators
- ☐ telephone and emergency phone numbers
- ☐ other materials available

2. Identify the location of each item above on the map of the laboratory.

3. When appropriate, read the instructions for the use of each piece of equipment to familiarize yourself with its operation and attempt to operate it correctly, if this is reasonable. For example:

- ☐ apparel: put on any goggles, masks, or laboratory aprons that you can and verify that you are wearing them correctly
- ☐ eyewash: activate the eyewash and verify the water is tepid
- ☐ fire escape map: exit the laboratory using the directions on the map so that you will evacuate correctly in case of a fire and identify the assembly point
- ☐ fire hose: locate the mechanism to activate the fire hose
- ☐ fire extinguisher: find the operating mechanism; note for which class(es) of fires it is used
- ☐ fume hoods/biological safety cabinets: be able to turn the air flow on and off, open and shut the sash; verify that air flow is adequate
- ☐ gas and electricity: make sure you understand how to shut off the gas and electricity in an emergency and identify emergency sources of power
- ☐ sharps disposal: practice discarding sharps and be sure you can correctly operate safety engineered devices
- ☐ safety shower: locate the ring that activates the water flow

Do not operate any piece of equipment with which you are unfamiliar or without permission of the laboratory supervisor in the area. Notify laboratory personnel if you find that any piece of equipment is not operating correctly. If the equipment is broken, it must be repaired; however, if you are operating it incorrectly, it may not be broken, and it is important that you learn the correct technique.

11

4. Locate inspection tags to see if the equipment has been recently serviced. Report any out of date devices and/or devices that do not have inspection tags.
5. As a participant in developing a safety culture in your laboratory (Unit 14), list any pieces of equipment that you did *not* find but you feel should be included. Beside each item, list the procedure(s) that justifies its inclusion. If possible, discuss your ideas with the supervisor of the area and together develop a strategy to eliminate the hazard or change protocols.

Accidents, emergencies & disasters

Laboratories should expend all possible effort toward accident prevention and emergency planning, but training in proper emergency response should not be neglected as it will prove necessary in the event that prevention efforts fail or unforeseen disasters strike. If an unfortunate incident occurs during an accident or emergencies, laboratories should be sure to learn from it and take action to prevent similar occurrences.

Accident reports

Any accidents involving personal injuries, even minor ones, must be immediately reported to a supervisor. When an accident occurs on the job, the supervisor must complete a written accident report, and the employer must take responsibility for the cost of medical care for the worker involved. Some incidents may seem minor at the time, yet can still result in serious repercussions. For example, a minor cut from a piece of glass could cause serious consequences if the glass was later found to be contaminated with an infectious agent. Failure to report the initial incident could then delay appropriate treatment, which could then result in the employee having difficulty receiving worker's compensation. Therefore, an accurate and detailed report for each and every accident is needed for proper medical care, insurance, worker's compensation, and other legal purposes, such as verification of regulatory compliance.

In the US, accident reports for sharps injury must comply with requirements of the Needlestick Safety and Prevention Act. Since these reports are tracked nationally, each report must include identification numbers instead of the victim's name to protect his or her privacy, as well highly detailed information about how the injury occurred, including the brand name and model of safety engineered sharp devices used. Regulators are increasingly interested in details for the devices involved in all types of accidents so that those attributed to device defects can be differentiated from those attributed to operator error. More detail regarding accident reports from sharps is beyond the scope of this text, but additional information can be found here: www.cdc.gov/niosh/stopsticks/sharpsinjuries.html (accessed January 22, 2016).

Accident reports are also needed to prevent accidents from recurring. Documentation of "near misses" is not legally required, but analysis of these incidents can be valuable for prevention. Analysis to determine the "root cause" (not the superficial or immediate cause) of an accident is an excellent way to identify procedure errors, areas in need of improvement, or the need for new policies. Safety protocols from regulatory and professional bodies cannot cover every situation, and examination of individual incidents can help prevent accidents due to unique situations. Accident reports are also useful to document whether correct procedures are being followed and to determine whether staff and/or equipment responded correctly to the emergency. For example, an accident may be caused by incorrectly performed first aid, a broken emergency exhaust fan, or a rescuer forgetting how to use the respirator. In addition to analyzing accident reports, examination of training records may be required to identify deficits in training. When the analysis process is complete, an action plan must be created with a timeline and accountability for completion. This action plan may include employee retraining in standard safety protocols, improving emergency procedures, and/or repairing and replacement of equipment. Several months after the incident, the safety committee should also evaluate the effects of these action plans to ensure that nothing has been overlooked and no new incidents have occurred.

Careful attention to detail on accident reports is critical. Follow-up medical care and accident analysis are more effective with complete, accurate information. Details can be easily forgotten over time, so prompt completion of documentation is important. The Occupational Safety and Health Administration (OSHA) maintains sample accident report forms but permits employers to create their own as long as it is easy to understand and the information collected is complete. In this accident report, processes that went well should be noted, and the people who responded with proficiency should be recognized. **f12.1** is a sample form showing the minimum information required.

All deaths and occupational injuries not considered "minor" must be reported to OSHA. OSHA's Occupational Injury and Reporting regulation 29 CFR parts 1904 and 1952 states that minor injuries or illnesses "do not involve death, loss of consciousness, days away from work, restriction of work or motion, transfer to another job, medical treatment other than first aid, or diagnosis of a significant injury or illness by a physician or other licensed health care professional."

For all employers with >10 workers in jobs with inherent risk, OSHA provides specific forms on which to record and track illnesses and injuries. Most laboratories have significant risks associated with them and thus are not exempt from such recordkeeping. However, small laboratories such as those performing near patient or "point of care" testing in clinics and doctors' offices may be exempt or partially exempt at the federal level. Additional detail can be found here: www.osha.gov/recordkeeping2014/OSHA3746.pdf (accessed November 9, 2015). Laboratories located in areas where state regulations prevail still need to consult their own occupational safety bodies regarding exemption status.

Employers must also complete OSHA Log Form 300, a log of all events that have occurred at an institution over one year. Form 300 does not require employee names, only the incidents themselves.

OSHA requires that injury records be retained for 5 years. No matter how many employees are at a site, employers must report any work related fatalities within 8 hours and must report within 24 hours in patient hospitalization, amputation, or eye loss, which is a change from the previous requirement (24 hours to report accidents that resulted in hospitalization of 3 or more people). Additional details and other reporting requirements are located at www.osha.gov/recordkeeping2014. An online form can be found at www.osha.gov/pls/ser/serform.html (accessed January 22, 2016).

 ISBN 978-089189-6463

Incident report number:			
Facility name:		Facility telephone:	
Facility address:			
Name and title of person filling out report:		Date and time of report:	
Date and time of incident:		Exact location of incident:	
Person involved in incident - full name:		Employee identification number:	
Home address and telephone:			
Job title:		Age:	Gender:
List other people injured in this incident with separate accident reports:			
Description of incident (provide enough detail so that the cause and manner of the incident is clear):			
Describe exact injury(s) present, if any:			
Describe exact treatment(s) given if injury present and by whom:			
Describe any other action taken, including emergency response measures:			
How could this incident have been prevented?			
Signature of Safety Officer:			
Print name	Title:	Date:	Time:
This completed form should be given to the Safety Officer for OSHA reporting and permanent storage			
Incident analysis & correction plan			
Causes of incident, including all contributory factors and probable root cause:			
Deficits noted, eg, prior training, compliance with protocols, incorrect performance of protocols, insufficient protocols, equipment failure, communication/sign failure, previously unidentified hazard:			
Person(s) to be commended and elements of emergency response that were well done:			
Correction plan, including all details to correct each deficit noted above. Beside each element of the correction plan, note person responsible and due date:			
Person responsible for verifying completion of correction plan, commending those with good performance and due dates:			
Analysis of ongoing problems from incident and correction plan effects 6 months after completion:			
Print name:	Title:	Date:	Time:
This completed form should be given to the Safety Officer for OSHA reporting and permanent storage			

f12.1 incident report form

12

Emergency & disaster planning

Laboratories need emergency and disaster plans. Chemical spill emergencies were discussed in Unit 3 related to "HAZWOPER" requirements, and accident response has been discussed throughout other units. Safety managers should develop plans with a broad perspective that includes weather emergencies, prolonged loss of electrical power, bioterrorist attacks, active shooters, bomb threats, and the like. Staff may need to react to multiple problems simultaneously (eg, fire, injury, chemical release), so planning should be as comprehensive as possible. Basic training for lockdowns is useful in almost every site and aids staff in choosing between attempting evacuation and "sheltering in place" during emergency events.

Training scenarios for wide ranging catastrophes can be difficult to create, so many facilities use "tabletop" drills in which crucial supervisors and staff members gather to talk about how they would respond to a particular scenario. Proposed responses can be discussed, analyzed, and used to identify weaknesses in the current plan and/or identify opportunities for additional training. These training scenarios, although artificial, can be very beneficial; when staff have had the opportunity, in the absence of the fear and stress that accompany an actual event, to think about proper response, they are more likely to make better choices during the confusion that accompanies real-life disasters.

Detailed emergency and disaster planning is more than can be presented here, especially since protocols are driven by the specific hazards of each site. The reader is referred to the Clinical and Laboratory Standards Institute (CLSI) document "Planning for Laboratory Operations During a Disaster; Approved Guideline" (GP36-A) to use as a basic framework. Another useful resource is the website maintained by the Department of Homeland Security, www.ready.gov. Appendix 12.1 contains checklists for bomb threats and suspicious packages, examples of the kind of training that may be required beyond basic laboratory safety. Although by no means comprehensive, **t12.1** lists many of the key elements needed in an emergency plan.

t12.1 Planning for emergencies, example considerations

Command and communication
People assigned to be in charge and reporting lines
Responsibilities of designated leaders
Communication methods, including those that can be used if power is off or cell towers not operational (such as walkie talkies)
Process to routinely verify that phone numbers, emails and methods of communication are current and functional
ICE or "in case of emergency" phone number designations
Alarms, sirens and building wide instructions
Electrical power
Prioritizing outlets powered by generators
Uninterruptible power supplies and battery power
Instrument protection from power surges or irregularities
Temperature control and ventilation
Emergency lighting and flashlights, particularly along evacuation routes
Shelter and escape
Main and alternate evacuation routes; methods for mobility impaired, escape masks
Access roads, provisions for driving out or in, full gas tanks in cars
When/how to shelter in place or move to designated shelters
Protocols for lockdown
Survival supplies, eg, food, water, medications, extra clothes, toiletries, wet wipes, air mattress, pillow, blanket
First aid supplies
Laboratory safety, function, and protection
Control of hazards, eg, turning off gas, stowing chemicals, putting biohazards away
Safe shutdown of operations
Preservation of fragile supplies and specimens
Ability and means to perform urgent testing
Transport of urgent specimens to other sites
Meeting medical testing demands during pandemics, mass casualties, emergencies
Alternate water sources for instrumentation, handwashing, reagent preparation
Hand hygiene without water
Backup and protection of vital information, both paper and electronic

12

 ISBN 978-089189-6463

First aid & emergency supplies

Emergency supplies with limited shelf lives must be routinely replaced, so laboratories must assign this responsibility to a staff member and document dates of replacement. Many emergency items that do not have expiration dates, eg, kitty litter, bandages, and fire blankets, also must be inspected annually to make sure they have not been removed and are still usable. Emergency supplies of all sorts have been discussed in other units, so the reader is referred to previous discussions for guidance in choosing appropriate emergency equipment for a given laboratory. Recommended first aid supplies vary depending on the hazards present and laboratory location. Laboratories in hospitals likely already have minimal supplies, while laboratories in more remote locations may have more varied supplies in greater quantities. The following items should be included in a laboratory first aid kit to meet both CLSI and American National Standards Institute (ANSI) Z308.1-2009 guidelines:

- absorbent compress (33 in^2)
- adhesive bandages (1 × 3 in, at least 16)
- adhesive tape (3/8 in × 5 yd)
- antibiotic topical (at least 6 individual applications of 0.14 oz or more)
- antiseptic swabs (at least 10 individual applications of 0.14 oz or more)
- antiseptic wipes (1 × 1 in, at least 10)
- antiseptic towelettes (24 in^2, at least 10)
- bandage compresses (3 sizes: 2 in, 3 in, and 4 in)
- burn dressing (4 × 4 in)
- burn treatment (at least 6 individual applications of 0.03 oz or more)
- cardiopulmonary resuscitation (CPR) barriers (1 or 2)
- cold pack (4 × 5 in)
- eye covering (2.9 in^2, at least 2 with attachments)
- eye/skin wash (at least 4 fl oz)
- gloves of medical examination quality
- roller bandages (2 in and 4 in sizes, at least 2)
- sterile pads (3 × 3 in each, at least 4)
- triangular bandage (40 × 40 × 56 in)

In addition, blankets to treat shock, splints, and an automatic external defibrillator (AED) are highly recommended. AEDs require routine inspection, and their use is described below. Staff education should include training in the proper use of all available materials and devices.

General first aid principles

Employers are responsible for making provisions for emergency healthcare no matter what the work shift. During the time between the injury occurring and treatment by healthcare professionals, laboratory staff may need to give first aid. First aid is temporary assistance given to a victim of a sudden illness or accident, and must be considered a stopgap measure, taken before professional assistance can be obtained. It is impossible in this Unit to cover anything except the most rudimentary elements of first aid. In addition, recommendations from professional bodies vary slightly, and the protocols of community and professional healthcare providers can differ. The information below is an amalgam of recommendations for the public, current at publication time, from the American Red Cross (ARC) and the American Heart Association (AHA). Every laboratory worker is strongly urged to take a basic first aid course and a CPR course, and to maintain certification and knowledge of the most recent guidelines.

So called "Good Samaritan" laws in all 50 states protect the general public from legal prosecution over outcomes from giving first aid, provided that the rescuer took "reasonable and prudent" actions without financial compensation. In light of these laws, fear of being sued should not stop a potential rescuer from attempting to help in an emergency situation; however, one should only render aid at one's skill level and try to "first, do no harm."

In the face of an emergency, the rescuer must stay calm, if only to help soothe a panicking accident victim. Also, time must be taken to make an accurate, albeit rapid, analysis of the situation so that the correct action is taken to save a life or prevent further injury. The basic steps of this analysis are also known as the "4 Cs": check, call, consent, care.

- **Check.** A rapid assessment of the scene and the victim(s) is necessary to determine whether the scene is unsafe or if the victim(s) can remain in place. The nature of the incident should be determined, and other people who can help should be quickly identified. Rescuers should also determine the condition of the victim and whether he or she is wearing a medical alert bracelet or necklace. Rescuers should shout loudly for help to bring as many people to the scene as possible who might be able to help. If the scene is unsafe, it may be necessary to forgo care and move the victim(s); however, a victim should only be moved if
  - the immediate area presents a life threatening hazard
  - the rescuer must reach another victim in worse condition
  - the rescuer must reposition the victim so that proper care can be given

If it is necessary to move an unconscious victim, he or she must be pulled by the long axis of the body. The rescuer should grasp the victim beneath both shoulders and support the victim's head with his or her arms. Avoid pulling sharply as one sharp pull can cause additional injuries. If possible, place the victim on a blanket or large coat and drag him or her, as this will provide extra head support and make movement easier. To escort conscious and ambulatory victims, place one of the victim's arms around your neck and wrap your arm around the victim's waist for support neck. If a wheeled chair is available, seat the victim in it and wheel him or her out.

- **Call.** A call should be placed to "911" or other appropriate emergency personnel as quickly as possible; if no phone is nearby, someone must be sent to get help. If the victim has an adult cardiac emergency, the call should be made before rendering any care. In breathing emergencies, it may be necessary to delay the call to provide a minute or 2 of care, as brain damage begins 4 minutes after oxygen deprivation. Whether or not one calls or cares for the victim first must be determined on a case by case basis ("call first" vs "call fast"). The exact location and details of the incident are important to convey calmly and clearly to emergency dispatch operators. An open telephone line should be maintained. "SAMPLE" is a useful acronym to remember crucial information to obtain about the victim:
  - **S**igns/symptoms in victim
  - **A**llergies or alert jewelry
  - **M**edications
  - **P**revious problems
  - **L**ast food or drink
  - **E**vents that led to the current situation

ISBN 978-089189-6463

- **Consent.** Even people with no knowledge of first aid can assist victims by giving them comfort and reassurance to help them remain calm. Beyond that, if the victim is conscious, rescuers must always obtain permission to give him or her first aid (in the case of minor children, parents or guardians must give their permission unless the condition is life threatening). The ARC directs rescuers to do the following to obtain consent: 1) identify yourself and state your level of training, 2) ask the victim if you may help, and 3) explain what you see and what you want to do. Under the law, when victims are unresponsive, unconscious, or otherwise impaired, rescuers do not need the victim to communicate consent, and can assist him or her on the assumption of implied consent.
- **Care.** Once care is begun, it must continue until another person takes over, the scene becomes unsafe, or the rescuer is too exhausted to continue. Life threatening conditions must be treated first to stabilize the victim until help arrives. When help comes or upon arrival at a medical facility, the person(s) rendering first aid must give complete and detailed information to the medical professional team, which may include the type of chemical/organism/radiation involved, the signs observed, the type of equipment involved, the duration of any harmful contact, and the first aid measures already taken. Effective treatment will greatly depend on the information provided by witnesses of the accident, and it is important for those who provide first aid to ensure that any available information is conveyed to the medical professionals treating the victim.

First aid procedures

The classic first 3 steps of first aid can be remembered with the acronym "ABC":

1. Airway: position the victim to maximize the airway opening
2. Breathing: assess breathing and airway obstruction
3. Circulation: assess the need for CPR or treatment with an AED

Consulting the "ABCs" is the first priority in treating any accident victim, as breathing and heart emergencies are life threatening. If breathing and circulation are stable, the next 3 priorities are as follows:

4. Stop all bleeding
5. Treat for shock
6. Treat the injury

The following is a more detailed explanation of each of the 6 major steps. Ideally, first aid providers should first perform hand hygiene and don protective equipment, but this is not always possible. Since the victim's allergies may be unknown, latex should be avoided. Improvising hand coverings with plastic wrap or bags is usually better than nothing.

Open the airway

Unresponsiveness can be established by gently tapping the victim and shouting, "Are you okay?" When someone is unconscious, his or her throat muscles can relax, causing the tongue to fall back and block the airway. The airway opening can be maximized by

- performing the "head tilt/chin lift, during which the victim is placed flat on his or her back and the rescuer lifts the victim's chin upward with his or her fingers while pushing gently down on the forehead
- using the jaw thrust maneuver, unless the victim is prone or has a head injury

12

Assess breathing

At one time, rescuers were told to take <10 seconds to "look, listen, and feel" for normal breathing in a victim. However, the absence of breathing proved difficult to detect, so more recent recommendations eliminate this step because it can delay treatment. For example, though many heart attack victims gasp intermittently, this breathing is not normal. Yet, unless the situation is clearly a breathing emergency, efforts to restore breathing are inappropriate and rapid initiation of cardiac treatment is of greater benefit. If the victim is an adult, chances are high that the situation is a cardiac emergency, so rescuers should move on to CPR or AED treatment as described below.

On the other hand, breathing emergencies are more likely in children than adults, so it is important to consider them in younger victims. Coughing, choking, turning blue, or failure of the chest to rise and fall with rescue breaths are signs of an obstructed airway. Conscious victims should be asked, "Can you speak?" as speech requires that some amount of air get into the lungs. Victims with a strong, forceful cough can probably clear obstructions on their own, and immediate intervention is not appropriate. If a conscious victim cannot speak or cough, the rescuer can clear obstructions by

- removing any foreign objects from the mouth that can be grasped and pulled easily; this step is recommended with great caution as objects as could become even more firmly embedded in the throat; "digging" for objects should be avoided
- bending the victim forward at the waist over one of the rescuer's arms, with the victim's mouth below chest level, and giving 5 back blows between the shoulder blades with the heel of one hand, using the strength from the elbow down rather than the whole arm
- performing abdominal thrusts (aka the "Heimlich maneuver"), in which the rescuer stands behind the victim and places a fist, thumb side facing inward, just above the victim's navel and below the ribs, while supporting the victim with a leg between the victim's legs; he or she should grab the fist with the opposite hand and give 5 sharp upward jerks, which should force air out of the lungs and dislodge any objects blocking the airway—if the victim is obese or pregnant, the rescuer should use chest compressions
- repeating back blows and abdominal thrusts until the object is forced out, or the victim can breathe/cough or becomes unconscious; brain cell death begins ~4 minutes after the brain stops receiving oxygen; if an airway obstruction cannot be removed within 2 minutes, the rescuer should begin administering CPR, during which the chest compressions may help dislodge the obstruction and deliver oxygenated blood to the brain

A victim who is wheezing and/or having trouble breathing may have asthma or severe allergies, and could simply need assistance with an inhaler or epinephrine pen. Most victims should be asked about known conditions, if possible.

Emergencies that require restoration of breathing include drowning and airway blockage. Ideally, rescue breathing is given with a mask or barrier, but this is not always possible. Direct contact with the victim's mouth must also be avoided if there was a chemical or biological splash in the area. Rescuers who do not want to perform rescue breathing without a barrier should perform hands only CPR as described below. The procedure for rescue breathing is:

1. Clear the airway as described above.

 ISBN 978-089189-6463

2. Once the airway is clear, give 2 slow breaths: pinch the victim's nostrils tightly shut, place your mouth over the victim's mouth, and make an airtight seal, either directly on the skin or through a barrier; if a mask is used, press it tightly against the victim's face to make a good seal. If the mouth is injured, close it by lifting the chin, and perform rescue breathing via the victim's nose. Each breath should last no >1-2 seconds, which helps to ensure that the lungs are inflated but not the stomach, as excess air in the stomach may cause vomiting. Excessive ventilation can also create unfavorable pressure and diminish cardiac output.
3. As you give each breath, check to see if the chest rises and falls. If the chest does not rise and fall, the airway is likely blocked; attempt to clear the airway again.
4. If breathing is not restored immediately, CPR with rescue breathing should be initiated as discussed below.

Assess & restore circulation

Current thinking about cardiac emergencies is that chest compressions are more important than rescue breathing, and that for the public, calling for help should take priority over initiating compressions ("call first"). The AHA advocates "hands only" CPR (www.heart.org/HandsOnlyCPR, accessed January 22, 2016) as a simple, effective procedure that can be performed by anyone. In the first few minutes after cardiac arrest, "hands only" CPR can be as effective as conventional CPR since the blood should still contain some oxygen. The AHA has even suggested that the traditional "ABC" acronym be changed to "CAB" for "compressions, airway, and breathing" because the evidence of positive outcomes for early chest compressions is so powerful. To perform "hands only" CPR

1. Call 911 or send someone for help
2. Place the victim flat on a hard surface
3. Remove clothing from the victim's chest area
4. Initiate chest compressions in a "push hard, push fast" mode. The AHA recommends a speed that matches the beat of the song "Stayin' Alive" by the Bee Gees. Compress the chest directly between the nipples using interlocked hands and straight arms at a 90° angle to the chest, which should allow you to use your body weight for the compressions, making CPR less tiring
5. For adults, compress the chest at least 2 inches at a rate of 100 per minute. Allow the chest to come back to normal height after each compression so that the heart can refill. (For children, compress the chest ~2 inches; for infants, compress ~1.5 inches using 2 fingers rather than the whole hands)
6. Continue compressions until the victim is revived, help arrives, an AED arrives, or you are too exhausted to continue

While rescue breathing is a secondary priority compared to chest compressions, it increases their benefit. Every 30 compressions should be followed by 2 breaths, delivered as described above. If an AED is available, it should be used as described below, but CPR should not be delayed while searching for and activating an AED.

The CPR techniques outlined here should be studied and practiced under qualified instructors. The increased availability of cell phones, AEDs, and advanced cardiac life support have changed the priorities during emergency situations, so CPR training for the public has been simplified to encourage action during an emergency; for example, community providers are no longer taught to check for a pulse before initiating chest compressions. Protocols are different for community providers and healthcare providers. The more advanced protocols for healthcare providers are not described here, but the reader is encouraged to learn more advanced CPR whenever possible.

Each emergency situation will be different, and the reader is left to simply make the best judgment for each occasion. The important message is that time matters. No one should take too long checking for breathing or a pulse. Calls for help and initiation of CPR must occur as soon as possible.

Stop any bleeding

A rescuer's first action should be to apply direct pressure on the wound with the cleanest cloth available. While additional dressings can be applied over the first dressings, the first dressings should not yet be removed as that could cause the bleeding to start again. If no other injury is present (eg, a broken bone), elevating the wound above the heart should slow the bleeding because the heart has to pump against gravity. Once a bandage has been applied, it should not be so tight that it cuts off circulation. The injured area should have feeling, warmth, and color if circulation has been maintained. Victims with nosebleeds should pinch their nostrils together and lean forward.

Only the most severe life threatening emergencies, coupled with an extreme delay in the arrival of medical help, would ever justify the use of a tourniquet. Because the victim could most likely lose the limb to which the tourniquet has been applied, one should be used only in cases of life threatening bleeding, when the loss of the limb can preclude the loss of life. As a last resort only, a tourniquet can be applied between the limb and the heart, as close as possible to the wound without being on or over it, and just tight enough to stop the bleeding. Once it has been applied, it should not be removed except by medical professionals because blood clots from the tourniquet site could enter the circulatory system. The exact time at which the tourniquet was applied should be written directly on the dressing, if possible. Otherwise it should be noted and communicated to the medical professional team.

No first aid measures for internal bleeding exist other than keeping the victim comfortable and warm. Immediate medical help is essential for internal bleeding. Signs and symptoms of internal bleeding include the following:

- areas of the body that are swollen, hard, and sore
- a faint rapid pulse
- skin that is pale, blue, clammy, or cold
- vomiting or coughing blood
- extreme thirst
- confusion, sleepiness, or loss of consciousness

If a body part has been severed, wrap it in a clean rag or sterile gauze, place it in a bag, and place the bag in cold water or ice. Do not place it directly on ice or in water, as this will damage the tissue. Handle lost teeth by their chewing edge, gently remove debris, and place them in saliva or in milk; give the victim a clean cloth or gauze to hold or bite down on over the empty tooth socket.

Treat for shock

Victims of virtually any serious injury may develop shock, but it is particularly associated with significant blood and/or fluid loss. In a serious injury, the body

 ISBN 978-089189-6463

attempts to send blood to vital organs first (brain, heart, lungs) and diverts it from the limbs. When the limbs are oxygen deprived and need blood, it is then diverted from vital organs. If the body tissues cannot compensate for this compromised circulatory pattern or if medical intervention does not arrive, death can occur. Signs and symptoms of shock may include any or all of the following:

- shallow, rapid breathing
- a rapid, weak pulse
- nausea and/or vomiting
- pale, cold, clammy skin
- thirst
- dilated pupils
- restlessness, irritability, or confusion

Shock is life threatening, and victims of serious injury should be treated for shock even if signs and symptoms have not yet developed. Treatment is as follows:

1. The victim should lie flat. If there are no injuries that preclude it, the feet can be elevated 12 inches or less to maintain blood supply to the brain; >12 inches may adversely affect breathing. If injuries of the legs, hips, head, spine, or neck are suspected, legs should not be elevated.
2. Do not give the victim any food or drink; if he or she is vomiting, place the victim on his or her side.
3. Keep the victim covered and as warm and calm as possible until help arrives, as pain and anxiety can hasten shock.

In cases of electrical shock, separate the victim from the electrical source with an insulator, not via direct contact with your hands or another conductor. Electrical shock should be suspected if burns are visible, and treatment should be identical to that for other causes of shock.

Treat the injury

Thermal & chemical burns

First degree thermal burns damage only the outer layer of skin and resemble sunburn. Second degree burns are slightly more severe because underlying tissue is damaged and blisters may be present. Jewelry and clothing must be removed from the burn area immediately as later swelling may make their removal difficult. First and second degree burns can be treated by immersing them in cool water or covering them with cool, moist dressings. Burns should not be treated with ice or very cold water as they can cause hypothermia and further damage tissue. While treatment in safety shower can be acceptable, if the water is very cold, the victim should not be allowed to spend longer than 15 minutes in the shower. Treatment for shock may be also necessary when severe second degree burns are sustained. Blisters should not be disturbed, and ointments, butter, and other treatments not designed for burns should never be applied because they can retain heat.

In third degree thermal burns, the underlying tissue is destroyed and the skin appears dark and charred. Loose clothing should be cut away, but any cloth embedded in the burn should be left alone as its removal may cause further damage. Third degree burns should be covered with cool, moist dressings, and victims should always be treated for shock. Ointments or ice should never be used, and medical help must be summoned as soon as possible.

Chemical burns should be flushed with copious amounts of water, and the affected clothing should be removed as soon as possible. If affected clothing cannot be removed without contaminating other body sites, it must be cut off. Strong neutralizing agents should not be used because the neutralization reaction produces heat. Acids and caustic chemicals may be neutralized with weak solutions, such as 2% bicarbonate for acids and 2% acetic acid for bases. Such solutions should be used only if they are already prepared and immediately available in a burn station or kit; rescuers must not take the time to prepare these solutions at the time of an incident, as it is far more effective and important to use water on the burn as soon as possible. Medical attention is always necessary, and severe cases may need treatment for shock. Medical personnel must always be informed of the exact chemical that caused the burn, because treatment can vary. For example, stating that "acid" caused the burn is not sufficient because hydrofluoric acid penetrates tissues deeply and can cause more damage even after flushing with water than other types of acids. If possible, medical personnel should be provided with the safety data sheet for the chemical.

Dry ice is extremely cold and can cause "burns" or frostbite. Damaged tissue should be treated with tepid, not hot, water.

Bone, muscle & joint injuries

Signs of bone, muscle, and joint injuries include deformity or swelling of the area, inability to move or use the body part normally, and coldness and/or numbness of the body part. When possible, a good way to check for swelling and deformity is to compare the injured body part with the uninjured part.

The acronym "RICE" is useful in remembering the basic steps in treating thesc injuries:

- **Rest.** The victim should not attempt to move or use the injured body part.
- **Immobilization.** The body part should be bandaged or splinted to prevent movement. Splints should extend above and below the injury, and the splint should be applied to the injury as it was found; no attempt should be made to restore a deformity to its original position. One option is to use anatomic splinting, a technique in which injured body parts are immobilized against uninjured parts of the body, eg, tying an injured leg to an uninjured leg, binding an arm to the chest, or bandaging 2 fingers together. If nothing else, the limb can be rested against the ground. If splinting is not possible, keeping the victim immobile and still is also effective.
- **Cold.** Ice or cold packs can be placed against the injured area to reduce pain and swelling.
- **Elevation.** If elevating the injured body part does not cause more pain, this can be beneficial. Propping up an injured leg is a good example.

After the bleeding is stopped, bone or joint injuries should be immobilized and the victim made as comfortable as possible until help arrives. Impaled objects should be left alone, and the bandages applied around them. Once bandaged, injured areas should be checked for feeling, warmth, and color to be sure that bandages have not been applied too tightly. Only professionals should attempt to set or treat broken bones, and victims should not be moved by nonemergency personnel unless there is danger in the immediate area. If the injury and/or bleeding is severe, treatment for shock may be necessary.

Injuries of the head, back, and neck are special cases, as any damage to the spinal cord can cause permanent paralysis. Therefore, it is particularly important to immobilize anyone with a suspected head, neck, or back injury, especially if a deformity is observed. Important

 ISBN 978-089189-6463

neurologic signs include seizures, nausea, visual disturbances, paralysis, numbness or tingling, loss of balance, confusion, or altered consciousness. Severe pain and clear and bloody fluids coming from the nose and ears are also important signs and constitute an emergency. The head should be immobilized in line with the body, but if the victim cannot keep his head straight, it should never be forced.

Eye injuries

Chemicals and biological splashes should be flushed out of the eyes with water for at least 15 minutes. The victim should be encouraged to roll his or her eyes around to expose all areas to the flow. Every attempt should be made to remove contact lenses, if any, while flushing the eyes, unless this delays treatment, eg, if the victim is not cooperative or if the contact lens is "stuck" to the eye by tissue damage. Contact lenses should never be forcibly removed from an eye injury.

A foreign object should be left in the eye unless it can be easily removed with a wet (never dry) piece of sterile gauze or cotton swab. Only superficial objects should be removed as a first aid measure, since rubbing injured eyes can cause more damage or embed the object more deeply. Objects that are difficult to remove or that impale the eye should be covered and left alone until medical help arrives. Because the eyes move simultaneously, both eyes should be covered to prevent movement of the injured eye, and the victim should be kept from rubbing the eye or attempting self treatment. Eye injuries are very uncomfortable and frightening, and every effort must be made to keep the victim calm. The victim should never be left alone, and medical help must be obtained as soon as possible.

Cuts/punctures

Serious bleeding must be stopped as described above. If chemicals or biohazards are involved and the wound is not bleeding profusely, it must be thoroughly flushed with water to neutralize the hazards. If the wound is minor and the tetanus vaccine is current (within 10 years for adults), the wound can be cleaned, dressed with antibiotic ointment, and bandaged with sterile gauze. Victims with wounds larger than 1 inch or wounds in which the edges of the skin will not touch should receive medical attention because they may require stitches. Wounds that contain objects or will not stop bleeding must also be treated professionally; medical attention may be necessary to remove objects from the wound and administer a tetanus shot. Wound dressings should be placed around rather than over embedded objects to stabilize them without pushing them in farther.

Heart attacks, strokes & sudden illness

Heart attacks, strokes, and sudden illness (eg, diabetic emergency or seizure) can happen anywhere and anytime, including in a laboratory. It is important to distinguish such conditions from those caused by laboratory accidents so that appropriate treatment can be given. If staff members have conditions such as diabetes or seizure disorders, training in first aid for those conditions may be advisable.

The acronym "FAST" can be used to recognize a stroke as follows:

- **Face:** weakness in the face, particularly if it is confined to 1 side, is a sign of stroke; ask the victim to smile to determine whether it's symmetrical; if not, suspect stroke

- **Arm:** weakness in the arms, particularly confined to 1 side, is a sign of stroke; ask the victim to raise both arms over his or her head; if he or she cannot or cannot do so symmetrically, stroke should be suspected
- **Speech:** slurred or unintelligible speech is a sign of stroke; ask the victim's name or other similarly simple questions to judge his or her ability to respond
- **Time:** the time should be noted if any of the above occurs, and emergency responders should be notified

Signs and symptoms of heart attacks are chest pain; arm, neck, or jaw pain; shortness of breath; nausea; and sweating. In recent years, emergency treatment for heart attacks in the form of AEDs has become more common; in many cases, AED treatment is superior. Many public places have AEDs placed in them, so knowing how to use one is important, even though AEDs have been designed to be intuitive even for untrained people. The diagrams for electrical pad placement are clear, and the instrument itself tells the user exactly what to do. The sequence is as follows:

1. Once a rescuer determines that a victim is possibly having heart trouble or a heart attack, he or she exposes the victim's skin on the upper torso, which must be bare and dry. Many AED kits contain a razor to shave off hair at the point of pad contact.

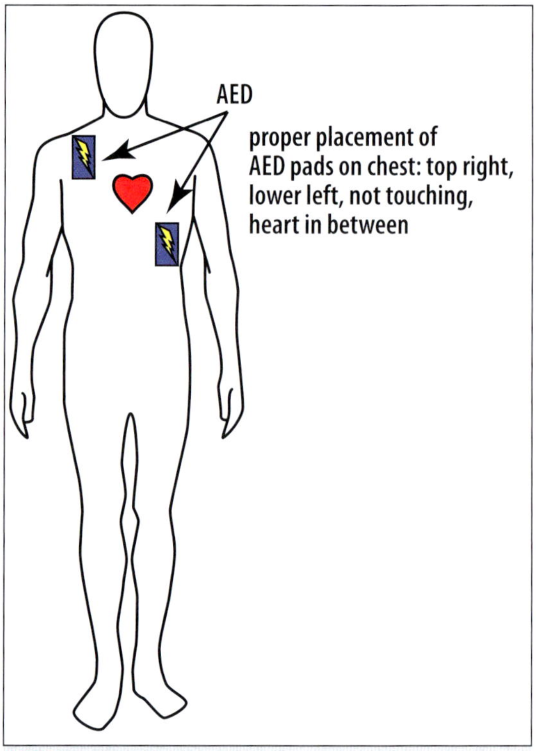
AED
proper placement of AED pads on chest: top right, lower left, not touching, heart in between

f12.2 automated external defibrillator schematic placement

2. The rescuer turns on the AED, attaches the disposable electrical pads to the device, then places one electrical pad on the victim's upper right torso, the other on the lower left torso **f12.2**. The pads must not be placed directly on jewelry or medication patches; skin patches such as nitroglycerin or nicotine should be removed. An AED can be used when the victim has a visible implanted device such as a pacemaker, but do not place the pads directly over the implanted devices and do not delay treatment to determine whether the victim has implanted devices that are not visible. The pads also cannot touch each other or anything metal. If the victim is small, eg, a child, a pad can be put on the chest and the other on the back, or pediatric pads can be used. Note that while victims <55 pounds or 8 years of age are better treated with pediatric pads, treatment should not be delayed in order to find them.

 ISBN 978-089189-6463

3. Once the pads are placed, the rescuer should stand back; no one should touch the victim.
4. The AED analyzes the victim's heart rhythm to determine if a shock is necessary.
5. If a shock is necessary, the AED will advise the rescuer to stand clear and then deliver a shock. Anyone touching the victim will also be shocked, so it is important to keep everyone clear.
6. The AED will again analyze the heart's rhythm to see if additional shocks are necessary. If heartbeat has not been restored, CPR should be continued with the AED pads remaining on the victim.

Because electricity is involved, AEDs cannot be used around flammables. Victims should be removed from water or conductive surfaces, as it is dangerous to use an AED on wet skin or metal. Cell phones and other devices should be moved at least 6 feet away or the heart rhythm analysis could be disrupted.

Good Samaritan laws in some states only permit nonprofessionals to administer 1 AED shock, but this varies. Rescuers should not hesitate to use these devices and should simply take care to follow the given directions carefully. Once an AED has been used, it is important to restock its supplies before putting it back in storage. Everyone should know the location of the nearest AEDs in their workplace, which can be recognized by the AED symbol shown in **f12.3**.

f12.3 automated external defibrillator symbol

Immediately administering aspirin to the victim has also been advocated for heart attacks. One uncoated adult aspirin can be given to a suspected heart attack victim unless the victim is allergic to aspirin, has stomach ulcers, is taking blood anticoagulants, or has been advised by a physician not to take aspirin. Aspirin should not be administered if a stroke is suspected.

Only the basics of first aid have been covered in this Unit. The reader is strongly encouraged to take both a CPR and a first aid course to practice the procedures outlined here and to learn the material in additional detail.

Summary table: accidents & first aid principles

Topic	Comments
Accidents	1. record details of all accidents, no matter how minor 2. analyze accidents and "near misses" for root causes, ways to improve policies and ways to prevent it from happening again 3. fill out OSHA forms as required; deaths must be reported to OSHA within 8 hours, and hospitalizations, amputations, or eye loss within 24 hours 4. develop comprehensive emergency and disaster plans
"Check, call, consent, care"	1. check the scene for safety, the nature of the incident, who can help, who should be treated first, who as a last resort should be moved 2. call 911 or other emergency 3. get victim's permission to render appropriate first aid; unconsciousness is implied consent; check for medic alert bracelets 4. don't move victim unless in immediate danger or position not conducive to first aid 5. move unconscious victims by grabbing under shoulders, supporting head with arms, and pulling on long axis of body 6. support ambulatory victims or place on wheeled chair to remove them 7. provide comfort, reassurance, and care at your appropriate skill level 8. establish unresponsiveness by tapping victim and shouting, "Are you okay?"
Airway: clear the airway	1. signs of airway obstruction: coughing, turning blue, choking, chest not rising 2. ask "Can you speak?"; victims who can speak are getting air 3. actions: a. remove visible foreign objects from mouth and throat ; do *not* dig for objects as they could be lodged even more deeply b. perform abdominal thrusts and/or back blows c. "chin lift, head tilt": lift chin, press forehead down gently; victims with strong coughs can usually clear their own airways
Breathing: assess breathing	1. determine whether breathing emergency or cardiac emergency 2. give 2 rescue breaths; watch chest rise and fall 3. initiate CPR if continued unresponsive
Circulation: assess & restore circulation	1. CPR chest compressions: "push hard, push fast" 2. chest compressions at rate 100 per minute 3. 2 breaths per 30 compressions if alone 4. don't delay CPR to wait for AED; initiate AED in cardiac emergencies when it is ready
Stop bleeding	1. apply direct pressure by covering wound with cleanest material available 2. apply additional dressings as necessary, but leave original dressing in place; check for feeling, warmth and color to be sure bandage not too tight 3. elevate the wound above the heart if no other injuries exist 4. only if bleeding is life threatening, apply a tourniquet between the wound and the heart, as close as possible to the wound; use of a tourniquet almost guarantees loss of the limb 5. internal bleeding: treat as shock, get immediate help 6. severed body part: wash, put in bag, put bag in cool water or ice 7. lost tooth: hold by top not by root, gently clean, put in saliva or milk; victim holds or bites on gauze over empty socket
Treat for shock	1. keep victim covered, warm, and flat; elevate feet less than 12 inches to maintain blood flow to brain unless injury contraindicates 2. do not give anything by mouth; place vomiting victims on their sides 3. burn marks: suspect electrical shock; separate victim with insulator

Summary table: accidents & first aid principles (continued)

Topic	Comments
Treat the injury	1. first and second degree thermal burns a. remove jewelry and clothing b. cover with cool water or cool moist dressings c. don't disturb blisters d. don't use ice; caution with shower if water cold e. treat for shock if burns severe 2. third degree burns a. cut away loose clothing; do not remove materials embedded in the burn b. cover with cool, moist dressing; don't use ice c. treat for shock and get immediate medical attention 3. chemical burns a. flush with water for 15 minutes b. remove affected clothing c. do not use strong neutralizing agents d. treat for shock and get medical attention 4. dry ice "burns" a. treat with tepid, not hot, water b. get medical attention
Bone & joint injuries	1. "RICE" a. REST affected limb b. IMMOBILIZE the affected limb; bandage around any impaled objects c. COLD to reduce pain and swelling d. ELEVATE if no injuries 2. move victim only if in immediate danger; move victims by pulling on long axis of body (under both shoulders and support head) 3. head/neck/spinal injuries a. immobilize victim with head "in line" with body if possible, but do not move body parts to do so b. neurological impairment signs or fluids coming from mouth and nose need urgent medical care
Cuts/punctures	1. stop serious bleeding 2. use water to flush out chemicals and biohazards. 3. seek medical attention to remove embedded objects, administer tetanus shots and get professional help for wounds >1 inch/skin edges that won't touch or won't stop bleeding 4. tetanus shots good for 10 years; if tetanus shot current, clean and dress minor wounds with antibiotic ointment and sterile gauze
Eye injuries	1. flush eyes for 15 minutes with water; roll eyes around to flush all 2. remove contact lenses if possible; do not force 3. if easily done, remove foreign objects with wet sterile gauze or swab 4. if foreign objects cannot be removed, cover both eyes to prevent movement 5. keep victim calm and get immediate medical attention
Stroke	1. Evaluate "FAST"; do not give aspirin 2. FACE: facial weakness, asymmetrical smile 3. ARMS: arm weakness, asymmetrical lift of arms over head 4. SPEECH: slurred or difficult speech; can't answer simple questions 5. TIME: note time symptoms started; time to call 911
Heart attacks	1. give 1 adult uncoated aspirin if not contraindications; start CPR and get AED 2. AED: follow directions from device; don't use on wet victim, on wet surface or in presence of flammables and electrical devices such as cell phones within 6 feet 3. don't apply AED pads over hairy skin (use razor), wet skin, medication patches, visible implants, jewelry 4. know location(s) of nearest AED devices and AED symbol
First aid kits	Gauze, bandages, antibiotic ointments, splints, chemical buffers, blankets, resuscitation barriers; see text & other units for comprehensive lists

Self evaluation questions

1 A laboratory worker goes into anaphylactic allergic shock and dies after inhaling powder from latex gloves. Which statement below is true?

- **a** this incident needs to be reported to OSHA within 8 hours
- **b** this incident needs to be reported to OSHA within 24 hours
- **c** this incident is only reportable if the facility has >10 employees
- **d** this incident is only reportable if the worker acquired the allergy occupationally

2 Which of the following accidents would NOT require an accident report?

- **a** serum specimen splashed in eye
- **b** first degree burn to the arm less severe than a sunburn
- **c** acid spilled in lap and immediately flushed under shower
- **d** needlestick injury with current tetanus shot and no bleeding
- **e** none of the above—all accidents, no matter how minor, must be reported

3 If a user made an error with a safety engineered device, the error must be included in the federal accident report.

- **a** true
- **b** false

4 Which of the following is (are) true?

- **a** accident reports should be completed as soon as possible after an incident
- **b** only the victim in an incident is authorized to complete an accident report
- **c** if an incident meets the OSHA definition of "minor," it does not need to be recorded at all
- **d** all of the above

5 You believe that a victim of a laboratory accident is going into shock. You should

- **a** lay the victim down and elevate his head
- **b** apply a cool cloth to the victim's forehead
- **c** gently give him as many sips of water as he will tolerate
- **d** cover the victim with a blanket or in some manner attempt to keep him warm
- **e** all of the above

6 You spilled concentrated sulfuric acid on your hand while making up a laboratory reagent. The correct sequence of actions to take is to

i. fill out an accident report
ii. report the injury to your supervisor
iii. neutralize the spill with 2% bicarbonate
iv. flush your hand with copious amounts of water

- **a** i, ii, iii, iv
- **b** iv, iii, ii, i
- **c** iii, iv, i, ii
- **d** iv, iii, i, ii
- **e** ii, iii, iv, i

 ISBN 978-089189-6463

7 What are the "4 Cs" that must be performed when an accident situation occurs?

8 You've responded to an emergency situation to provide first aid to an unconscious victim. Of the tasks below, which do you perform first?

- **a** stop the bleeding of a head wound
- **b** try to arouse the victim by tapping his shoulder and yelling, "Are you okay?"
- **c** perform a head tilt/chin lift and then "look, listen, and feel" for breathing
- **d** check the pulse and initiate chest compressions

9 All of the following could be performed to clear the victim's airway and restore breathing EXCEPT:

- **a** removing any foreign objects
- **b** performing the jaw thrust maneuver
- **c** tilting the head so that the chin is touching the chest
- **d** performing abdominal thrusts (the Heimlich maneuver)
- **e** performing mouth to mouth or mouth to nose resuscitation

10 You must stop the bleeding of a cut to the arm. No bones are broken, and the bleeding is not life threatening. All of the following could be performed EXCEPT:

- **a** elevating the arm
- **b** applying a pressure bandage
- **c** applying direct pressure with a clean cloth
- **d** applying a tourniquet between the wound and the heart but not over the wound

11 You must treat a third degree burn. Which of the following should you do?

i. apply ice
ii. treat for shock
iii. apply a cool, moist dressing
iv. apply burn ointment

- **a** i & iii
- **b** ii & iv
- **c** ii & iii
- **d** i, ii & iii
- **e** i, ii, iii & iv

12

12 Bone and joint injuries should be _________ until help arrives.

13 Foreign objects in the eye that are difficult to remove [should/should not (choose one)] be left in the eye, and then the eye should be covered. When covering an eye injury, both eyes should be covered because _______________.Chemicals should be flushed from the eyes using large amounts of _________, and if they are being worn, one should attempt to remove __________ from the eyes.

14 Accident victims should be moved

- **a** if injury is visible
- **b** if they are the most severely injured in a group
- **c** if they are in immediate physical danger at the accident site
- **d** if they can comfortably be moved by grasping 1 arm and 1 leg
- **e** all of the above

15 AEDs must not be used in the presence of or in direct contact with __________, __________, or __________.

16 Each of the following is CORRECT regarding adult CPR technique EXCEPT:

- **a** chest should be compressed at least 2 inches deep between the nipples
- **b** chest should be compressed at a rate of 100 compressions per minute
- **c** if an AED is available, an attempt to should be made to use it before CPR is engaged
- **d** single rescuers should give 2 breaths for every 30 compressions

17 AED pads can be placed over

- **a** implanted pacemakers that are visible under the skin
- **b** jewelry
- **c** tattoos
- **d** nicotine skin patches
- **e** all of the above
- **f** none of the above

18 When should you legally attempt first aid?

- **a** when the victim is unconscious
- **b** when you think a child's injury could be life threatening
- **c** when you have explained who you are and what you know, and the victim consents to care
- **d** all of the above

 ISBN 978-089189-6463

19 Which action(s) below is (are) correct?

- **a** putting a lost tooth in milk
- **b** placing a severed finger directly in ice
- **c** removing and replacing a bandage when it is soaked through with blood
- **d** administering aspirin to a victim with signs consistent with stroke
- **e** all of the above

20 Which of the following is correct?

- **a** using chest compressions on a choking pregnant woman
- **b** having assistants hold down a combative victim so AED can be used
- **c** using AED on a victim with cardiac arrest while dialing 911 on a cell phone
- **d** applying a pressure bandage to an area of severe bleeding until skin goes numb
- **e** none of the above

Answers

1 **a**

2 **e**

3 **a**

4 **a**

5 **d**

6 **b**

7 **check, call, consent, care**

8 **b**

9 **c**

10 **d**

11 **c**

12 **immobilized**

13 **should; the eyes move simultaneously, and movement of the injured eye should be avoided; water; contact lenses**

14 **c**

15 **flammables, water, or metal**

16 **c**

17 **c**

18 **d**

19 **a**

20 **a**

ISBN 978-089189-6463

Appendix 12.1: bomb threat & suspicious packages checklists

If you receive a telephoned bomb threat, you should do the following:

- Get as much information from the caller as possible. Try to ask the following questions:
  - When is the bomb going to explode?
  - Where is it right now?
  - What does it look like?
  - What kind of bomb is it?
  - What will cause it to explode?
  - Did you place the bomb?
- Keep the caller on the line and record everything that is said.
- Notify the police and building management immediately.

Be wary of suspicious packages and letters. They can contain explosives, chemical or biological agents. Be particularly cautious at your place of employment. Some typical characteristics postal inspectors have detected over the years, which ought to trigger suspicion, include parcels that:

- Are unexpected or from someone unfamiliar to you.
- Have no return address or a return address that can't be verified as legitimate.
- Are marked with restrictive endorsements such as "Personal," "Confidential," or "Do not X ray."
- Have protruding wires or aluminum foil, strange odors or stains.
- Show a city or state in the postmark that doesn't match the return address.
- Are of unusual weight given their size or are lopsided or oddly shaped.
- Are marked with threatening language.
- Have inappropriate or unusual labeling.
- Have excessive postage or packaging material, such as masking tape and string.
- Have misspellings of common words.
- Are addressed to someone no longer with your organization or are otherwise outdated.
- Have incorrect titles or titles without a name.
- Are not addressed to a specific person.
- Have hand written or poorly typed addresses.

With suspicious envelopes and packages other than those that might contain explosives, take these additional steps against possible biological and chemical agents.

- Refrain from eating or drinking in a designated mail handling area.
- Place suspicious envelopes or packages in a plastic bag or some other type of container to prevent leakage of contents. Never sniff or smell suspect mail.
- If you do not have a container, then cover the envelope or package with anything available (e.g., clothing, paper, trash can, etc.) and do not remove the cover.
- Leave the room and close the door or section off the area to prevent others from entering.
- Wash your hands with soap and water to prevent spreading any powder to your face.
- If you are at work, report the incident to your building security official or an available supervisor, who should notify police and other authorities without delay.
- List all people who were in the room or area when this suspicious letter or package was recognized. Give a copy of this list to both the local public health authorities and law enforcement officials for follow-up investigations and advice.
- If you are at home, report the incident to local police.

Source: www.ready.gov/explosions, accessed 11/2015

 ISBN 978-089189-6463

Unit 13

Accident situations

Scenario A

During a toxicology extraction, Phillip was evaporating an organic solvent in a hood atop a hot plate purchased at a discount store. Suddenly, a spark from the hot plate ignited a burst of flames across the top of the beaker. Phillip immediately grabbed a carbon dioxide (CO_2) extinguisher and snuffed out the flames. Disgusted over the mishap that ruined his toxicology extraction, Phillip dropped the fire extinguisher on the floor and walked out of the laboratory for a coffee break. When he came back to his work station, Phillip found a fire squad actively fighting a large blaze coming from the hood.

1 Number these steps from 1 to 5, putting them in the proper sequence of actions Phillip should have followed after the ignition of the first blaze.

__ Evacuate all other personnel
__ Put the fire out with the extinguisher
__ Pull the fire alarm and call the proper extension
__ Close the glass door of the hood and turn it off
__ Watch for rekindling of the blaze

2 How could the first fire have been prevented?

a By performing the procedure with the hood turned off
b By carrying out the procedure in the sink rather than the hood
c By using a procedure that does not require the evaporation of an organic solvent
d By using nonsparking, explosion proof electrical equipment designed for use with flammable solvents

3 Was a CO_2 extinguisher acceptable to use in this situation?

a no, and that is why the fire started up again
b no, a water extinguisher should have been used
c no, a dry chemical extinguisher should have been used
d yes, although a water extinguisher could also have been used
e yes, although a dry chemical extinguisher could also have been used

4 There is still some CO_2 left in the extinguisher Phillip used. What should be done?

a the CO_2 extinguisher should be refilled no matter how much CO_2 is left
b the CO_2 extinguisher should be refilled if less than half of the CO_2 is left
c the CO_2 extinguisher should be returned to its storage site and the remainder of the CO_2 can be used on the next fire

5 Was this the type of fire that is appropriate for an extinguisher?

a yes
b no, because of the size of the fire
c no, because of the location of the fire
d no, because of the hazardous gases produced by the fire

Scenario B

Lauren was becoming discouraged because her small volumetric flasks kept coming back from the utility room dirty. One day she decided to clean the flasks herself and proceeded to put together a cleaning solution of concentrated sulfuric acid and potassium dichromate at her workstation in the chemistry department. After Lauren had poured a small amount of cleaning solution in each of the small volumetric flasks, she set the beaker containing cleaning solution on a small shelf above the workbench. Because the shelf was too narrow to hold the beaker adequately, a slight jarring of the shelf caused the beaker to overturn and pour acid over the workbench, wetting Lauren from the waist down. Sizing up her situation, Lauren decided not to use the nearby safety shower but to go the restroom to remove her clothing. Another scientist, Kelly, saw her going to the restroom in a hurry and went to see what was wrong. Kelly quickly rushed Lauren to the safety shower and then to the emergency department. Later in the emergency department, the physician evaluated the burns on Lauren's hip and thigh to be first and second degree burns.

6 What action(s) should Lauren have taken to prevent this accident?

- **a** wear an acid protective apron
- **b** prepare strong acid solutions over the sink
- **c** never put acid on a high shelf or in an open beaker
- **d** all of the above

7 What is the first thing Lauren should have done to minimize her injury?

- **a** report to the supervisor for help and advice
- **b** neutralize the acid with 20% sodium hydroxide
- **c** wash off in a safety shower and remove her contaminated clothing
- **d** go immediately to the emergency department so that medical treatment could be initiated as soon as possible

8 If you were the supervisor evaluating the report of this accident, what improvements would you make in your laboratory to prevent this from occurring again?

- **a** increase the number of safety showers and eyewashes
- **b** retrain laboratory personnel in safety procedures, particularly the handling of acids
- **c** set up burn stations with 20% sodium hydroxide and 20% hydrochloric acid
- **d** remove all strong acids from the laboratory and place them into locked storage

9 How should the acid spill at Lauren's workstation be cleaned?

- **a** use gloves and body protection to sweep up the waste
- **b** create a "dike" around the spill with kitty litter to prevent it from spreading
- **c** sprinkle it with a weak neutralizer such as sodium bicarbonate
- **d** all of the above

 ISBN 978-089189-6463

10 What else should Kelly have done when she realized what had happened?

- **a** she should have notified a supervisor before she treated or transported Lauren anywhere
- **b** she should have asked someone to check the scene to see if anyone else was hurt or in danger
- **c** she should have checked the SDSs for concentrated sulfuric acid and potassium dichromate to be sure they could be flushed down the safety shower before she put Lauren inside it
- **d** all of the above

Scenario C

Cooper was setting up a suction flask. Using his bare hands, he attempted to put a piece of glass tubing through a rubber stopper. The tubing was very difficult to insert, so when Cooper increased the force on the tubing, the glass broke and caused a deep cut in his left palm.

11 What steps should Cooper have taken to prevent the accident?

i. heat the glass tube
ii. use a chemically compatible lubricant on the glass tubing
iii. never use rubber stoppers; instead use cork
iv. use a towel or some other protection for his hands

- **a** iv only
- **b** i & iii
- **c** ii & iv
- **d** i, ii & iii

12 Cooper was able to stop the bleeding in his palm without difficulty. His next course of action should be to

- **a** fill out an incident report form and give it to the supervisor
- **b** go to the emergency department and ask for a tetanus shot and an antibiotic
- **c** report the accident to the supervisor immediately and get medical attention
- **d** apply a waterproof bandage to the cut, finish his work, and report the accident to the supervisor later

13 The laboratory has the following choices for disposal. Where should the glass that punctured Cooper's palm be disposed?

- **a** ordinary trash can
- **b** segregated chemical waste containers
- **c** needle disposal box, biohazardous waste
- **d** broken glass container, nonbiohazardous waste

Scenario D

Alexander needed to change the hydrogen tank of the gas chromatograph because the gauge indicated that the cylinder was almost empty, and he knew that it should never be emptied completely. He took a wrench and began to disconnect the fitting. When the fitting was slightly loose, a sudden, violent release of gas emitted from the fitting. Alexander was not hurt, so he proceeded to change the cylinders as usual. Within a few minutes, Kevin entered the room that housed the gas chromatograph to put some specimens for the gas chromatograph on an old rotator. When he turned on the rotator, a spark caused the entire room to burst into flames.

14 How could this accident have been prevented?

- **a** only nonsparking equipment should have been used
- **b** Alexander should have turned off the valve of the hydrogen tank before removing the fitting
- **c** Alexander should have ventilated the laboratory and notified a supervisor that there had been a hydrogen gas leak
- **d** all of the above

15 If Alexander had turned off the valve the hydrogen tank, how could he tell if it was safe to remove the fitting?

Scenario E

Ethan needed 0.5 McFarland bacterial suspensions for antibiotic sensitivity testing and wanted to vortex mix them before verifying correct turbidity. He turned the vortex mixer to "HIGH" and placed the first tube into the vortex. Because it was not covered, it splashed his entire face, including his eyes and mouth. Ethan grabbed a paper towel and wiped off his face with soap and water. Within a few days, Ethan had conjunctivitis in his eye and a positive blood culture.

16 What action should Ethan have taken immediately after the accident?

- **a** he should have flushed his entire face, eyes, and mouth for at least 15 minutes with the eye wash
- **b** he should have washed his face and flushed his eyes with a disinfectant such as iodine or phenol
- **c** he should have reported to his supervisor so that he or she could advise him based on the organism he was handling
- **d** he should have reported to the emergency department for prophylactic antibiotics and a vaccination against the organism he was handling

17 How could this accident have been prevented?

- **a** by mixing suspensions with a stir bar
- **b** by vortexing only tightly covered samples
- **c** by vortexing only inside a biological safety cabinet
- **d** by eliminating antibiotic sensitivity testing with bacterial suspensions

 ISBN 978-089189-6463

Scenario F

Adalyn plugged in a vortex mixer beside the sink where she was working. When she spilled some water on the electrical cord, she noticed that a portion of the cord was frayed and that it was giving off sparks. She tried to unplug the cord and received a severe shock.

18 To prevent the shock that she received, the best thing Adalyn could have done is to

- **a** turn the vortex off before doing anything else
- **b** put on latex gloves to pull out the plug
- **c** put on nitrile gloves to pull out the plug
- **d** turn off power to the outlet from the breaker box
- **e** create a dike around the water spill with kitty litter and then absorb all of it with paper towels before doing anything else

19 What unsafe practice(s) contributed to Adalyn's situation?

- **a** using a frayed electrical cord
- **b** using electrical equipment around a source of water
- **c** using an electrical receptacle near a source of water
- **d** all of the above

20 This scenario indicates that the electrical receptacle was probably not________, and the laboratory should consider replacing it if proximity to water cannot be eliminated.

- **a** polarized
- **b** grounded
- **c** a ground fault circuit interrupt
- **d** a surge protected wet receptacle

Scenario G

Laurel had been processing blood specimens when she was told she had a phone call. She took the call in the laboratory and did not remove her gloves. Later another worker, Anna, washed her hands so that she could eat. She used the same phone to let her supervisor know that she was going to lunch. 12 weeks later, Anna had an acute case of hepatitis B.

21 What policy would you recommend to this laboratory?

- **a** equipment should be designated for use by either gloved or nongloved personnel
- **b** staff should not receive phone calls while processing blood or performing other laboratory functions
- **c** equipment such as telephones and computer keyboards should be disinfected daily with a mid level disinfectant
- **d** iodine hand scrubs capable of killing the hepatitis B virus should be used for routine handwashing

22 Which of the following is true regarding Anna's case of hepatitis?

- **a** Anna's employer is financially responsible for all medical care if Anna's infection is job related
- **b** it is Anna's fault for using a laboratory telephone without gloves, so she must pay for any medical costs
- **c** Laurel is responsible for Anna's medical expenses because she is the one who contaminated the telephone
- **d** if Anna had gotten the hepatitis B vaccine, she would not have hepatitis; therefore, Anna is responsible for the medical costs

Scenario H

David carefully packed a dozen whole blood samples in a cardboard box and surrounded them with Styrofoam "peanuts" to prevent breakage. He labeled the package with the name and address of the reference laboratory, and took the package to the US Post Office. The next day, an upset post office supervisor called the laboratory. The package had leaked over a mail sorter, and he wanted to know the contents of the package.

23 What did David do wrong?

i. he did not use triple packaging
ii. he used a cardboard box rather than plastic
iii. he failed to use absorbent material in the package to contain leakage
iv. he sent the blood through the US mail, and federal mail sorters do not know how to properly handle blood

- **a** iv only
- **b** i & iii
- **c** ii & iv
- **d** i, ii & iii

 ISBN 978-089189-6463

24 What else should David have put on the exterior of the package?

- **a** a clear plastic bag
- **b** no more than 2 layers of plastic cling wrap so that package markings are still visible
- **c** the UN 3373 label with the words "Biological Substances, Category A" next to the diamond
- **d** the UN 3373 label with the words "Biological Substances, Category B" next to the diamond
- **e** not enough information is given to answer, but David should not have been shipping a Category A substance; if the specimens met Category B criteria, he should have used the proper label

Scenario I

For many months, Rachel had been disposing of a manufacturer's pre-prepared reagent mixture down the sink. When the sink got clogged, a plumber was called to examine the sink. The sink was an old one, and the drain was composed of metal pipes. When the plumber tried to unscrew the pipes with a wrench, they exploded.

25 What was in the pre-prepared mixture that most likely caused this problem?

- **a** mercury
- **b** picric acid
- **c** nitroglycerin
- **d** sodium azide
- **e** perchloric acid

26 What could be done to prevent this from occurring again?

- **a** install plastic pipes
- **b** use a commercial drain cleaner such as Drano to unclog pipes
- **c** use a plunger to unclog pipes
- **d** use a plumber's "snake" to unclog pipes

27 Where should Rachel have been able to find the information telling her about the disposal of this reagent?

- **a** in the procedure for the test using that product
- **b** in the chemical hygiene plan for the laboratory
- **c** in the safety data sheet for the product
- **d** all of the above

Scenario J

Andrew, a recent hire, had never worked in a laboratory before. He was given 7 tubes of blood to spin in a centrifuge. He placed the tubes in random carriers, closed the centrifuge, and turned it on. A few minutes later, he heard a spectacular crash and saw that the centrifuge had fallen off the edge of the bench and onto the floor.

28 What probably caused the centrifuge to fall off the bench?

- **a** a short circuit in the motor
- **b** defective suction cups on the bottom
- **c** unbalanced contents with regard to weight
- **d** someone knocking against it and normal motor vibrations

29 What should Andrew do first to respond to this incident?

- **a** notify a supervisor and fill out an incident report
- **b** if tube breakage is visible, clear the immediate vicinity to let aerosols settle
- **c** unplug the centrifuge and create a dike around the immediate area with kitty litter and spray it with 10% bleach
- **d** check the sources of the blood for infectious disease

Scenario K

Patrick noticed that when he removes his gloves at the end of each work day, his hands are itchy, and he seems to be developing a mild red rash. His wife lent him some of her hand cream and suggested that he use it during the day under his gloves. This did not improve the situation.

30 What should Patrick do?

- **a** use latex free gloves
- **b** use powderless gloves
- **c** use gloves that are both latex and powder free
- **d** wear glove liners underneath the gloves and wash with a more gentle soap
- **e** report the problem to his supervisor so that the exact cause of the rash can be investigated

31 What else has Patrick done that has put him at additional risk?

____

____

____

____

____

Scenario L

Emily got a thermal burn on her arm from accidently leaning on a hot plate. She ran her arm under cold water and noticed several blisters. She used some antibiotic ointment on the burn, wrapped it in clean gauze, and resumed her work. Within a few minutes, the pain in her arm was so intense that she felt woozy.

32 What did Emily do wrong?

- **a** she used ointment on a burn
- **b** she did not seek medical attention
- **c** she did not report the incident to her supervisor
- **d** all of the above

33 The blisters on the burn indicate that Emily probably has __________ degree burns. Because she is feeling "woozy," staff should treat her for __________.

Scenario M

Evan finished working on the laboratory computer, removed his gloves, and put some alcohol gel on his hands to decontaminate them. As he left the laboratory, he turned off the light switch on the wall. The light switch sparked and ignited the alcohol gel.

34 What did Evan do wrong?

- **a** he used alcohol gel instead of washing his hands
- **b** he did not let the gel dry before touching anything else
- **c** he turned off the light; laboratories must always be illuminated for safety
- **d** all of the above

35 Is alcohol hand gel ever appropriate to use after glove removal when working with biohazards?

- **a** yes, if hands are not visibly soiled
- **b** yes, if the laboratory is BSL 1
- **c** yes, if the laboratory is not BSL 4
- **d** yes, if the laboratory is BSL 1 or 2
- **e** no

36 How should Evan's supervisor respond to this incident?

- **a** reprimand Evan and note the incident in his annual evaluation
- **b** remove all alcohol hand gel from the laboratory
- **c** initiate an education campaign to remind workers to wait at least 15 seconds after gel application before touching anything
- **d** all of the above

 ISBN 978-089189-6463

Scenario N

William was working on the third floor of the laboratory in the east wing. A "Code Red" was announced for the second floor east wing. William took the elevator in the east wing to evacuate. The elevator shaft is adjacent to the compressed gas storage room, so that when the compressed gas room exploded, it damaged the elevator mechanism and stopped the elevator. William was trapped in the elevator for 1 hour until the fire rescue team could release him.

37 Each of the following is an error that William made EXCEPT:

- **a** he should have walked to the west wing first and then used the stairs
- **b** he should not have evacuated using an elevator
- **c** he should not have evacuated toward a high risk area
- **d** he should have waited until a supervisor told him to evacuate

38 If you were William's supervisor, what would you do?

____

____

____

Scenario O

The explosion proof refrigerator needed defrosting, so Benjamin loaded all the flammable chemicals from the refrigerator into the chemical fume hood. The number of chemicals crowded the interior so that the sash would not fit all the way down. When the supervisor entered the laboratory, she smelled a strong odor of organic solvent and immediately tried to identify the source before there was an explosion.

39 What is the likely source of the odor?

- **a** the chemical fume hood drawing air poorly
- **b** the inside of the explosion proof refrigerator releasing its fumes
- **c** the frost inside the explosion proof refrigerator releasing its fumes

40 What should be done now?

____

____

____

 ISBN 978-089189-6463

Scenario P

A laboratory had an isolated room with controlled airflow to handle *Mycobacterium tuberculosis* cultures. Brad, the building janitor, has a compromised immune system, which he has not disclosed. He developed a terrible cough and was diagnosed with tuberculosis. Shortly thereafter, annual tuberculosis skin testing of all staff revealed that office workers down the hall had converted from negative to positive results. Workers in the laboratory were unaffected.

41 What is the first thing that should be investigated in this situation?

- **a** the community contacts of the office staff
- **b** the air handling system responsible for the exhaust from the tuberculosis laboratory
- **c** whether the tuberculosis laboratory is still at a positive pressure relative to the rest of the building
- **d** whether the tuberculosis laboratory is still at a negative pressure relative to the rest of the building

42 Does this incident have to be reported to OSHA?

- **a** no, OSHA no longer has a tuberculosis standard
- **b** no, the worker's immune system was compromised
- **c** yes, it should be recorded and submitted in the annual OSHA report if it is determined that the TB was acquired through occupational exposure
- **d** it depends on how many employees work in the facility

43 Did Brad make a mistake in not disclosing his condition? Explain your answer.

____

____

____

44 Brad died of tuberculosis several weeks after he was diagnosed. A leak in the HEPA filtration system of the biological safety cabinets was discovered that allowed unfiltered exhaust to enter the general air circulation of the building. Does Brad's death have to be reported to OSHA?

- **a** Yes, within 8 hours of determining that this was an occupationally caused death.
- **b** Yes, within 24 hours of determining that this was an occupationally caused dcath.
- **c** No, because Brad was immunocompromised and did not disclose this information.
- **d** No, because the death occurred so long after the incident.

13

Scenario Q

Lynn was the laboratory supervisor on duty when a safety engineered device failed and an employee was injured. In talking to several of the other employees, Lynn discovered that many of them had experienced "close calls" with these devices, but none had been previously injured. Lynn contacted several vendors to ask for samples of other devices and allowed all the employees to try them. She then discarded all old devices and purchased the one preferred by the staff. Feeling that she had solved the problem, she sat down to write the incident report, but could not remember details such as the model number of the device involved and the exact time of the incident.

45 What did Lynn do right?

46 What did Lynn do wrong?

47 What did this incident reveal about the "safety culture" in this laboratory?

13

 ISBN 978-089189-6463

Answers

Scenario A

1 1, 4, 2, 3, 5

2 d

3 e

4 a

5 a

Scenario B

6 d

7 c

8 b

9 d

10 b

Scenario C

11 c

12 c

13 c

Scenario D

14 d

15 both the internal pressure gauge and flow gauge on the regulator would read 0

Scenario E

16 a

17 b

Scenario F

18 d

19 d

20 c

Scenario G

21 a

22 a

Scenario H

23 b

24 e

Scenario I

25 d

26 a

27 d

Scenario J

28 c

29 b

Scenario K

30 e

31 only nonpetroleum based lotions should be used under gloves; if the lotion Patrick is using is petroleum based, the gloves could be compromised, as certain petroleum based hand creams can dissolve glove material, leaving small holes

Scenario L

32 d

33 second; shock

Scenario M

34 b

35 a

36 c

Scenario N

37 d

38 Review procedures related to fire and fire evacuation routes with all employees immediately. When the review is complete, require every employee to participate in a fire drill. Repeat fire training at least annually.

Scenario O

39 a

40 ventilate the room immediately and place chemicals in an alternate storage site such as a vented flammable cabinet

Scenario P

41 b

42 c

43 Yes. While workers are allowed to maintain their privacy, generally they should disclose conditions that put them at additional risk. Employers are required to keep health information confidential and to provide reasonable extra measures of protection.

44 a

Scenario Q

45 She was thorough in getting details about the defective device from everyone and involved the employees in the selection of the best device. She acted quickly to permanently remove the hazardous device.

46 She waited too long to complete the incident report and may be missing some vital information. Accident reports must be completed as soon as reasonably possible after an incident.

47 The "close calls" regarding this device suggested an accident waiting to happen, yet the people involved did not report it. Employees must receive a clear message that safety is everyone's job and that their observations and input are welcome.

 ISBN 978-089189-6463

Unit 14

A culture of safety

Sound rules and protocols alone do not make a laboratory truly safe. The attitudes of all personnel in an organization create an atmosphere regarding safety that is either positive or negative. An institutional culture that prioritizes and values safety, promotes cooperation among all workgroups, and optimizes work conditions for every employee, takes safety management beyond mere rules to a place where workers are both safe and satisfied in their jobs.

Lab safety, only part of the whole picture

In the very first Unit of this book, 2 things are emphasized: the first is that each individual is responsible for contributing to the safety of the whole laboratory; the second is that the process of keeping an environment safe is never ending, as it includes complying with evolving regulations, evaluating incidents, improving procedures, and staff training. The left half of **f14.1** is also used as **f1.1** in Unit 1 to illustrate the overall process of safety management in the laboratory.

This Unit expands concepts of safety management to include the laboratory's role within the larger organization. Few laboratories exist in isolation; they are usually part of a larger entity with its own mission. For example, research laboratories are funded to provide new and accurate information for many purposes. Educational laboratories support a curriculum designed to train competent

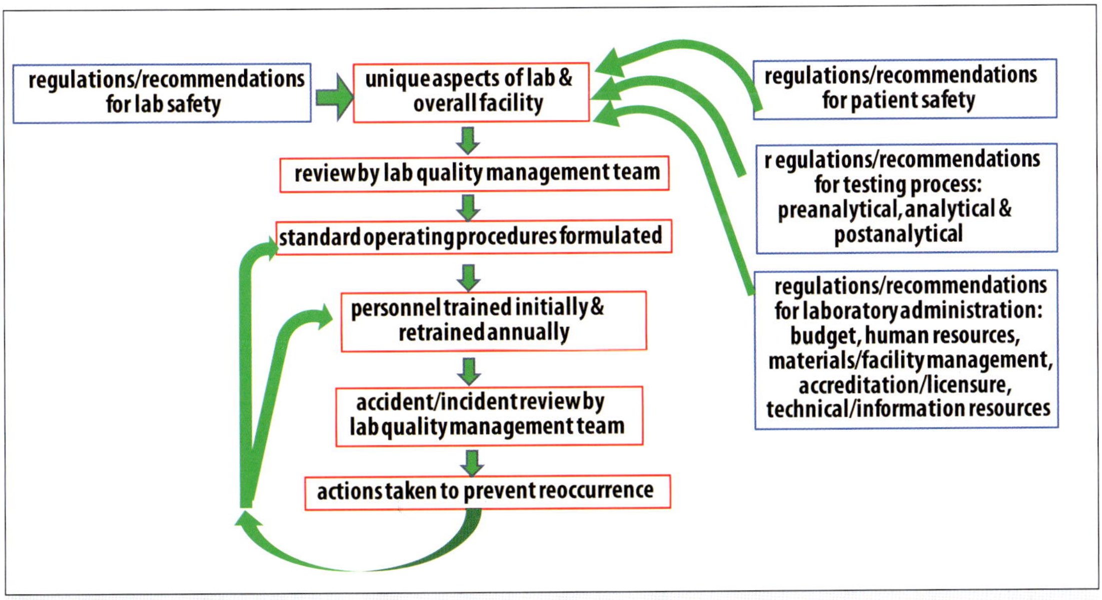

f14.1 diagram of safety process

14

practitioners. Medical laboratories exist in all manner of healthcare operations to provide timely testing that directly impacts patient care. A laboratory, therefore, must also create policies and protocols that support the expectations, limitations, and desired outcomes of the larger organization in which it resides. The right half of f14.1 illustrates some of the competing priorities for medical laboratories, which must be considered in conjunction with laboratory safety. In this part of the figure, the lab quality management team has replaced the safety officer/committee as the guiding entity. Global quality management ensures not just the safety but also the quality of the laboratory's performance relative to the overall institution.

Incorporating safety into organizational culture

Ideally, a culture of safety permeates the entire institution. New employees are introduced to a group that expects and values safe behavior, in which resources and policies for safety are not in competition with other endeavors. Upper management supports safety initiatives and commits resources to permit best practices. Safety considerations are deliberated in every activity and initiative, and each person in the organization recognizes safety as a priority that is equal to other priorities. All of this is easier to say than it is to do; the safest way to do something may not be the fastest, easiest, or cheapest way. Therefore, everyone in the organization must be willing participants in a culture that seeks out ways to keep everyone safe while appropriately balancing all other priorities and resources.

Also discussed in previous units is the concept of using accident analysis to prevent recurrence of incidents. Reliable analysis requires accurate and truthful accident reports. To facilitate this, managers must create an atmosphere that is positive and constructive rather than negative, accusatory, and punitive. To be sure, workers who repeatedly and willfully engage in unsafe behavior may need discipline and termination, but these actions should be the very last resort. In a safety culture, incidents are valued as learning opportunities, which helps workers feel secure in voicing their concerns about unsafe conditions and telling the truth in incident reports, even if they are at fault. If workers feel that telling the truth could have negative consequences, they may not admit their own mistakes or report their colleagues' unsafe behavior. For example, someone afraid of reprisal might claim that a device failed rather than admit he or she used it incorrectly; such a claim could then incur the unneeded expense of replacing of a nonfaulty device rather than initiating a necessary retraining program.

A reactive approach to incidents that have already occurred is necessary, but far less desirable than a proactive approach. In a safety culture, this means that possible accidents are anticipated, and at risk conditions are corrected to prevent incidents from occurring. For example, in so called "failure mode analysis," tasks are analyzed step by step for problems that could arise, and protocols are adjusted to prevent any accident determined to be reasonably possible.

In a safety culture, every person's input is solicited and valued, and communication between supervisors and staff is frequent and honest. A general rule in assessing procedures is that the best way to ensure compliance is to make the process as easy and comfortable as possible. The people

 ISBN 978-089189-6463

who actually perform the task in question usually have the best insight into the issues that must be considered and may often have the best solutions to those issues. When supervisors design protocols or make purchases in isolation, they might not have all the information necessary for the best outcome. Therefore, workers themselves should be asked to provide their opinion and concerns, and be encouraged to give honest feedback that will not be judged or dismissed. For example, a worker caught in a back room without eye protection could be scolded and disciplined, but in a safety culture, the worker instead would be asked why he or she wasn't using the goggles provided, and that answer might be more revealing: eg, "The goggles hurt or don't fit," "The room is too hot and the goggles fog up," or "The goggles don't work with prescription glasses." These honest answers can then lead to the root cause for the incident, which can then be corrected. Further, if this worker was involved in the design of the task and the purchase of the safety equipment, he or she likely would have complied with safety protocol in the first place.

Education and training, not punitive measures, are the first and best strategies for correcting unsafe behaviors. Since most people are highly motivated by self preservation, compliance with safety protocols is higher when workers believe that they are at risk, that the risk is significant, and that the protocols work and are worth the effort, so education about hazards helps workers to take them seriously. Well organized and thorough training on protocols ensures that workers know what to do in an incident, and why it needs to be done. Training should also be cooperative rather than dictatorial, as workers who are encouraged to give suggestions for making procedures less onerous are more likely to comply. Still, while some safety measures are not negotiable, the difficulty or disruption that is associated with them can be minimized.

Clinical Laboratory Standards Institute
Document GP17-A3
Safety Focus by Month

January: specimen transport
February: hoods and environmental issues
March: personal protective equipment
April: ergonomics
May: bloodborne pathogens
June: general safety
July: compressed gases
August: safe work practices
September: chemical hygiene
October: fire safety
November: electrical safety
December: waste management

f14.2 CLSI monthly safety focus suggestions

An atmosphere that is constructive and formative rather than hostile and accusatory not only invites input, but also rewards it. Regular commendations of people who perform safety measures well or take the time to report hazards increase the whole staff's comfort with doing the same. Supervisors and managers should be evaluated not just on how problems were solved but also on how employees were coached and recognized.

At all levels, a team approach to solving problems, in which representatives from multiple disciplines participate, is more likely recommend measures that answer the needs of all stakeholders, not just the laboratory. A formal process should be initiated to schedule safety reminders and activities, which needs not be elaborate or time consuming to be effective. The Clinical and Laboratory Standards Institute recommends focusing on a different safety topic each month **f14.2**, as well multiple simple ways exist to promote safety. One idea is to give a small reward to for the employee with the best hazard-reducing idea related to the month's topic. Computer screen savers could display safety tips relevant to the monthly theme. Posters with safety messages on the current topic could be displayed in the bathroom. At the management level, each month's topic could, independent of changes in regulations or incidents, trigger targeted reviews, risk assessments, and inspections. All of these activities contribute to an atmosphere in which the group culture gently guides everyone toward the same behavioral standards.

Safety priorities in a medical lab

Medical laboratories are a good illustration of how the overall mission of the institution can compete with laboratory safety. Without question, the primary goal of all healthcare facilities is excellent patient outcomes. To this end, the laboratory's contribution is test results that are not only accurate but also delivered in time to effect good patient care. Medical laboratories in hospitals must provide this service 24 hours a day and do not have the luxury of delaying procedures until everything is ready and safe. Medical laboratory safety management includes having enough staff to perform requested services without anyone being overburdened or rushed. However, the demand for medical care is unpredictable. Understaffing the laboratory is a hazard for both workers and patients, but overstaffing the laboratory is wasteful and costly. Further, the entire healthcare industry is under significant pressure to reduce costs, and this impacts not just staff levels but also budgets for safety initiatives. Maintaining laboratory safety and patient safety simultaneously is difficult, but essential.

Institute of Medicine

Quality Aims to Improve Healthcare Delivery

Patient safety
Patient centeredness
Efficiency
Effectiveness
Equity
Timeliness

f14.3 IOM 6 quality goals for healthcare

In 1999, the Institute of Medicine (IOM) published *To Err Is Human*, a document that outlined the scope and extent of preventable medical errors in the US, and called for aggressive changes in the healthcare system. The following year, 6 quality aims were published as a framework to address some of these concerns **f14.3**.

Patient safety and laboratory safety overlap nicely in some areas. As stated in *To Err Is Human*, "commonly, errors are caused by faulty systems, processes, and conditions that lead people to make mistakes or fail to prevent them," which is unquestionably true for both patient and laboratory safety, and supports the notion in this book that robust protocols and procedures are a necessary starting point. The statement also supports the creation of a safety culture in which flawed processes are to blame, not individuals. Areas of potential conflict exist, however. While healthcare workers understand that they have entered a profession with hazards, in extreme circumstances the desire for self protection may conflict with delivering care. A good example is the significant risk posed by Ebola patients and their specimens; it would be overly disruptive

 ISBN 978-089189-6463

and costly to treat every specimen as if it had Ebola, yet some provision must be made for these unusual events, which must strike a balance between workers' physical safety and effective patient care. A far more common conflict comes with the goal of timeliness. The pressure to collect specimens, transport them to the laboratory, and perform testing as quickly as possible can be extreme; "turnaround time" is one of the most common metrics on which a laboratory or workers are evaluated. Medical laboratory workers often feel rushed and perform multiple tasks at once, not just due to the pressure of performance evaluations but also due to their personal commitment to patient care. Doing too many things at top speed is both a lab safety and a patient safety concern, an inherent conflict that is recognized and mitigated as much as possible in a safety culture.

In 1996, the Joint Commission, which accredits hospitals, created the formal Sentinel Event Policy to help hospitals that experience serious adverse events improve safety procedures and learn from sentinel events. The term "sentinel" is used because the incidents are sufficiently serious that immediate investigation and correction of processes are necessary. Sentinel events are those that result in death, permanent harm, or severe temporary harm and/or require intervention to sustain life. The 2015 policy related to laboratories (www.jointcommission.org/assets/1/6/CAMLAB_19_SE_all_CURRENT.pdf, accessed January 26, 2016) calls for the following actions after a sentinel event:

- formalized response by a team to stabilize the patient, disclose the event to the patient and his or her family, and provide support to the family and staff involved in the event
- notification of the leadership of the laboratory
- immediate investigation
- completing a comprehensive and systematic analysis to identify the event's causal and contributory factors
- corrective actions
- implementation timeline for the corrective actions
- systemic improvement

Additional details are beyond the scope of this text but can be readily located in the document cited above and on The Joint Commission's website. The Institute for Healthcare Improvement also has many resources related to patient safety initiatives (www.ihi.org/IHI/Topics/PatientSafety, accessed January 26, 2016).

Regardless of a laboratory's location and mission, the creation of a safety culture is the best means to ensure prevention of and appropriate response to accidents. An atmosphere in which everyone regards safety as part of the job and values his or her own role in contributing to a safe environment results in ongoing vigilance for hazards. Everyone is held individually accountable for their actions, but also believes that improving job processes, rather than punishing the people involved, is the best response to incidents. A positive climate that includes safety is a vital part of completing any laboratory's mission.

Summary table: culture of safety

Topic	Comment
Principles of a safety culture	1. laboratory must fulfill mission of organization without compromising safety 2. laboratory quality management team better model to include all priorities of lab instead of just a safety committee 3. constructive and formative atmosphere, not negative and punitive 4. workers learn from incidents rather than get punished; promotes truthful reporting 5. proactive response to prevent incidents in addition to reactive response 6. workers feel safe to give input to all decisions, protocols, and policies 7. highest compliance is when the right way is also the easiest and most comfortable 8. recognition of workers who perform well or give good suggestions regarding hazards 9. monthly focus on a safety topic 10. correct flawed processes before disciplining people 11. all workers believe promoting safety in every activity is part of their job.
Unique aspects of medical laboratories	1. difficult to staff appropriately since demand for medical care not predictable 2. pressure to produce test results as fast as possible 3. insufficient staffing leads to unsafe multitasking and rushing 4. need to provide healthcare even when patients are very high risk
Institute of Medicine's quality goals	1. patient safety 2. patient centeredness 3. efficiency 4. effectiveness 5. equity 6. timeliness
The Joint Commission	1. sentinel event is so severe that immediate analysis and correction required 2. sentinel events defined as incidents that result in death, permanent harm, or severe temporary harm/intervention required to sustain life 3. process for handling sentinel events includes analysis for root cause and contributory factors, plan of correction, timeline for correction, and analysis of the system

Self evaluation questions

1 A laboratory manager allowed the entire staff to try several brands of protective gloves. By a unanimous vote, the staff chose a brand of gloves that was comfortable and fit well. The manager knew that all the gloves had the same level of barrier protection, but the brand chosen by the staff was double the price of the others. What should the manager do?

- **a** buy the gloves that the staff wants
- **b** buy the staff's second choice, which is a cheaper brand
- **c** tell the staff the price difference and ask them if it is worth it
- **d** try to find a cheaper brand of gloves that the staff likes just as well
- **e** c and d, and if they don't work out, a

 ISBN 978-089189-6463

2 A night shift worker was caught not wearing a lab coat. He said the lab was too hot, so the temperature was adjusted to his liking. He was caught not wearing a lab coat again, and because he said the one he had was uncomfortable, a lab coat that he liked was purchased. Then he was caught not wearing a lab coat again and finally admitted that he just doesn't like wearing them. What should the manager do?

- **a** move him to a job that doesn't require a lab coat
- **b** buy other brands of lab coats to see if he likes any of them
- **c** terminate him; when all efforts fail to incur compliance with a mandatory protective measure, the lab has a duty to not tolerate unsafe practices

3 The Emergency Department (ED) sends specimens to the laboratory via pneumatic tube. Lately, many specimens were not sufficiently padded or were not enclosed in sealed bags. How should this problem be solved?

- **a** the lab manager should notify the ED manager that the person(s) who did not follow procedure need to be identified and written up
- **b** the lab manager should refuse to accept any specimens from the ED until its staff have all been retrained
- **c** the lab manager should keep a log listing days, times, and problems observed, and then meet with the ED manager to analyze cause(s) for the breaches in protocol and develop the best solution
- **d** all of the above

4 A lab manager decides a procedure is too difficult to perform safely in her lab. Although the medical staff insists they need this test to be available 24 hours a day, she elects to send the test to a reference lab. Which IOM quality aims is she not supporting? (Choose as many as apply.)

- **a** patient safety
- **b** patient centeredness
- **c** efficiency
- **d** effectiveness
- **e** equity
- **f** timeliness

5 A laboratory test was performed incorrectly and the patient's discharge had to be delayed by 3 days. This is a sentinel event.

- **a** true
- **b** false

6 Each of the following is required as a response to a sentinel event EXCEPT:

- **a** identification of root cause of the event
- **b** identification of conditions that contributed to the event, even though they didn't directly cause it
- **c** notification of OSHA
- **d** implementation due dates for items in the correction plan
- **e** verification that the patient and family received all necessary care and support

Answers

1 **e**

2 **c**

3 **c**

4 **a, b, f**

5 **b**

6 **c**

 ISBN 978-089189-6463

Posttest

Self evaluation questions

1 Each organization below is paired CORRECTLY with one of its functions EXCEPT:

- **a** CDC: standards for handling biohazards
- **b** EPA: regulation of hazardous waste disposal
- **c** OSHA: certifying personnel for laboratory work
- **d** NRC: radiation and radioactive waste management
- **e** DOT: regulation of hazardous material transportation

2 Which OSHA standard below is primarily a performance standard?

- **a** Formaldehyde
- **b** Bloodborne Pathogens
- **c** Hazard Communication
- **d** Hazardous Chemicals in Laboratories

3 In order to reduce biohazard contamination, workers must handle objects in the direction of "clean to dirty." In other words, once a clean object enters a contaminated zone, it never returns to the clean zone. OSHA would classify this as a(n)

- **a** engineering control
- **b** work practice control
- **c** personal protective equipment control
- **d** administrative control

4 Which of the following is (are) always on an SDS for a chemical?

- **a** methods of disposal
- **b** methods for safe use
- **c** description of hazards
- **d** all of the above

5 Hazards should be shipped

- **a** in properly labeled, triple layer containers
- **b** in sturdy, unbreakable containers so that padding is not necessary
- **c** with enough absorbent material for a container prone to leaking
- **d** with enough padding that fragile containers can be shipped
- **e** all of the above

6 If the EPA has granted reciprocity or equivalency to a state department of the environment, then

- **a** the state can have regulatory authority in place of the EPA
- **b** the state's regulations are as strict or stricter than the EPA's
- **c** facilities should follow state requirements rather than EPA requirements
- **d** all of the above

7 Which of the following is a best practice to avoid ergonomic injury?

- **a** bend at the waist to lift a heavy load
- **b** when operating a computer mouse, use a pad under the wrist
- **c** position a computer screen >30 inches from the eyes
- **d** use a chair with a straight, rigid back
- **e** construct laboratories with hard, durable floors

8 Which of the types of fire below could you use water on?

- **a** Class A
- **b** Class B
- **c** Class C
- **d** Class D
- **e** all of the above
- **f** it is never acceptable to use water on a fire in the laboratory

9 Name the 4 components required to start a fire (the fire "quadrahedron"): (2 points)

__________ __________ __________ __________

10 A hydrogen gas fire would be a _____fire and correspond to _____ shown below.

- **a** Class A; symbol 2
- **b** Class B; symbol 3
- **c** Class C; symbol 1
- **d** Class D; symbol 2
- **e** Class E; symbol 3

1

2

3

11 The first priority in the event of a fire is

- **a** locating the nearest fire extinguisher
- **b** turning off current and shutting the doors
- **c** ensuring the safety of all people by initiating an evacuation
- **d** pulling the fire alarm and calling the appropriate extension

 ISBN 978-089189-6463

12 The best way to evacuate a fire scene is to

a use the nearest stairway or elevator
b leave doorways open as you exit so that the fire is clearly visible
c cover your mouth and nose with a damp cloth as you run toward the door
d exit via a route that moves from areas of high danger to areas of low danger
e all of the above

13 When using a CO_2 fire extinguisher you should

a aim at the top of the fire and work your way down
b hold the discharge horn firmly to direct the stream accurately
c activate the extinguisher by pulling out the locking pin and firmly squeezing the double handle
d all of the above

14 Which sketch below represents a dry chemical extinguisher?

a 
b 
c 

15 A shipment of paper towels is on fire, and a silver fire extinguisher is nearby. Is it okay to use this extinguisher on the fire?

a yes, because water can be used on a Class A fire
b no, because a Class K extinguisher cannot be used on a Class A fire
c yes, but only if there is a clear evacuation route
d not enough information is given; extinguisher type and size of fire must be known

16 A first floor laboratory, which is 20 feet by 35 feet, has one door that leads directly to an exit corridor and one window opposite the door that leads directly outside. Should another door be added for evacuation?

a not necessarily because the laboratory is <1000 square feet
b not necessarily because in this case the window can serve as a secondary evacuation route
c a & b
d yes, OSHA regulations always require primary and secondary exits

17 When should automatic fire suppression systems such as sprinklers be used in a laboratory?

- **a** in areas with unattended, high risk operations
- **b** in areas with solvent recycling devices
- **c** in areas that lack water reactive chemicals
- **d** all of the above
- **e** never

18 The most significant hazard of a chemical with the NFPA label shown is

- **a** health
- **b** instability
- **c** corrosivity
- **d** flammability

19 Matching

______ environmental hazard

______ health hazard

______ poison

______ flammable

20 The HMIS label shown would be most appropriate for which chemical listed below?

- **a** mercury
- **b** formaldehyde
- **c** hydrogen cyanide
- **d** compressed nitrogen gas

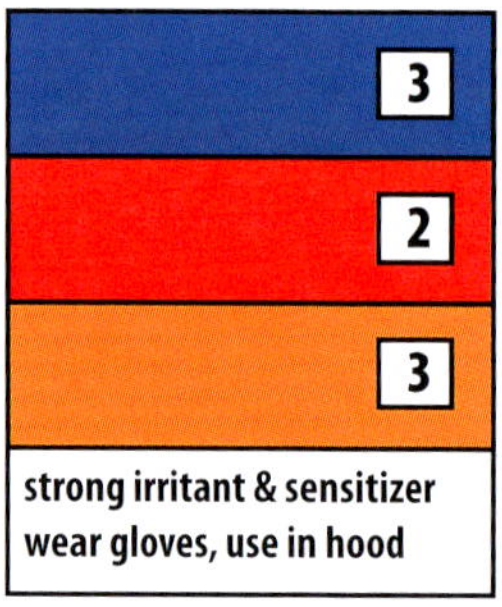

21 A secondary container chemical label is shown. Which of the following is (are) true?

15% hydrochloric acid
strong corrosive
use in the hood wearing gloves, face protection & goggles

- **a** it would be GHS compliant if a signal word were added
- **b** it would be OSHA compliant if it had a GHS pictogram added
- **c** it needs a date (prepared, received, expiration) and the initials of the person involved
- **d** all of the above

pt

 ISBN 978-089189-6463

22 Match the hazard below to its correct method of storage, handling or disposal.

___ combustibles

___ cyanide compounds

___ corrosives

___ mercury

___ oxidizers

___ ether

___ azides

___ water soluble radioactive waste

a do not flush in sinks with copper or lead plumbing
b isolate these from contact with hydrocarbons
c isolate these from contact with acids
d form explosive peroxides with oxygen; dispose after 1 year
e wear safety goggles and add these slowly to water to minimize splashing
f store in a vented safety cabinet
g flush with copious volumes of water down the sink if local regulations permit
h extreme health hazard; clean spills with special kit and wear respiratory protection

23 A properly installed and maintained chemical fume hood has

a a clean HEPA filter
b an airflow of 100 linear feet per minute
c the hood exhaust pipe connected to the building's air handler
d all of the above

24 Formaldehyde exposure can be minimized by

a using a backdraft hood since formaldehyde is heavier than air
b performing environmental monitoring to ensure PELs are not exceeded
c requiring work practices that ensure that no odor of formaldehyde can be detected
d all of the above

25 Which of the following should not be mixed with bleach or chlorine containing compounds?

a xylene
b paraffin
c picric acid
d formaldehyde

pt

26 If a hazardous spill occurs, how should absorbent material be used?

- **a** first in the middle of the spill, then working outward
- **b** first at the largest part of the spill, then working outward
- **c** first around the entire perimeter of the spill, then working inward
- **d** absorbent material should not be used for hazardous materials since it will then become hazardous too

27 Matching

_____ crystallized picric acid

_____ hydrogen cyanide

_____ xylene

_____ glutaradehyde

_____ frozen carbon dioxide

- **a** irritant; causes asthma
- **b** fatal in small concentrations
- **c** physical and chemical hazard
- **d** very flammable and toxic
- **e** explosive

28 An strong oxidizing acid should not be mixed with a(n)

- **a** organic acid
- **b** cyanide compound
- **c** strong base
- **d** any of the above

29 Which chemical bottle(s) below must be transported in a plastic bucket?

- **a** 50 mL of xylene
- **b** 1000 mL of phosphate buffer, pH 7.5
- **c** 500 mL of concentrated hydrochloric acid
- **d** all of the above

30 When is an electrical consumer product acceptable for lab use around hazardous chemicals?

- **a** when it has a 3 prong plug
- **b** when it has a polarized plug
- **c** when it is certified, by an OSHA Nationally Recognized Testing Laboratory
- **d** all of the above
- **e** none of thc above; electrical equipment that is safe to use around lab hazards is usually manufactured differently

 ISBN 978-089189-6463

31 Which of the following is (are) acceptable way(s) to use electrical equipment in the lab?

- **a** using permanent extension cords if rated for the amount of electricity in use
- **b** operating equipment with wet hands if it is plugged into a GFCI
- **c** unplugging an instrument by pulling on the cord if the plug is 3 pronged
- **d** using 2 prong mating receptacles with grounding tails for 3 prong plugs
- **e** marking malfunctioning equipment with "OUT OF SERVICE"
- **f** all of the above

32 Electrical shock occurs when

- **a** the electricity has a high enough voltage
- **b** the electricity has a high enough amperage
- **c** a person contacts an electrical circuit with an insulator
- **d** a person becomes part of a completed electrical circuit
- **e** all of the above

33 Electric shock in the laboratory is unlikely because most fuses will prevent the level of current needed for shock from entering an instrument.

- **a** true
- **b** false

34 Emergency generators should not be connected directly to power lines because

- **a** the generator could be damaged
- **b** laboratory equipment connected to the generators could be damaged
- **c** electrical "backfeed" from the generator could harm workers trying to restore power to the normal power lines
- **d** electrical "backfeed" from the generator could harm the laboratory workers trying to use equipment connected to it

35 OSHA requires 5 general categories of electrical hazard management. Name them and give an example of each.

______________ ____

______________ ____

______________ ____

______________ ____

______________ ____

36 Each of the following is a sign of electrical malfunction EXCEPT:

- **a** the same fuse blowing twice in one week
- **b** a circuit breaker being tripped during a thunderstorm
- **c** the cord plugged into an outlet feeling warm
- **d** an instrument giving the user a "tingle" when it is touched

37 Each of the following is paired CORRECTLY with an accepted method of disposal EXCEPT:

- **a** human blood: incineration
- **b** broken glass: puncture resistant container
- **c** needles: recap and discard in biohazard bag
- **d** microbiological cultures, biosafety level 2: autoclaving
- **e** scalpels: discard entire instrument into puncture resistant container

38 Identify which practice below is unsafe:

- **a** carrying a volumetric flask by the "neck"
- **b** lubricating glass connections with glycerin
- **c** operating centrifuge with a fixed cover over the moving parts
- **d** leaving a biological safety cabinet on after a spill has occurred inside it
- **e** opening an autoclave after the pressure and temperature have dropped to ambient
- **f** all of the above

39 Which object below is paired correctly with the hazard it presents?

- **a** coverslipped glass slide of tissue that has been preserved and stained: biohazard
- **b** brain tissue frozen in a cryostat for processing: biohazard
- **c** flask of *Staph aureus* culture loosely capped for autoclave processing: explosion hazard
- **d** plastic tube of sputum suspected of respiratory pathogen tightly capped for centrifugation: explosion hazard

40 The "one handed scoop" technique for covering a needle meets OSHA specifications.

- **a** true
- **b** false

41 Which safety engineered sharps device below is acceptable to OSHA?

- **a** a device that can be opened so the sharp can be released into a puncture resistant container and the device reused
- **b** an accessory added at the completion of work to automatically sheathe the sharp
- **c** a device a user can activate with both hands behind the sharp
- **d** all of the above

42 Observe the centrifuge shown here.

- **a** this centrifuge is loaded incorrectly
- **b** this centrifuge could make odd or excessive noise
- **c** the centrifuge may "walk" (move across the bench) when it is turned on
- **d** all of the above

 ISBN 978-089189-6463

43 What do you do if the heat sensitive tape doesn't change color after an autoclave cycle? (Choose more than one answer.)

- **a** assume the contents have been sterilized; this is what is supposed to happen
- **b** assume the contents have not been sterilized; this is not what is supposed to happen
- **c** check that the autoclave was set up correctly and repeat the cycle
- **d** check to see if the spore test was performed in the last week; if the contents passed, assume the contents were sterilized

44 A broth culture of a dangerous multidrug resistant organism is mixed with bleach and then autoclaved. What will happen?

- **a** the heat sensitive autoclave tape will not turn the proper color
- **b** if spores are in the autoclave, the spore test will fail
- **c** the container will explode
- **d** the container will boil over

45 Each statement below is CORRECT EXCEPT:

- **a** mouth pipetting is never acceptable under any circumstances
- **b** biosafety level 4 is for handling nonpathogens in a teaching laboratory
- **c** the 3 broad classes of disinfectants are heat, chemicals, and radiation
- **d** aerosols can form when a hot inoculating loop is placed in broth or when open tubes are centrifuged
- **e** objects contaminated with biohazards such as computer keyboards, telephones, and pencils are called fomites

46 What is wrong with this biosafety cabinet?

- **a** sash is not low enough
- **b** closed heat source should not be in cabinet
- **c** objects should not be on air intake grill
- **d** vortex should not be in cabinet
- **e** all of the above

47 Choose the CORRECT statement below:

- **a** infectious molds should be handled at BSL 1
- **b** bulk volatile organic solvents can be handled under a biological safety cabinet
- **c** the minimum biological safety cabinet classification for handling BSL 3 organisms is Class 3
- **d** centrifuges should be opened as soon as you hear breakage so that you can remove the broken glass and minimize damage to the centrifuge
- **e** none of the above is correct

48 If no aerosols or splashing are expected, specimens from patients with viral hepatitis can be handled at BSL

a 1
b 2
c 3
d 4

49 All of following are acceptable when handling animals in a laboratory EXCEPT:

a wearing heavy, "biteproof" gloves
b autopsying animals in biological safety cabinets
c inoculating animals using a press cage or sedation
d autoclaving all cages after you have cleaned them out
e assuming that even the control animals are infectious

50 The World Health Organization organism classification of "high individual and community risk" best correlates to CDC biosafety level

a 1
b 2
c 3
d 4

51 Each protocol below is a CDC "standard microbiological practice" EXCEPT

a decontaminating work surfaces after completion of work
b an ongoing insect and rodent control plan
c handwashing after completion of work
d hepatitis B and BCG vaccination of workers
e work practices that minimize aerosols

52 All of the following are components of an exposure control plan EXCEPT

a disposal protocol for chemicals
b provisions for hepatitis B vaccine
c treatment protocols for blood and body fluid exposure
d task assessments for all staff who contact blood and body fluids
e disposal protocols for contaminated materials

53 A mid level disinfectant is the minimum requirement to chemically decontaminate which of the following organism(s)?

a Creutzfeldt-Jakob agent
b *Mycobacterium tuberculosis*
c *Bacillus anthracis* and *Clostridium difficile* spores
d all of the above

pt

54 If a patient has a condition that warrants "Droplet Precautions" under the Transmission Based Precautions, then staff who contact that patient must

- **a** wear protective equipment whenever in the patient's room
- **b** wear protective equipment when it's necessary to be within ~6 feet of the patient
- **c** be vaccinated against HBV
- **d** be vaccinated with the BCG vaccine

55 Choose the CORRECT statement(s).

- **a** autoclaving procedures which inactivate biosafety level 4 organisms should deactivate infectious prions
- **b** once the cryostat freezes a tissue, it is no longer infectious
- **c** once tissues have been chemically fixed and stained, they are no longer infectious
- **d** high level disinfectants and sterilants are the same thing
- **e** all of the above

56 A 10% solution of household bleach made 30 days ago is considered a

- **a** low level disinfectant
- **b** mid level disinfectant
- **c** high level disinfectant
- **d** none of the above

57 A laboratory that can provide confirmatory but not definitive testing in the US Laboratory Response Network is considered a

- **a** reference lab
- **b** national lab
- **c** sentinel lab
- **d** BSL 2 lab
- **e** BSL 3 lab

58 Can blood or tissue specimens ever be transported by pneumatic tube systems?

- **a** yes, if they are bagged so that spills would be contained
- **b** yes, if they are in sufficient padding to prevent breakage
- **c** yes, as long as dangerous organisms like Ebola are not suspected
- **d** all of the above
- **e** no

59 Autopsies can generate hazardous aerosols. Which protocol below is NOT CORRECT?

- **a** remove mask as soon as procedures are complete and replace with a fresh one
- **b** wet bone saws before use and collect dust in vacuum bags
- **c** bag the head when cutting, especially if rabies or prions are suspected
- **d** place specimens into preservatives carefully to avoid splashing

60 When you are beginning work in a biological safety cabinet you should

- **a** turn it on and let it run for at least 4 minutes before starting work
- **b** load all the supplies needed for the entire procedure into the cabinet
- **c** verify airflow is ~75 lfpm
- **d** all of the above

61 Which of the following would be effective against a nonspore forming bacterium?

- **a** ultraviolet light at 265 nm
- **b** ethanol
- **c** 10% household bleach
- **d** all of the above

62 Which organism below is the most hazardous if it is spilled?

- **a** *Staphylococcus aureus*
- **b** *Streptococcus pyogenes*
- **c** *Neisseria meningitides*
- **d** *Clostridium difficile*

63 If a high level disinfectant is used, it should penetrate every spill adequately, and no additional disinfection is necessary.

- **a** true
- **b** false

64 The disposal method of choice for infectious prions is

- **a** NaOH decontamination followed by autoclaving
- **b** autoclaving followed by bleach decontamination
- **c** ethanol decontamination
- **d** incineration

65 A piece of equipment needs to be sent out for repair. It is used for biohazards, but is not labeled as such since the entire laboratory is considered biohazardous. What needs to be done to send this instrument out?

- **a** decontaminate the instrument entirely
- **b** label the instrument as biohazardous and package it to meet biohazard shipping standards
- **c** either of the above is acceptable
- **d** none of the above is acceptable; repair must be done inside the laboratory, and the repair person must receive proper orientation and protection

 ISBN 978-089189-6463

66 A broth culture of a BSL 3 organism was spilled outside of a biosafety cabinet. Which is (are) correct?

- **a** all staff in the location should try to evacuate while holding their breath
- **b** the room should remain empty for 60 minutes
- **c** the nearest person should put absorbent material around the spill as quickly as possible
- **d** the nearest person should spray a mid level disinfectant on the spill as quickly as possible
- **e** a & b
- **f** c & d

67 Fill in the blanks. Compressed gas should be stored in a(n) ____________ position and kept ________________ to the wall at all times so that it will be confined if there is a sudden release of pressure. Gas cylinders should never be used until they are completely ______________, or negative pressure could suction materials into the cylinder. Inventory management for cylinders should keep full and empty cylinders _______________, and should ensure that the ________________ cylinders are the ones used first. You should only use the __________________ for the particular type of gas which you are using since they are designed to prevent incompatible chemicals from mixing.

68 If you have to move a compressed gas cylinder

- **a** use a hand truck if it's large
- **b** use a hand truck no matter the size
- **c** lift it by the valve or regulator
- **d** lift it by the valve cover
- **e** a & c
- **f** a & d

69 You need an oxygen gas tank for procedure, so before you connect it to your instrument, you should verify that it is green.

- **a** true
- **b** false

70 Gas leaks of any type can cause fatal anoxic atmospheres.

- **a** true
- **b** false

pt

71 Compressed gas cylinders

- **a** can never be reused because of the stress caused by the high pressures
- **b** can be reused, but must be checked every year
- **c** can be reused, but users are still responsible for checking for cracks, dents, rust, etc before using the gas
- **d** b & c

72 On a compressed gas regulator, the pressure gauge reads >0, but the flow gauge reads 0. What does this mean?

- **a** the cylinder valve is open
- **b** the valve regulating flow is open
- **c** a & b
- **d** cannot tell from just this information

73 Which of the following is (are) true?

- **a** gas leaks in basements are more hazardous if the gas is lighter than air
- **b** gas leaks in basements are more hazardous if the gas is heavier than air
- **c** pressure release rupture disks will burst if the internal pressure of a gas gets too high
- **d** pressure release rupture disks will melt if the room temperature gets too hot
- **e** a & c
- **f** b & d

74 Small compressed gas leaks from cylinders and fittings are best detected by

- **a** visual inspection
- **b** smell
- **c** sound
- **d** soapy solution

75 If work has concluded with a gas for the day, what is (are) the best thing(s) to do?

- **a** turn off the flow regulator
- **b** turn off the valve
- **c** turn off the instrument to which it is connected
- **d** all of the above
- **e** leave everything as is if it is in working order

76 A gas cylinder is venting violently. What is the best thing to do?

- **a** turn off the flow regulator
- **b** turn off the valve
- **c** turn on the chemical fume hoods and all ventilation
- **d** evacuate

 ISBN 978-089189-6463

77 Which of the following is (are) true?

- **a** "cryogloves" are made to provide sufficient protection for hands immersed in liquid gases
- **b** since liquid cryogen gases are not in the gaseous state, they do not present an asphyxiant hazard
- **c** when filling a Dewar flask, the liquid gas should be dispensed toward the top of the flask and allowed to run gently down the sides of the flask to the bottom
- **d** if skin is exposed to a cryogen, it should be flushed for 15 minutes using slightly warm (not hot) water
- **e** all of the above

78 Which term below refers to the amount of radiation to which a person was exposed?

- **a** rem
- **b** Curie
- **c** Becquerel
- **d** half-life

79 Identify which isotope is being handled properly.

- **a** ^{32}P being used behind lead shielding
- **b** ^{51}Cr and ^{14}C wastes being discarded into the same receptacle
- **c** ^{103}Pd held in lead receptacle for 170 days before being disposed of into ordinary waste stream
- **d** ^{57}Co being chosen for a procedure instead of ^{125}I

80 All of the following are common negative consequences of radiation exposure EXCEPT:

- **a** fetal damage
- **b** DNA mutations
- **c** malignancies
- **d** hearing loss
- **e** stem cell death

81 A laboratory worker manipulates radioactive materials using tongs rather the hands. Which "ALARA" principle does this illustrate?

- **a** time
- **b** distance
- **c** shielding
- **d** decay in storage

pt

82 When work with radioisotopes is complete, they are immediately returned to shielded storage rather than being left on the workbench. Which "ALARA" principle(s) does this illustrate?

a time
b distance
c shielding
d all of the above

83 Radioactive seeds implanted into tumors are usually _____ emitters.

a α
b β
c γ
d δ

84 Radioactive emissions can be

a mutagenic
b carcinogenic
c lethal
d all of the above

85 Which radiation dosimeter(s) below is (are) being worn correctly?

a ring worn over the glove
b badge in pocket inside the lab coat
c badge clipped to lab coat lapel
d all of the above

86 Which is most dangerous?

a inhaled α particles
b handling test tube containing β particles
c γ emitter stored in a refrigerator
d X ray emitter stored in a cabinet

87 In order to handle radioactivity, a laboratory must have an NRC issued license.

a true
b false

88 If possible, which of the following should be kept separated?

a pipettes used for radioisotopes and regular pipettes
b wet and dry radioactive trash
c trash of different isotopes
d sinks used for radioisotopes and regular sinks
e all of the above

 ISBN 978-089189-6463

89 Which of the following is proper procedure?

- **a** shipment of radioisotopes is left unattended on the loading dock
- **b** since few radioisotopes are used in a laboratory, an inventory and disposal log is not maintained
- **c** since a wipe test has never showed contamination in the laboratory, the procedure is discontinued
- **d** records related to radioisotopes are kept 1 year and then shredded
- **e** none of the above

90 One of the most critical documents for demonstrating compliance with EPA and RCRA regulations is the

- **a** EPA license to treat waste
- **b** EPA license to store waste
- **c** EPA license to generate waste
- **d** completed Hazardous Waste Manifest

91 An EPA regulated characteristic waste is defined as a waste that

- **a** cannot be discarded in EPA administered landfills
- **b** appears on an EPA regulatory list because of its characteristics
- **c** appears on an EPA, state, or local regulatory list because of its characteristics
- **d** does not appear on an EPA regulatory list but has 1 of the following characteristics: toxicity, reactivity, flammability or corrosivity

92 The Resource Recovery and Conservation Act says

- **a** that facilities generating <1 Ci of radioactive waste don't need to recycle it
- **b** that facilities are responsible for their chemical waste until its ultimate disposal
- **c** that all chemical waste must be recycled by EPA methods that have low environmental impact
- **d** that chemical waste must be reduced by 10% annually in every facility until a 50% reduction has been achieved

93 All of the following are TRUE EXCEPT:

- **a** since chemical waste will no longer be used in the lab, labeling requirements are less stringent
- **b** the EPA does not have an infectious waste/medical waste category, but most state and local governments do
- **c** the EPA and DOT have developed a hazardous waste manifest that is the base form now required in all 50 states
- **d** based on the issued permit, the EPA can fine labs for either having too much hazardous waste on site or holding it too long without disposal

 ISBN 978-089189-6463

94 A fluorescent light bulb

- **a** is not classified as hazardous waste
- **b** is classified as EPA listed waste
- **c** is classified as EPA characteristic waste
- **d** is classified as EPA universal waste

95 A cup of urine labeled with a patient's name is discarded into a trash can in a medical lab. This trash is taken to the public landfill.

- **a** this is unacceptable because the name was not removed
- **b** this is unacceptable because urine is better disposed in the sanitary sewer
- **c** this would be acceptable if the trash can is biohazardous
- **d** this is acceptable
- **e** a & b

96 Which of the following is (are) never acceptable?

- **a** flushing chemicals down the drain when it is connected to a septic tank that discharges into ground water
- **b** being an EPA large quantity generator and holding waste for pick up for a year
- **c** having a satellite waste accumulation area across the street from the laboratory
- **d** all of the above

97 Who of the following need(s) waste management training?

- **a** administrative assistant who tracks the hazardous waste manifests
- **b** stockroom clerk who inspects the central waste accumulation site weekly
- **c** laboratory staff who discard the waste
- **d** all of the above

98 A licensed waste handler did not provide a completed waste manifest within 30 days. What should be done?

- **a** give him/her another 30 days
- **b** contact the waste handler to follow-up
- **c** send an exception report to the EPA
- **d** b & c

99 Compliance with HAZWOPER regulation is not complete unless

- **a** all employees at a site are trained on HAZWOPER regulations
- **b** all employees at a site know to call the EPA National Emergency Response Center
- **c** an emergency response coordinator is appointed and trained
- **d** all employees at a site know how to contain a hazardous chemical spill
- **e** a & d
- **f** b & c

 ISBN 978-089189-6463

100 Onsite waste recycling devices are designed to have no hazards and are almost always the most cost effective means to treat chemical waste.

- **a** true
- **b** false

101 Each type of equipment below is paired CORRECTLY to one of its functions EXCEPT:

- **a** heat resistant gloves: handling hot glassware
- **b** face shields: protection against hazardous fumes
- **c** goggles: eye protection against chemical splashes
- **d** safety shower: wash off and dilute chemical spills on the body
- **e** latex or vinyl gloves: protection against some chemicals and biohazards

102 Which of the following may be permitted in some labs?

- **a** smoking
- **b** eating and drinking
- **c** applying cosmetics
- **d** using nonpetroleum based hand lotion
- **e** replacing a contact lens that has fallen out

103 Which of the following can cause increased carriage of microbes on the hands?

- **a** chipped nail polish
- **b** rings and other jewelry
- **c** long natural nails and artificial nails
- **d** all of the above

104 Workers should be monitored carefully for reactions to gloves because

- **a** some people have latex allergies
- **b** some people have allergies to chemicals used to process gloves
- **c** broken and irritated skin puts workers at additional risk
- **d** all of the above

105 The barrier function of latex gloves can be compromised by all of the following EXCEPT:

- **a** exposure to light
- **b** exposure to ozone
- **c** storage in an unheated room
- **d** contact with certain hand lotions
- **e** prolonged contact with sweaty hands

106 All of the following is (are) important limitations of alcohol based hand gels EXCEPT:

- **a** the gels are flammable
- **b** more gel must be used to kill spore forming organisms
- **c** some viruses are not killed by the action of the alcohol
- **d** work cannot continue until the gel has dried on the hands

107 Which situation(s) below is (are) acceptable?

- **a** using gloves to operate a tablet device that never leaves the biohazard laboratory
- **b** using a cell phone in the laboratory as long as you deglove and wash hands first
- **c** using ear buds to listen to music, which is turned down low enough that emergency sounds can be heard
- **d** all of the above
- **e** none of the above

108 Signs communicating that caution is needed should be

- **a** red, black, and white
- **b** yellow and black
- **c** green and white
- **d** orange
- **e** yellow and purple

109 An eyewash is inspected in a chemical laboratory that handles flammables, corrosives, and oxidizers. Which finding(s) below does (do) not meet ANSI standards?

- **a** water pressure: 20 psi
- **b** water temperature: 107° F
- **c** shower 30 feet from workers accessible through a door that opens toward the eyewash
- **d** water flow: 30 gallons per minute
- **e** all of the above

110 Which protocol below is INCORRECT?

- **a** bleaching eyewash covers and nozzles weekly
- **b** verifying shower water temperature annually
- **c** laundering lab coats in cold water with antiseptic detergent
- **d** wearing ear plugs if the device you are using interferes with hearing ordinary conversation
- **e** wearing identification badges under lab coats when working with hazards

111 Of the personal protective equipment below, which is usually put on last and removed first?

- **a** mask
- **b** face shield
- **c** gloves
- **d** lab coat

 ISBN 978-089189-6463

112 What kind of protection does a face mask offer? (Choose more than one answer.)

- **a** prevents infectious patients from spreading respiratory pathogens through large droplets
- **b** prevents health care workers from transmitting respiratory pathogens to patients through large droplets
- **c** blocks aerosolized micro-organisms from being inhaled
- **d** blocks chemical vapors from being inhaled

113 You can look directly at a UV light source to see if it's working if you are wearing the proper eye protection.

- **a** true
- **b** false

114 A worker with chronic obstructive pulmonary disease (COPD) claims that she cannot breathe when wearing an N-95 respirator. What should be done?

- **a** switch her to a surgical mask
- **b** try another brand of respirator
- **c** switch her to a chemical cartridge respirator
- **d** don't allow her to work in situations that require a respirator

115 What should you do before you put on gloves to work with hazards?

- **a** wash your hands
- **b** sanitize your hands with alcohol
- **c** verify that the gloves have powder inside
- **d** inspect the gloves for holes or defects

116 Which of the following should be avoided when working in the laboratory?

- **a** wearing a "breakaway" lanyard
- **b** wearing shorts
- **c** wearing long skirts that are only an inch above the floor
- **d** all of the above

117 All of the following are functions of accident reports EXCEPT:

- **a** documenting worker's compensation claims
- **b** documenting cause for employee termination
- **c** documenting appropriateness of actions taken
- **d** analyzing the incident to prevent similar ones from occurring

118 Every laboratory that has paid employees must track injuries on OSHA Form 300

- **a** true
- **b** false

119 To comply with the Needlestick Prevention and Safety Act accident reporting provisions, accident reports must include

- **a** the brand name of the safety engineered device used
- **b** the exact model of the safety engineered device used
- **c** all of the user's actions that led to the injury
- **d** all of the above

120 Which of the following must be inspected at least annually?

- **a** general purpose absorbent material for spills that has no expiration date
- **b** respirator chemical cartridges that have expiration dates
- **c** contents of first aid kits
- **d** all of the above

121 According to the American Red Cross, the first step in first aid is to

- **a** make sure the airway is open
- **b** restore breathing and heartbeat
- **c** stop any bleeding
- **d** treat for shock
- **e** treat the wound

122 Each condition below is paired CORRECTLY with its appropriate first aid treatment EXCEPT:

- **a** airway blocked by an object: attempt to remove the object with the fingers
- **b** shock: keep victim warm and elevate the feet
- **c** first degree burn: cover with cool water or a cool, moist dressing
- **d** object impaled in eye: cover both eyes and do not attempt to remove the object
- **e** slurred speech and lopsided smile: transport victim to hospital

123 After an accident has occurred, the appropriate reason(s) that a victim could be moved is (are)

- **a** the immediate area is life threatening
- **b** the rescuer needs access to a more seriously injured person
- **c** the rescuer cannot give appropriate care with the victim in his current position
- **d** all of the above

 ISBN 978-089189-6463

124 Matching.

_____ shock
_____ stroke
_____ electrcial shock
_____ second degree burn

- **a** skin is redlike a sunburn
- **b** skin is red and has blisters
- **c** skin is black, charred and clothing is embedded in the burn
- **d** skin shows evidence of a burn and victim has no breathing or heart beat
- **e** victim can raise one arm over head but not the other
- **f** victim is pale, listless, and thirsty with rapid breathing and heartbeat

125 A "call fast" instead of a "call first" situation is a(n)

- **a** child who appears to be choking
- **b** adult who has sudden chest pain
- **c** adult who suddenly can't remember her name and has slurred speech
- **d** child with heart condition gasping for breaths

126 When should you refrain from giving first aid?

- **a** when a parent refuses consent for a child with a nonlife threatening condition
- **b** when an adult refuses care because she finds out you are not a doctor
- **c** when the accident scene is so unsafe you feel that your own life is in danger
- **d** when an adult's medical alert bracelet says, "Do not resuscitate"
- **e** all of the above

127 You bandaged a wound but you see blood seeping through the bandage. Which of the following should you do?

- **a** remove the bandage and replace it with a thicker one
- **b** put ice over the wound
- **c** put pressure on the wound with another bandage
- **d** elevate the wound above the heart (assuming no other injuries)
- **e** a & b
- **f** c & d

128 You are eating lunch with a coworker and she suddenly has trouble breathing. Her lips and tongue are swollen, and she asks you to fetch her purse so she can get her epinephrine pen. What should you do?

- **a** dial 911
- **b** perform the head tilt/chin lift
- **c** give 2 rescue breaths and begin CPR
- **d** do as she asks and get her purse

129 To dislodge an airway obstruction, chest compressions may be more effective than abdominal thrusts in

- **a** pregnant victims
- **b** obese victims
- **c** victims who are lying flat and difficult to lift
- **d** all of the above

130 Match each to the appropriate symbol:

_____ explosive symbol

_____ eyewash symbol

_____ compressed gas symbol

_____ laser symbol

_____ fire blanket symbol

_____ radiation symbol

_____ deluge shower symbol

_____ biohazard symbol

_____ electrical hazard symbol

_____ assembly point symbol

_____ no smoking symbol

_____ AED symbol

a

b

c

d

e

f

g

h

i

j

k

l

pt

 ISBN 978-089189-6463

131 The safest laboratory is the one that

- **a** has all the recommended safety equipment
- **b** has trained all of its workers thoroughly in safety
- **c** has supervisors who are aware of all safety standards
- **d** has all written policies in compliance with OSHA standards

132 A laboratory with a healthy "safety culture"

- **a** ensures that workers who do not comply with safety policies are terminated
- **b** solicits input from all workers regarding hazards they have observed
- **c** hires outside consultants to recommend that best work practices and safety equipment to be used
- **d** all of the above

Posttest answers

1 c

2 b

3 b

4 d

5 a

6 d

7 b

8 a

9 fuel, oxygen or oxidizing agent, heat source, independently sustainable chemical chain reaction

10 b

11 c

12 d

13 c

14 c

15 d

16 c

17 d

18 b

19 environmental hazard: c; health hazard: b; poison: d; flammable: a

20 b

21 c

22 combustibles: f
cyanide compounds: c
corrosives: e
mercury: h
oxidizers: b
ether: d
azides: a
liquid radioactive waste: g

23 b

24 d

25 d

26 c

27 crystallized picric acid: e
hydrogen cyanide: b
xylene: d
glutaraldehyde: a
frozen carbon dioxide: c

28 d

29 c

30 e

31 e

32 d

33 b

34 c

35 grounding (ie, low resistance connection of electrical device to the earth to disperse charge)
guarding (ie, separation of electrical sources from the public, eg, using locked doors)
insulation (ie, nonconductors surrounding sources of electricity, eg, plastic covers)
circuit protection devices (ie, devices that protect against surges of current, eg, fuses, circuit breakers, GFCI)
safe work practices (ie, techniques that reduce electrical hazard, eg, separation of equipment from water, not using extension cords, following maintenance procedures)

36 b

37 c

38 a

39 b

 ISBN 978-089189-6463

40	**b**
41	**c**
42	**d**
43	**b & c**
44	**c**
45	**b**
46	**c**
47	**e**
48	**b**
49	**d**
50	**d**
51	**d**
52	**a**
53	**b**
54	**b**
55	**c**
56	**d**
57	**a**
58	**d**
59	**a**
60	**d**
61	**d**
62	**c**
63	**b**
64	**d**
65	**c**
66	**e**
67	**vertical; chained; empty; separated; oldest; regulator**
68	**b**
69	**b**
70	**a**
71	**c**
72	**a**
73	**f**
74	**d**
75	**d**
76	**d**
77	**d**
78	**a**
79	**c**
80	**d**
81	**b**
82	**d**
83	**c**
84	**d**
85	**c**
86	**a**
87	**b**
88	**e**
89	**e**
90	**d**
91	**d**
92	**b**
93	**a**
94	**d**
95	**e**
96	**d**
97	**d**
98	**a**
99	**f**
100	**b**

101 b

102 d

103 d

104 d

105 c

106 b

107 a

108 b

109 e

110 c

111 c

112 a & b

113 b

114 d

115 d

116 b

117 b

118 b

119 d

120 d

121 a

122 a

123 d

124 shock: f
stroke: e
electrical shock: d
second degree burn: b

125 a

126 e

127 f

128 d

129 d

130 explosive symbol: i
eyewash symbol: b
compressed gas symbol: h
laser symbol: e
fire blanket symbol: a
radiation symbol: d
deluge shower symbol: c
biohazard symbol: f
electrical hazard symbol: l
assembly point symbol: k
no smoking symbol: g
AED symbol: j

131 b

132 b

 ISBN 978-089189-6463

References

All websites accessed on November 23, 2015 or later.

American Heart Association [2012] *Heartsaver First Aid/CPR/AED Student Workbook*. Dallas, TX. 2011.

American National Standard Institute [2009] ANSI Z308.1-2009 Minimum Requirements for Workplace First Aid Kits and Supplies

American National Standard Institute [2014] ANSI Z358.1-2014 Emergency Eyewash and Shower Equipment Compliance Guide. Available at www.eyewashdirect.com/v/vspfiles/pdf-new/eyewash-ansi-2015.pdf

American Red Cross [2011] *Adult First Aid/CPR/AED Ready Reference* Available at www.redcross.org/images/MEDIA_CustomProductCatalog/m4240170_Adult_ready_reference.pdf

American Society of Microbiology [2010] Sentinel level clinical laboratory protocols for suspected biological threat agents and emerging infectious diseases. Available at www.asm.org/index.php/guidelines/sentinel-guidelines

American Society of Microbiology [2011] Packing and shipping infectious substances. Available at www.asm.org/images/pdf/Clinical/pack-ship-7-15-2011.pdf

Amr S, Bollinger M [2004] Latex allergy and occupational asthma in health care workers: adverse outcomes. *Environ Health Perspect* 112(3)

Camiener G [2004] Laboratory budget reduction through hazardous waste minimization. *Lab Med,* 35(1), 9-12. January 2004.

Centers for Disease Control and Prevention. Basic Infection Control and Prevention Plan for Outpatient Oncology Settings. Available at www.cdc.gov/flu/professionals/infectioncontrol/resphygiene.htm

Centers for Disease Control and Prevention. Biological Risk Assessment Worksheet. Available at www.cdc.gov/biosafety/publications/BiologicalRiskAssessmentWorksheet.pdf

References

Centers for Disease Control and Prevention. Bioterrorism Training and Education. Available at emergency.cdc.gov/bioterrorism/training.asp

Centers for Disease Control and Prevention [2003] Exposure to Blood. What Healthcare Personnel Need to Know. Available at www.cdc.gov/HAI/pdfs/bbp/Exp_to_Blood.pdf

Centers for Disease Control and Prevention [2007] Guidelines for Isolation Precautions: Preventing Transmission of Infectious Agents in Healthcare Settings. Available at www.cdc.gov/hicpac/2007IP/2007isolationPrecautions.html

Centers for Disease Control and Prevention [2011] Laundry: Washing Infected Material. Available at www.cdc.gov/HAI/prevent/laundry.html

Centers for Disease Control and Prevention [2012] Guide for Shipping Infectious Substances. Available at www.cdc.gov/od/eaipp/shipping

Centers for Disease Control and Prevention [2013] NIOSH Stop Sticks Campaign. www.cdc.gov/niosh/stopsticks/sharpsinjuries.html

Centers for Disease Control and Prevention [2014a] Laboratory Response Network. Available at emergency.cdc.gov/lrn/index.asp

Centers for Disease Control and Prevention [2014b] NIOSH-Approved Particulate Filtering Facepiece Respirators. www.cdc.gov/niosh/npptl/topics/respirators/disp_part

Centers for Disease Control and Prevention [2015a] CDC Laboratory Training. Available at www.cdc.gov/labtraining

Centers for Disease Control and Prevention [2015b] Preventing Healthcare Infections. Available at www.cdc.gov/HAI/prevent/prevention.html

Centers for Disease Control and Prevention [2015c] Workbook for Designing, Implementing, and Evaluating a Sharps Injury Prevention Program. Available at www.cdc.gov/sharpssafety/resources.html

Centers for Disease Control and Prevention. [2012] Respirator Fact Sheet. Available at www.cdc.gov/niosh/npptl/topics/respirators/factsheets/respfact.html

Clinical Laboratory Standards Institute [2007] QMS04-A2 (formerly GP18-A2) *Laboratory Design*, Approved Guideline

Clinical Laboratory Standards Institute [2011] GP05-A3. *Clinical Laboratory Waste Management*, Approved Guideline, 3e

Clinical Laboratory Standards Institute [2012] GP12-A3 *Clinical Laboratory Safety*, Approved Guideline, 3e

Clordisys [2014] Ultraviolet Light Disinfection Data Sheet. Available at www.cleanhospital.com/pdfs/uv_data_sheet.pdf

Dimenstein IB [2009] A pragmatic approach to formalin safety in anatomical pathology. *Lab Med* 40(12):740-746

Furr AK, ed [2000] *CRC Handbook of Laboratory Safety*, 5e. CRC Press

Gile TJ [2001] Ergonomics in the laboratory. *Lab Med* 32(5):263-267

ref

 ISBN 978-089189-6463

Healthcare Environmental Resource Center. Available at hercenter.org

Henderson D, Dembry L, Fishman N et al [2010] Society for Healthcare Epidemiology in America guideline for management of healthcare workers. *Infect Control Hosp Epidemiol* 31(3):203-32

Henry JB, ed [2001] *Clinical Diagnosis and Management by Laboratory Methods*, 20e. WB Saunders

Institute for Healthcare Improvement. Patient Safety. Available at www.ihi.org/IHI/Topics/PatientSafety

Institute of Medicine [2001] *Crossing the Quality Chasm: A New Health System for the 21st Century*. National Academy Press

Jensen PA, Lambert LA, Iademarco MF et al [2005] Guidelines for preventing transmission of *Mycobacterium tuberculosis* in health care settings. *MMWR* 54(RR17):1-141

Kohn LT, Corrigan JM, Donaldson MS, ed [1999] *To Err Is Human: Building a Safer Health System*. National Academy Press

Markenson D, Ferguson J et al [2010] 2010 American Heart Association and American Red Cross international consensus on first aid science with treatment recommendations. *Circulation* 122:S582-S605

Mazurek GH, Jereb J, Lobue P, Iademarco MF et al [2005] Guidelines for using the QuantiFERON-TB Gold test for detecting *Mycobacterium tuberculosis* infection, United States. *MMWR* 54(RR15):49-55

Miller JM, Astles R, Baszler T et al [2012] Guidelines for safe work practices in human and animal medical diagnostic laboratories: recommendations of a CDC-convened, Biosafety Blue Ribbon Panel. *MMWR* 61(01);1-101. Available at www.cdc.gov/mmwr/preview/mmwrhtml/su6101a1.htm

National Fire Protection Association [2015a] NFPA 2001 *Clean Agent Fire Extinguishing Systems*

National Fire Protection Association [2015b] NFPA 45 *Fire Protection for Laboratories Using Chemicals*

National Fire Protection Association [2015c] NFPA 99 *Health Care Facilities*

National Institutes of Health [2013] Guidelines for Research Involving Recombinant or Synthetic Nucleic Acid Molecules. Available at osp.od.nih.gov/sites/default/files/NIH_Guidelines_0.pdf

Neumar, R et al [2015] 2015 AHA guidelines update for cardiopulmonary resuscitation and emergency cardiovascular care. *Circulation* 132:18suppl2

Occupational Safety and Health Administration. Healthcare Industry Compliance Assistance Quick Start. Available at www.osha.gov/dcsp/compliance_assistance/quickstarts/health_care/index_hc.html

Occupational Safety and Health Administration [1987] *Federal Register* 29 CFR 1910.120. *Hazardous Waste Operations and Emergency Response*

Occupational Safety and Health Administration [1990a] *Federal Register* 29 CPR Part CFR 1910.147. *Control of Hazardous Energy*

References

Occupational Safety and Health Administration [1990b] *Federal Register* 29 CPR Part 1910.1450. *Occupational Exposure to Hazardous Chemicals in Laboratories*

Occupational Safety and Health Administration [1991] *Federal Register,* 29 CFR Part 1910.1030. *Occupational Exposure to Bloodborne Pathogens,* Final Rule

Occupational Safety and Health Administration [1992] *Federal Register* 29 CPR Part 1910.1048 *Occupational Exposure to Formaldehyde*

Occupational Safety and Health Administration [2001a] CPL 02-02-069 - CPL 2-2.69 - Enforcement procedures for the *Occupational Exposure to Bloodborne Pathogens*

Occupational Safety and Health Administration [2001b] Federal Register, 29 CFR Part 1910.1030. Occupational Exposure to Bloodborne Pathogens; Needlestick and Other Sharps Injuries, Final Rule 66:5317-5325

Occupational Safety and Health Administration [2002] Federal Register 29 CPR Part CFR 1910.157. Portable Fire Extinguishers

Occupational Safety and Health Administration [2011a] Laboratory Safety Ergonomics for the Prevention of Musculoskeletal Disorders OSHA FS-3462. Available at www.osha.gov/Publications/laboratory/OSHAfactsheet-laboratory-safety-ergonomics.pdf

Occupational Safety and Health Administration [2011b] Laboratory Safety Guidance OSHA 3404-11R. Available at www.osha.gov/Publications/laboratory/OSHA3404laboratory-safety-guidance.pdf

Occupational Safety and Health Administration [2014] Updates to OSHA's Recordkeeping Rule: Who Is Required To Keep Records and Who Is Exempt. DEA FS-3746. Available at www.osha.gov/recordkeeping2014/OSHA3746.pdf

Occupational Safety and Health Administration [2015a] Health Effects from Contaminated Water in Eyewash Stations. OSHA 3818. Availablle at www.osha.gov/Publications/OSHA3818.pdf

Occupational Safety and Health Administration [2015b] Hospital Respiratory Protection Program Toolkit 3767. Available at www.osha.gov/Publications/OSHA3767.pdf. May 2015

Occupational Safety and Health Administration [2015c] OSHA's Recordkeeping Rule, Available at www.osha.gov/recordkeeping2014

Occupational Safety and Health Administration [2015d] Permissible Exposure Limits¬—Annotated Tables. Available at www.osha.gov/dsg/annotated-pels/index.html

Ozanne G [2002] Latex, vinyl or nitrile? Characteristics of certain gloves used in the laboratory or in the field to reduce risk of skin exposure to biological agents. *Can J Med Lab Sci* 64:29-35

ref

 ISBN 978-089189-6463

The Joint Commission [2014] Sentinel Event Policy and Procedures. Available at www.jointcommission.org/sentinel_event_policy_and_procedures

Tosini W, Ciotti C, Goyer F et al [2010] Needlestick injury rates according to different types of safety-engineered devices: results of a French multicenter study. *Infect Control Hosp Epidemiol* 31:402-407

United Nations Economic Commission for Europe [2015] A Guide to the Globally Harmonized System of Classification and Labeling of Chemicals (GHS). Available at www.osha.gov/dsg/hazcom/ghsguideoct05.pdf

United States Department of Health and Human Services [2009] *Biosafety in Microbiological and Biomedical Laboratories,* 5e. US Government Printing Office

United States Department of Homeland Security [2007] DHS Chemicals of Interest. Available at www.dhs.gov/xlibrary/assets/chemsec_appendixa-chemicalofinterestlist.pdf

United States Department of Homeland Security [2015] Critical Infrastructure: Chemical Security. Available at www.dhs.gov/critical-infrastructure-chemical-security

United States Environmental Protection Agency [2012] Hazardous Waste Generator Regulations: A User-Friendly Reference Document Version 6. Available at www2.epa.gov/hwgenerators/hazardous-waste-generator-regulations-user-friendly-reference-document

United States Environmental Protection Agency [2015a] Hazardous Waste Generators. Available at www2.epa.gov/hazgen

United States Environmental Protection Agency [2015b] Hazardous Wastes. www3.epa.gov/epawaste/hazard/index.htm

United States Environmental Protection Agency [2015c] Table Noting Which States Have Hazardous Waste Generator Categories That Are the Same as the Federal Categories and Which Have Different Categories. Available at www.epa.gov/hwgenerators/table-noting-which-states-have-hazardous-waste-generator-categories-are-same-federal

United States Environmental Protection Agency [2015d] Uniform Hazardous Waste Manifest EPA Form 8700-22, Available at www3.epa.gov/epawaste/hazard/transportation/manifest/pdf/newform.pdf

United States Federal Select Agent Program [2014] Select Agents and Toxins List. Available at www.selectagents.gov/SelectAgentsandToxinsList.html

United States Food and Drug Administration [2006] 21CFR Part 800 [Docket No. 2003N-0056(formerly 03N-0056)]. Medical Devices; Patient Examination and Surgeons' Gloves; Test Procedures and Acceptance Criteria. Final Rule. *Fed Regist* 71(243):75865-75879.

ref

References

United States Food and Drug Administration [2015] FDA-Cleared Sterilants and High Level Disinfectants with General Claims for Processing Reusable Medical and Dental Devices. Available at www.fda.gov/ MedicalDevices/ DeviceRegulationand Guidance/ ReprocessingofReusable MedicalDevices/ucm437347.htm

United States Nuclear Regulatory Commission [2012] NRC Form 3 Notice to Employees. Available at www.nrc.gov/reading-rm/doc-collections/forms/nrc3.pdf

United States Nuclear Regulatory Commission [2015] 31.11 General license for use of byproduct material for certain in vitro clinical or laboratory testing. Available at www.nrc.gov/reading-rm/doc-collections/cfr/part031/part031-0011.html

United States Postal Service [2015] USPS Publication 52 Hazardous, Restricted, and Perishable Mail. Available at pe.usps.com/text/pub52/welcome.htm

Weinert, George [1996] Latex antigen, glutaraldehyde trigger complaints and solutions. *ADVANCE for Medical Laboratory Professionals* Feb 5

World Health Organization [2009] WHO Guidelines on Hand Hygiene in Health Care. Available at apps.who.int/iris/bitstream/10665/44102/1/9789241597906_eng.pdf

World Health Organization [2015] WHO: A Guide for Shippers of Infectious Substances. Available at www.who.int/ihr/infectious_substances/en/

ISBN 978-089189-6463

Index

Page numbers followed by t or f indicate tables or figures, respectively.

Index

 ISBN 978-089189-6463

D

Index

 ISBN 978-089189-6463

G

H

Index

 ISBN 978-089189-6463

M

N

O

 ISBN 978-089189-6463

P

R

S

 ISBN 978-089189-6463

T

U

V

W

X

 ISBN 978-089189-6463